Woven Fabrics

Woven Fabrics

Edited by **John Tegmeyer**

𝒞LANRYE
INTERNATIONAL

New Jersey

Published by Clanrye International,
55 Van Reypen Street,
Jersey City, NJ 07306, USA
www.clanryeinternational.com

Woven Fabrics
Edited by John Tegmeyer

International Standard Book Number: 978-1-63240-526-5 (Hardback)

This book contains information obtained from authentic and highly regarded sources. Copyright for all individual chapters remain with the respective authors as indicated. A wide variety of references are listed. Permission and sources are indicated; for detailed attributions, please refer to the permissions page. Reasonable efforts have been made to publish reliable data and information, but the authors, editors and publisher cannot assume any responsibility for the validity of all materials or the consequences of their use.

The publisher's policy is to use permanent paper from mills that operate a sustainable forestry policy. Furthermore, the publisher ensures that the text paper and cover boards used have met acceptable environmental accreditation standards.

Trademark Notice: Registered trademark of products or corporate names are used only for explanation and identification without intent to infringe.

Printed in the United States of America.

Contents

Preface

Every book is initially just a concept; it takes months of research and hard work to give it the final shape in which the readers receive it. In its early stages, this book also went through rigorous reviewing. The notable contributions made by experts from across the globe were first molded into patterned chapters and then arranged in a sensibly sequential manner to bring out the best results.

Woven fabrics are a popular variety of textiles. This book deals with the production and manufacturing of woven fabrics. Generally, woven fabrics are the traditional textile fabrics used for cloth production, and are used extensively in several fabric compositions as intermediate goods that affect human actions. The relative significance of woven fabrics as traditional textile materials is tremendously huge, and at present, application fields of woven fabrics as technical textiles is rapidly extended by utilizing its geometric features and advantages. For instance, this book covers systematic approaches to fabric design, micro and nano technology required to create woven fabrics, as well as the theories for industrial applications.

It has been my immense pleasure to be a part of this project and to contribute my years of learning in such a meaningful form. I would like to take this opportunity to thank all the people who have been associated with the completion of this book at any step.

Editor

Modeling of Woven Fabrics Geometry and Properties

B. K. Behera[1], Jiri Militky[2], Rajesh Mishra[2] and Dana Kremenakova[2]
[1]Department of Textile Technology, Indian Institute of Technology, Delhi,
[2]Faculty of Textile Engineering, Technical University of Liberec,
[1]India
[2]Czech Republic

1. Introduction

There are many ways of making fabrics from textile fibers. The most common and most complex category comprises fabrics made from interlaced yarns. These are the traditional methods of manufacturing textiles. The great scope lies in choosing fibers with particular properties, arranging fibers in the yarn in several ways and organizing in multiple ways, interlaced yarn within the fabric. This gives textile designer great freedom and variation for controlling and modifying the fabric. The most common form of interlacing is weaving, where two sets of threads cross and interweave with one another. The yarns are held in place due to the inter-yarn friction. Another form of interlacing where the thread in one set interlocks with the loops of neighboring thread by looping is called knitting. The interloping of yarns results in positive binding. Knitted fabrics are widely used in apparel, home furnishing and technical textiles. Lace, Crochet and different types of Net are other forms of interlaced yarn structures. Braiding is another way of thread interlacing for fabric formation. Braided fabric is formed by diagonal interlacing of yarns. Braided structures are mainly used for industrial composite materials.

Other forms of fabric manufacture use fibers or filaments laid down, without interlacing, in a web and bonded together mechanically or by using adhesive. The former are needle punched nonwovens and the later spun bonded. The resulting fabric after bonding normally produces a flexible and porous structure. These find use mostly in industrial and disposable applications. All these fabrics are broadly used in three major applications such as apparel, home furnishing and industrial.

The traditional methods of weaving and hand weaving will remain supreme for high cost fabrics with a rich design content. The woven structures provide a combination of strength with flexibility. The flexibility at small strains is achieved by yarn crimp due to freedom of yarn movement, whereas at high strains the threads take the load together giving high strength. A woven fabric is produced by interlacing two sets of yarns, the warp and the weft which are at right angles to each other in the plane of the cloth (Newton, 1993). The warp is along the length and the weft along the width of the fabric. Individual warp and weft yarns are called ends and picks. The interlacement of ends and picks with each other produces a coherent and stable structure. The repeating unit of interlacement is called the weave (Robinson & Marks, 1973). The structure and properties of a woven fabric are dependent upon the constructional parameters as thread density, yarn fineness, crimp, weave etc.

The present chapter establishes some interesting mathematical relationships between these constructional parameters so as to enable the fabric designer and researcher to have a clear understanding of the engineering aspects of woven fabrics. This is an attempt to transform from an experience based designing into an engineered approach to model woven fabric constructions.

1.1 Elements of fabric structure

Plain weave has the simplest repeating unit of interlacement. It also has the maximum possible frequency of interlacements. Plain weave fabrics are firm and resist yarn slippage. Figure 1 shows plain weave in plan view and in cross-section along warp and weft. The weave representation is shown by a grid in which vertical lines represent warp and horizontal lines represent weft. Each square represents the crossing of an end and a pick. A mark in a square indicates that the end is over the pick at the corresponding place in the fabric that is warp up. A blank square indicates that the pick is over the end that is weft up. One repeat of the weave is indicated by filled squares and the rest by crosses. The plain weave repeats on two ends and two picks.

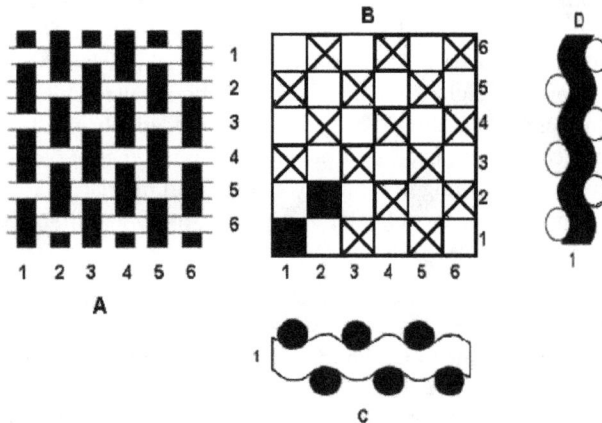

Fig. 1. Plan (A), Weave representation (B) Cross-sectional view along warp (D) Cross-sectional view along weft (C) for plain weave

1.2 Regular and irregular weaves

1.2.1 Regular weaves

Regular weaves (Grosicki, 1988) give a uniform and specific appearance to the fabric. The properties of the fabric for such weaves can be easily predicted. Examples of some of the common regular weaves are given in figure 2.

1/1 plain   2/2 matt   1/3 twill   1/4 sateen   2/2 warp rib   2/2 weft rib

1/3 sateen base crepe   3/1 sateen base crepe weave

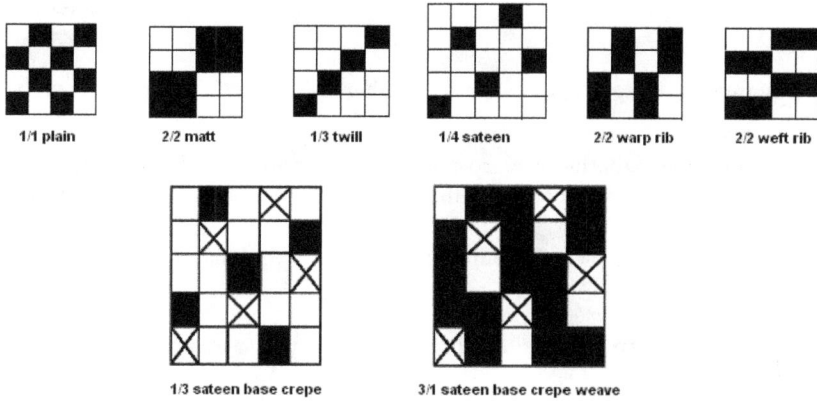

Fig. 2. Regular weaves

1.2.2 Irregular weaves

Irregular weaves are commonly employed when the effect of interlacement is masked by the coloured yarn in the fabric. Such weaves are common in furnishing fabric. In such structures the prediction of mechanical properties is difficult. Examples of some of the common irregular weaves are given in figure 3.

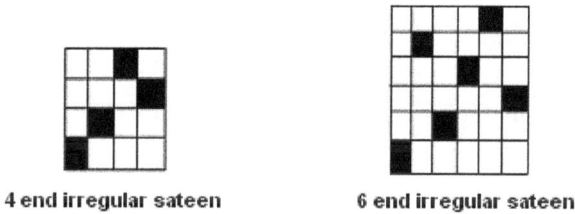

4 end irregular sateen   6 end irregular sateen

Fig. 3. Irregular weaves

1.3 Mathematical representation of different weaves

The firmness of a woven fabric depends on the density of threads and frequency of interlacements in a repeat. Fabrics made from different weaves cannot be compared easily with regard to their physical and mechanical properties unless the weave effect is normalized. The concept of average float has been in use since long, particularly for calculating maximum threads per cm. It is defined as the average ends per intersection in a unit repeat. Recently this ratio termed as weave factor (Seyam, 2002; Weiner, 1971) has been used to estimate tightness factor in fabric.

1.3.1 Weave factor

It is a number that accounts for the number of interlacements of warp and weft in a given repeat. It is also equal to average float and is expressed as:

$$M = \frac{E}{I} \qquad (1)$$

Where E is number of threads per repeat, I is number of intersections per repeat of the cross-thread.

The weave interlacing patterns of warp and weft yarns may be different. In such cases, weave factors are calculated separately with suffix1 and 2 for warp and weft respectively.

Therefore, $M_1 = \frac{E_1}{I_2}$; E_1 and I_2 can be found by observing individual pick in a repeat

and $M_2 = \frac{E_2}{I_1}$; E_2 and I_1 can be found by observing individual warp end in a repeat.

1.3.2 Calculation of weave factor

1.3.2.1 Regular weaves

Plain weave is represented as $\frac{1}{1}$; for this weave, E_1 the number of ends per repeat is equal to 1+1=2 and I_2 the number of intersections per repeat of weft yarn =1+ number of changes from up to down (vice versa) =1+1=2.

Table 1 gives the value of warp and weft weave factors for some typical weaves.

Weave	E_1	I_2	E_2	I_1	M_1	M_2
1/1 Plain	2	2	2	2	1	1
2/1 Twill	3	2	3	2	1.5	1.5
2/2 Warp Rib	2	2	4	2	1	2
2/2 Weft Rib	4	2	2	2	2	1

Table 1. Weave factor for standard weaves

E_1 and E_2 are the threads in warp and weft direction
I_2 and I_1 are intersections for weft and warp threads

1.3.2.2 Irregular weaves

In some weaves the number of intersections of each thread in the weave repeat is not equal. In such cases the weave factor is obtained as under:

$$M = \frac{\Sigma E}{\Sigma I} \qquad (2)$$

Using equation 2 the weave factors of a ten-end irregular huckaback weave shown in figure 4 is calculated below.

Weave factor, $M = \dfrac{10+10+10+10+10+10+10+10+10+10}{10+6+10+6+10+6+10+6+10+6} = \dfrac{100}{84} = 1.19$

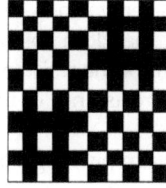

Fig. 4. Ten-end Huck-a-back weave

2. Geometrical model of woven structures

The properties of the fabric depend on the fabric structure. The formal structure of a woven fabric is defined by weave, thread density, crimp and yarn count. The interrelation between fabric parameters can be obtained by considering a geometrical model of the fabric. The model is not merely an exercise in mathematics. It is not only useful in determining the entire structure of a fabric from a few values given in technological terms but it also establishes a base for calculating various changes in fabric geometry when the fabric is subjected to known extensions in a given direction or known compressions or complete swelling in aqueous medium. It has been found useful for weaving of maximum sett structures and also in the analysis and interpretation of structure-property relationship of woven fabrics. Mathematical deductions obtained from simple geometrical form and physical characteristics of yarn combined together help in understanding various phenomena in fabrics.

2.1 Basic relationship between geometrical parameters

The geometrical model is mainly concerned with the shape taken up by the yarn in the warp or weft cross-section of the fabric. It helps to quantitatively describe the geometrical parameters. The basic model (Pierce, 1937) is shown in figure 5. It represents a unit cell interlacement in which the yarns are considered inextensible and flexible. The yarns have circular cross-section and consist of straight and curved segments. The main advantages in considering this simple geometry are:

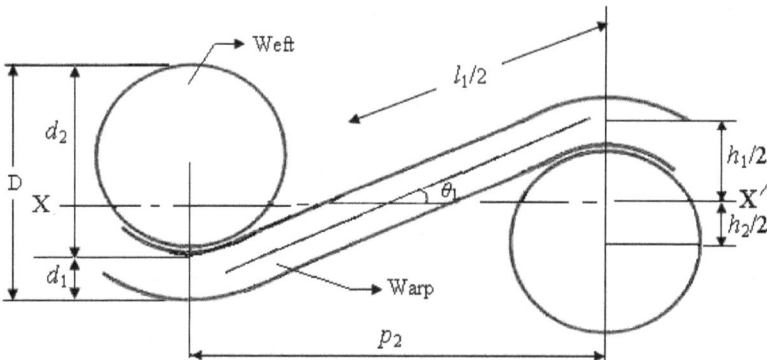

Fig. 5. Peirce's model of plain weave

1. Helps to establish relationship between various geometrical parameters
2. Able to calculate the resistance of the cloth to mechanical deformation such as initial extension, bending and shear in terms of the resistance to deformation of individual fibers.
3. Provide information on the relative resistance of the cloth to the passage of air, water or light.
4. Guide to the maximum density of yarn packing possible in the cloth.

From the two-dimensional unit cell of a plain woven fabric, geometrical parameters such as thread-spacing, weave angle, crimp and fabric thickness are related by deriving a set of equations. The symbols used to denote these parameters are listed below.

d - diameter of thread
p - thread spacing
h - maximum displacement of thread axis normal to the plane of cloth (crimp height)
θ - angle of thread axis to the plane of cloth (weave angle in radians)
l - length of thread axis between the planes through the axes of consecutive cross- threads (modular length)
c - crimp (fractional)
$D = d_1 + d_2$

Suffix 1 and 2 to the above parameters represent warp and weft threads respectively.

In the above figure projection of yarn axis parallel and normal to the cloth plane gives the following equations:

$$c_1 = \frac{l_1}{p_2} - 1 \tag{3}$$

$$p_2 = (l_1 - D\theta_1)\cos\theta_1 + D\sin\theta_1 \tag{4}$$

$$h_1 = (l_1 - D\theta_1)\sin\theta_1 + D(1 - \cos\theta_1) \tag{5}$$

Three similar equations are obtained for the weft direction by interchanging suffix from 1 to 2 or vice-versa as under:

$$c_2 = \frac{l_2}{p_1} - 1 \tag{6}$$

$$p_1 = (l_2 - D\theta_2)\cos\theta_2 + D\sin\theta_2 \tag{7}$$

$$h_2 = (l_2 - D\theta_2)\sin\theta_2 + D(1 - \cos\theta_2) \tag{8}$$

Also, $d_1 + d_2 = h_1 + h_2 = D \tag{9}$

In all there are seven equations connecting eleven variables. If any four variables are known then the equations can be solved and the remaining variables can be determined. Unfortunately, these equations are difficult to solve. Researchers have tried to solve these equations using various mathematical means to find new relationships and also some simplified useful equations.

2.1.1 Relation between weave composition and structural parameters

When the interlacement pattern is modified by changing the float length, the structure of the fabric changes dramatically. It has a profound effect on the geometry of the yarn interlacement and related properties in the woven fabric. The maximum weavability limit is predicted by extending the Peirce's geometrical model for non-plain weaves by soft computing. This information is helpful to the weavers in avoiding attempts to weave impossible constructions thus saving time and money. It also helps to anticipate difficulty of weaving and take necessary steps in warp preparations. The relationship between the cover factors in warp and weft direction is demonstrated for circular and racetrack cross-section for plain, twill, basket and satin weave in later part of this chapter. Non plain weave fabric affords further flexibility for increasing fabric mass and fabric cover. As such they enlarge scope of the fabric designer and researcher. Figure 6 shows the relationship between warp and weft thread spacing for different weaves for a given yarn.

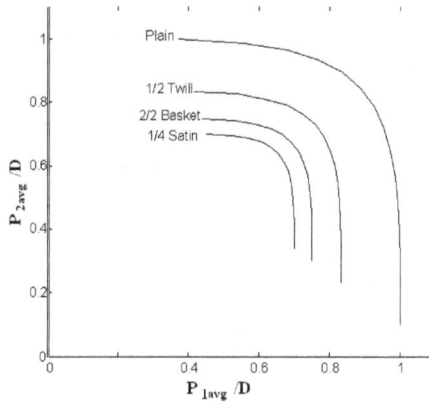

Fig. 6. Relation between average thread spacing in warp and weft for different weaves (Circular cross-section, yarn tex=30, Ø=0.6, ρ=1.52)

2.2 Some derivatives

2.2.1 Relation between p, h, θ and D

From equations 4 and 5 we get:

$$\left(l_1 - D\theta_1\right) = \frac{p_2 - D\sin\theta_1}{\cos\theta_1} = \frac{h_1 - D(1 - \cos\theta_1)}{\sin\theta_1}$$

or $D(\sec\theta_1 - 1) - p_2\tan\theta_1 + h_1 = 0$

Substituting $x_1 = \dfrac{\tan\theta_1}{2}$

we get, $x_1^2\left(D - \dfrac{h_1}{2}\right) - p_2 x_1 + \dfrac{h_1}{2} = 0$

For real fabrics $\quad x_1 = \dfrac{tan\theta_1}{2} = p_2 - \dfrac{\sqrt{p_2^2 - 2h_1(D - \frac{h_1}{2})}}{2D - h_1} = \dfrac{p_2 - \sqrt{p_2^2 - h_2^2 - D^2}}{D + h_2}$

Using value of x_1, one can calculate θ, l and c and also other parameters.

Similarly, using equation 7 and 8, and by eliminating l and substituting x_1 as above, we will arrive at a more complex equation as:

$$\frac{c_1}{2} + x_1 \frac{D}{p_2} - x_1^2 (1 + \frac{c_1}{2}) = \frac{D}{p_2}(1 - x_1^2)\tan^{-1}x_1$$

It is difficult to solve this equation algebraically for x_1. However one can substitute value of x_1 obtained earlier to solve this equation just for an academic interest.

These seven equations have been solved by soft computing in order to establish several useful relationships. However, at this stage, one can generalize the relationship as:

$$h_1 = f(p_2, c_1)$$

This function f can be obtained by plotting p and h for different values of c.

2.2.2 Functional relationship between p, h, c

Trigonometric expansion of equations 4 and 5 gives:

$$p_2 = l_1 - \frac{l_1\theta_1^2}{2} + \frac{D\theta_1^3}{3} + \frac{l_1\theta_1^4}{24} + ----$$

$$h_1 = l_1\theta_1 - \frac{D\theta_1^2}{2} - \frac{l_1\theta_1^3}{6} + \frac{D\theta_1^4}{8} + ----$$

When θ is small, higher power of θ can be neglected which gives:

$$h_1 = l_1\theta_1, \; p_2 = l_1, \; c_2 = \frac{\theta_1^2}{2}, \; h_1 = p_2\sqrt{2c_1}$$

and these equations reduce to:

$$\theta_1 = (2c_1)^{\frac{1}{2}} \tag{10}$$

$$\theta_2 = (2c_2)^{\frac{1}{2}} \tag{11}$$

$$h_1 = \frac{4}{3}p_2\sqrt{c_1} \tag{12}$$

$$h_2 = \frac{4}{3} p_1 \sqrt{c_2}$$ (13)

These four equations are not new equations in this exercise. They are derived from the previous seven original equations. However they give simple and direct relationships between four fabric parameters h, p, c and θ.

2.2.3 Jammed structures

A woven fabric in which warp and weft yarns do not have mobility within the structure as they are in intimate contact with each other are called jammed structures. In such a structure the warp and weft yarns will have minimum thread spacing. These are closely woven fabrics and find applications in wind-proof, water-proof and bullet-proof requirements.

During jamming the straight portion of the intersecting yarn in figure 5 will vanish so that in equation 4 and 5, $l_1 - D\theta_1 = 0$

$$\frac{l_1}{D} = \theta_1$$

Equations 4 and 5 will reduce to

$$h_1 = D(1 - \cos\theta_1)$$

$$p_2 = D\sin\theta_1$$

Similarly, for jamming in the weft direction $l_2 - D\theta_2 = 0$, equations 7 and 8 will reduce to the above equations with suffix interchanged from 1 to 2 and vice-versa.

For a fabric being jammed in both directions we have:

$$D = h_1 + h_2 = D(1 - \cos\theta_1) + D(1 - \cos\theta_2)$$

$$\text{or } \cos\theta_1 + \cos\theta_2 = 1$$ (14)

$$\sqrt{1 - \left(\frac{p_1}{D}\right)^2} + \sqrt{1 - \left(\frac{p_2}{D}\right)^2} = 1$$ (15)

This is an equation relating warp and weft spacing of a most closely woven fabric.

2.2.4 Cross threads pulled straight

If the weft yarn is pulled straight $h_2 = 0$ and $h_1 = D$,

Equation 5 will give $D = (l_1 - D\theta_1)\sin\theta_1 + D(1 - \cos\theta_1)$

$$\cos\theta_1 = \left(\frac{l_1}{D} - \theta_1\right)\sin\theta_1$$

$$\text{or } \theta_1 + \cot\theta_1 = \frac{l_1}{D} \tag{16}$$

This equation gives maximum value of θ_1 for a given value of l_1/D

The above equation will be valid for warp yarn being straight by interchange of suffix from 1 to 2.

However, the weft thread can be restricted in being pulled straight by the jamming of warp threads.

In such a case,

$$l_1 - D\theta_1 = 0$$

$$\text{or } \theta_1 = \frac{l_1}{D}$$

Equation 5 will become

$$h_2 = D - h_1 = D - D(1 - \cos\theta_1) = D\cos\frac{l_1}{D} \tag{17}$$

If the weft thread is pulled straight and warp is just jammed

$$\text{Then } \frac{l_1}{D} = \theta_1 = \frac{\pi}{2} \tag{18}$$

These are useful conditions for special fabric structure.

2.2.5 Non circular cross-section

So far, it is assumed that yarn cross-section is circular and yarn is incompressible. However, the actual cross-section of yarn in fabric is far from circular due to the system of forces acting between the warp and weft yarns after weaving and the yarn can never be incompressible. This inter-yarn pressure results in considerable yarn flattening normal to the plane of the cloth even in a highly twisted yarn. Therefore many researchers have tried to correct Peirce's original relationship by assuming various shapes for the cross-section of yarn. Two important cross-sectional shapes such as elliptical and race-track are discussed below.

2.2.5.1 Elliptical cross-section

Peirce's elliptical yarn cross-section is shown in figure 7; the flattening factor is defined as

$$e = \sqrt{\frac{b}{a}}$$

Where b = minor axis of ellipse, a = major axis of ellipse

The area of ellipse is $(\pi/4)ab$. If d is assumed as the diameter of the equivalent circular cross-section yarn, then

$$d = \sqrt{ab}$$

$$h_1 + h_2 = d_1 + d_2 = b_1 + b_2$$

$$b_1 + b_1 = h_1 + h_2 = \frac{4}{3}\left[p_1\sqrt{c_2} + p_2\sqrt{c_1} \right] \tag{19}$$

Yarn diameter is given by its specific volume, v and yarn count as under:

$$d_{mils} = 34.14\frac{\sqrt{v}}{\sqrt{N}} \text{ , } N \text{ is the English count.}$$

$$d_{cm} = \frac{\sqrt{Tex}}{280.2\sqrt{\varphi \rho_f}} = \frac{\sqrt{Tex}}{280} \text{ , assuming, } \varphi = 0.65, \rho_f = 1.52 \text{ for cotton fiber}$$

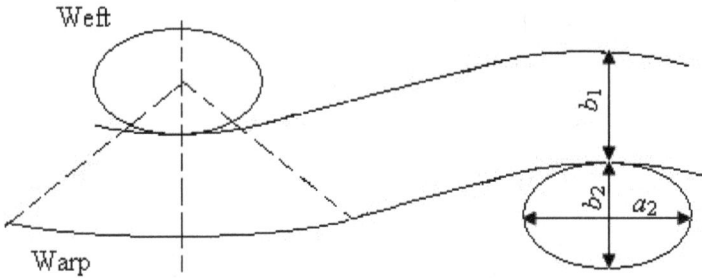

Fig. 7. Elliptical cross-section

This can be used to relate yarn diameter and crimp height by simply substituting in equation 19 to obtain:

$$h_1 + h_2 = d_1 + d_2 = D = 34.14\left(\sqrt{\frac{v_1}{N_1}} + \sqrt{\frac{v_2}{N_2}} \right) \tag{20}$$

$$h_1 + h_2 = d_1 + d_2 = \frac{1}{280.2}\left(\sqrt{\frac{T_1}{\varphi_1 \rho_{f1}}} + \sqrt{\frac{T_2}{\varphi_2 \rho_{f2}}} \right) = \frac{1}{280}\left(\sqrt{T_1} + \sqrt{T_2} \right)$$

assuming, $\varphi = 0.65, \rho_f = 1.52$ for cotton fiber

These are useful equation to be used subsequently in the crimp interchange derivation.

2.2.5.2 Race track cross-section

In race track model (Kemp, 1958; Love, 1954) given in figure 8, a and b are maximum and minimum diameters of the cross-section. The fabric parameters with superscript refer to the zone AB, which is analogous to the circular thread geometry; the parameters without superscript refer to the race track geometry, a repeat of this between CD. Then the basic equations will be modified as under:

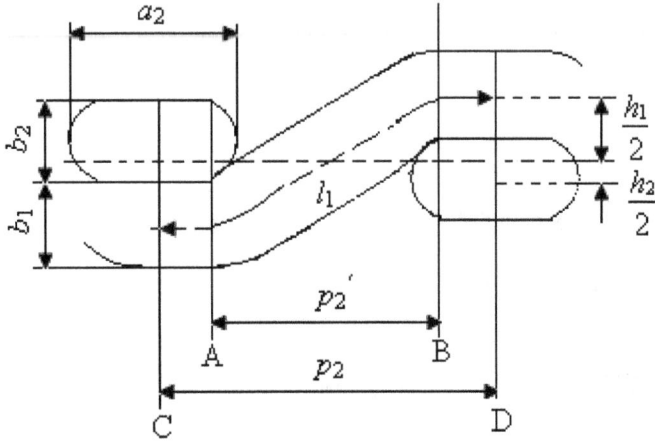

Fig. 8. Race track cross-section

$$p_2' = p_2 - (a_2 - b_2) \tag{21}$$

$$l_1' = l_1 - (a_2 - b_2) \tag{22}$$

$$c_1' = \frac{l_1 - p_2'}{p_2'} = \frac{c_1 p_2}{p_2 - (a_2 - b_2)} \tag{23}$$

Similarly,

$$c_2' = \frac{c_2 p_1}{p_1 - (a_1 - b_1)} \tag{24}$$

$$h_1 = \frac{4}{3} p_2' \sqrt{c_1'} \tag{25}$$

$$h_2 = \frac{4}{3} p_1' \sqrt{c_2'} \tag{26}$$

$h_1 + h_2 = B = b_1 + b_2$

And also if both warp and weft threads are jammed, the relationship becomes

$$\sqrt{B^2 - \left(p_1'\right)^2} + \sqrt{B^2 - \left(p_2'\right)^2} = B \tag{27}$$

2.3 Prediction of fabric properties

Using the fabric parameters discussed in the previous section it is possible to calculate the *Fabric thickness, Fabric cover, Fabric mass* and *Fabric specific volume.*

2.3.1 Fabric thickness

Fabric thickness for a circular yarn cross-section is given by

$$h_1 + d_1 \text{ or } h_2 + d_2, \text{ whichever is greater.}$$

When the two threads project equally, then $h_1 + d_1 = h_2 + d_2$
In this case the fabric gives *minimum thickness* $= 1/2(h_1 + d_1 + h_2 + d_2) = D; h_1 = D - d_1$

Such a fabric produces a smooth surface and ensures uniform abrasive wear.

In a fabric with coarse and fine threads in the two directions and by stretching the fine thread straight, maximum crimp is obtained for the coarse thread. In this case the fabric gives *maximum thickness* as under;

$$\text{Maximum Thickness} = D + d_{coarse}, \text{ since } h_{coarse} = D$$

When yarn cross-section is flattened, the fabric thickness can be expressed as $h_1 + b_1$ or $h_2 + b_2$, whichever is greater

2.3.2 Fabric cover

In fabric, cover is considered as fraction of the total fabric area covered by the component yarns. For a circular cross-section cover factor is given as:

$$\frac{d}{p} = \frac{E\sqrt{T}}{280.2\sqrt{\varphi \rho_f}} = \frac{K}{28.02\sqrt{\varphi \rho_f}}$$

$$K = E\sqrt{T} \times 10^{-1} \; K \text{ is cover factor}$$

T is yarn tex, E is threads per cm $= 1/p$ suffix 1 and 2 will give warp and weft cover factors.

for $\dfrac{d}{p} = 1$, cover factor is maximum and given by,

$$K_{max} = 28.02\sqrt{\varphi \rho_f}$$

Fractional fabric cover is given by:

$$\frac{d_1}{p_1} + \frac{d_2}{p_2} - \frac{d_1 d_2}{p_1 p_2} = \frac{1}{28.02}\left[K_1 + K_2 - \frac{K_1 K_2}{28.02}\right]$$

Multiplying by 28.02 and taking $28.02 \approx 28$ we get fabric cover factor as under:

$$\text{Fabric cover factor} = K_1 + K_2 - \left(K_1 K_2 / 28\right) \tag{28}$$

For race track cross-section the equation will be

$$\frac{a}{p} = \frac{d}{ep\sqrt{1+\frac{4}{\pi}\left(\frac{1}{e}-1\right)} \times 28.02\sqrt{\varphi\rho_f}} \qquad \text{here } e = b/a$$

$$= \frac{K}{e\sqrt{1+\frac{4}{\pi}\left(\frac{1}{e}-1\right)} \times 28.02\sqrt{\varphi\rho_f}}$$

For elliptical cross-section the equation will be;

$$\frac{a}{p} = \frac{d}{ep} = \frac{E\sqrt{T}}{280.2\,e\,\sqrt{\varphi_{,f}}} = \frac{K}{28.02\,e\sqrt{\varphi_{,f}}} \qquad (29)$$

$$\text{Here } e = \sqrt{\frac{b}{a}} \text{ and } d = \sqrt{ab}$$

2.3.3 Fabric mass (Areal density)

$$\text{gsm} = [T_1 E_1(1+c_1)+T_2 E_2(1+c_2)] \times 10^{-1} \qquad (30)$$

$$\text{gsm} = \sqrt{T_1}\,[(1+c_1)\,K_1 + (1+c_2)\,K_2\beta] \qquad (31)$$

E_1, E_2 are ends and picks per cm.
T_1, T_2 are warp and weft yarn tex
Here K_1 and K_2 are the warp and weft cover factors, c is the fractional crimp and $d_2/d_1 = \beta$.

In practice the comparison between different fabrics is usually made in terms of gsm. The fabric engineer tries to optimize the fabric parameters for a given gsm. The relationship between the important fabric parameters such as cloth cover and areal density is warranted.

2.3.4 Fabric specific volume

The apparent specific volume of fabric, v_F is calculated by using the following formula:

$$v_F = \frac{\text{fabric thickness (cm)}}{\text{fabric mass (g/cm}^2)} \qquad (32)$$

Fabric mass (g/cm²) = 10^{-4} x gsm

Fabric packing factor, $\Phi = v_f/v_F$ $\qquad (33)$

Here v_f, v_F are respectively fiber and fabric specific volume.

A knowledge of fiber specific volume helps in calculating the packing of fibers in the fabric. Such studies are useful in evaluating the fabric properties such as warmth, permeability to air or liquid.

2.4 Maximum cover and its importance

Maximum cover in a jammed fabric is only possible by keeping the two consecutive yarns (say warp) in two planes so that their projections are touching each other and the cross thread (weft) interlaces between them. In this case the weft will be almost straight and maximum bending will be done by the warp.

$$d_1/p_1 = 1 \text{ will give } K_1 = K_{max}$$

and the spacing between the weft yarn, $p_2 = D \sin\theta_1 = D$ (for $\theta = 90^0$) , $p_2 = d_1 + d_2$

$$\frac{d_2}{p_2} = \frac{d_2}{d_1 + d_2} = \frac{2}{3} \quad \text{for } d_2 = 2d_1$$

This will give $K_2 = 2/3 \, K_{max}$,

If $d_1 = d_2$ then $d_1 = d_2/p_2 = 0.5$ and $K_2 = 0.5 \, K_{max}$

This is the logic for getting maximum cover in any fabric.

The principles are as under:

1. Use fine yarn in the direction where maximum cover is desired and keep them in two planes so that their projections touch each other and use coarse yarn in the cross direction.
2. As in (1) instead of coarse yarn insert two fine yarns in the same shed.

Both options will give maximum cover in warp and weft but first option will give more thickness than the second case.

The cover factor indicates the area covered by the projection of the thread. The ooziness of yarn, flattening in finishing and regularity further improves the cover of cloth. It also gives a basis of comparison of hardness, crimp, permeability, transparency. Higher cover factor can be obtained by the lateral compression of the threads. It is possible to get very high values only in one direction where threads have higher crimp. Fabrics differing in yarn counts and average yarn spacing can be compared based on the fabric cover. The degree of flattening for race track and elliptical cross-section can be estimated from fabric thickness measurements to evaluate b and a from microscopic measurement of the fabric surface.

The classical example in this case is that of a poplin cloth in which for warp threads

$p_1 = d_1$ and for $d_1 = d_2 = D/2$ and for jamming in both directions

$$p_1 = D \sin \theta_2$$

$$d = D/2 = D \sin \theta_2$$

$$\theta_2 = 30^\circ = 0.5236$$

$$\theta_1 = 82^0 18' 1.4364 (\text{using } \cos\theta_1 + \cos\theta_2 = 1)$$

$$p_2 = D\sin\theta_1 = 0.991D \approx 2p_1$$

$$l_1 = D\theta_1 = 1.14364$$

$$l_2 = D\theta_2 = 0.5236$$

$$c_1' = 0.45, \ c_2' = 0.0472$$

This is a specification of good quality poplin which has maximum cover and ends per cm is twice that of picks per cm.

3. Application of geometrical model

3.1 Computation of fabric parameters

The basic equations derived from the geometrical model are not easy to handle. Research workers (Nirwan & Sachdev, 2001; Weiner, 1971) obtained solutions in the form of graphs and tables. These are quite difficult to use in practice. It is possible to predict fabric parameters and their effect on the fabric properties by soft computing (Newton, 1995). This information is helpful in taking a decision regarding specific buyers need. A simplified algorithm is used to solve these equations and obtain relationships between useful fabric parameters such as thread spacing and crimp, fabric cover and crimp, warp and weft cover. Such relationships help in guiding the directions for moderating fabric parameters.

Peirce's geometrical relationships can be written as

$$\frac{p_2}{D} = (K_1 - \theta_1)\cos\theta_1 + \sin\theta_1 \tag{34}$$

$$\frac{h_1}{D} = (K_1 - \theta_1)\sin\theta_1 + (1 - \cos\theta_1) \tag{35}$$

Where $K_1 = l_1/D$ and two similar equations for the weft direction will be obtained by interchanging the suffix 1 with 2 and vice versa. The solution of p_2/D and h_1/D is obtained for different values of θ_1 (weave angle) ranging from 0.1– $\pi/2$ radians. Such a relationship is shown in figure 9.

It is a very useful relationship between fabric parameters for engineering desired fabric constructions. One can see its utility for the following three cases

1. Jammed structures
2. Non-jammed fabrics
3. Special case in which cross-threads are straight

3.1.1 Jammed structures

Figure 9 shows non linear relationship between the two fabric parameters p and h on the extreme left. In fact, this curve is for jamming in the warp direction. It can be seen that the jamming curve shows different values of p_2/D for increasing h_1/D, that is warp crimp. The theoretical range for p_2/D and h_1/D varies from 0-1. Interestingly this curve is a part of circle and its equation is:

$$\left(\frac{p_2}{D}\right)^2 + \left(\frac{h_1}{D} - 1\right)^2 = 1 \tag{36}$$

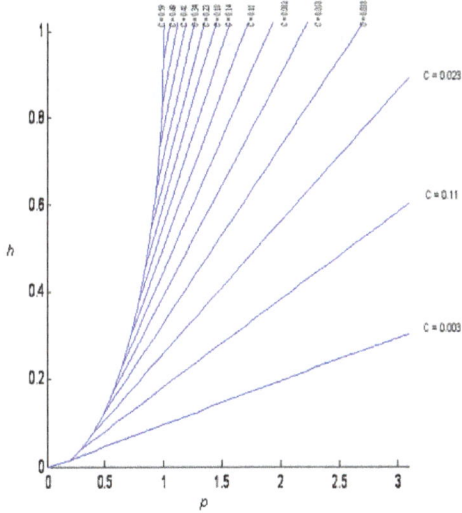

Fig. 9. Relation between thread spacing and crimp height

with centre at (0, 1) and radius equal to 1.

For jamming in the warp direction of the fabric the parameters p_2/D and corresponding h_1/D can be obtained either from this figure or from the above equation.

The relationship between the fabric parameters over the whole domain of structure being jammed in both directions can be obtained by using an algorithm involving equations from the previous section.

Another useful relationship between the crimps in the two directions is shown in figure 10. It indicates inverse non-linear relationship between c_1 and c_2. The intercepts on the X and Y axis gives maximum crimp values with zero crimp in the cross-direction. This is a fabric configuration in which cross-threads are straight and all the bending is being done by the intersecting threads.

Figure 11 shows the relation between h_1/p_2 and h_2/p_1. The figure shows inverse linearity between them except at the two extremes. This behavior is in fact a relationship between the square root of crimp in the two directions of the fabric.

Other practical relations are obtained between the warp and weft cover factor and between cloth cover factor and fabric mass (gsm).

Figure 12 gives the relation between warp and weft cover factor for different ratio of weft to warp yarn diameters (β). The relation between the cover factors in the two directions is sensitive only in a narrow range for all values of β. The relation between the cover factors in the two directions are inter- dependent for jammed structures. Maximum threads in the warp or weft direction depend on yarn count and weave. Maximum threads in one direction of the fabric will give unique maximum threads in the cross-direction. The change in the value of β causes a distinct shift in the curve. A comparatively coarse yarn in one direction with respect to the other direction helps in increasing the cover factor. For $\beta = 0.5$, the warp yarn is coarser than the weft, this increases the warp cover factor and decrease the weft cover factor. This is due to the coarse yarn bending less than the fine yarn. Similar effect can be noticed for β =2, in which the weft yarn is coarser than the warp yarn. These results are similar to earlier work reported by Newton (Newton, 1991 & 1995; Seyam, 2003).

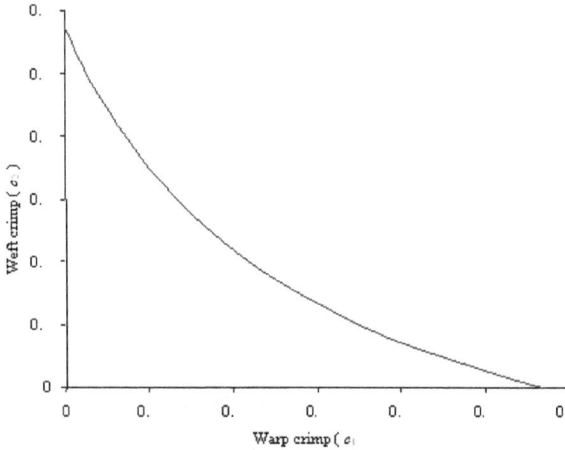

Fig. 10. Relation between warp and weft crimp for jammed fabric

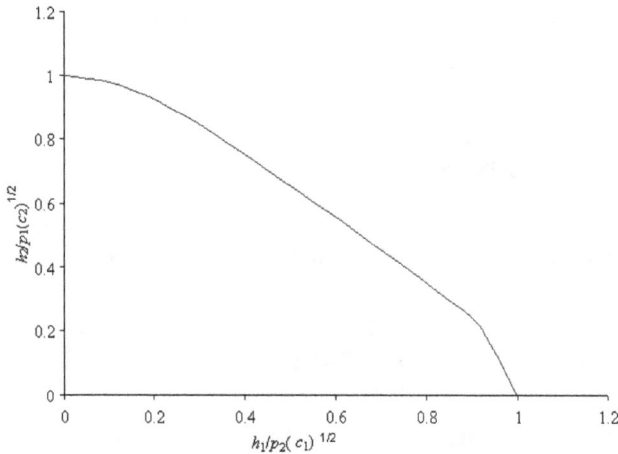

Fig. 11. Relation between warp and weft crimp in jammed fabric

The relation between fabric mass, (gsm) with the cloth cover (K_1+K_2) is positively linear (Singhal & Choudhury, 2008). The trend may appear to be self explanatory. Practically an increase in fabric mass and cloth cover factor for jammed fabrics can be achieved in several ways such as with zero crimp in the warp direction and maximum crimp in the weft direction; zero crimp in the weft direction and maximum crimp in the warp direction; equal or dissimilar crimp in both directions. This explanation can be understood by referring to the non-linear part of the curve in figure 11.

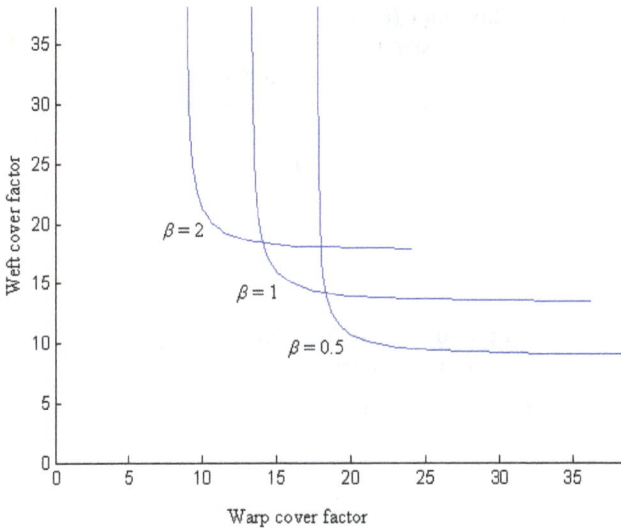

Fig. 12. Relation between warp and weft cover factor for different β in jammed fabric

3.1.2 Non-jammed structure

It can be seen that the relation between p_2/D and corresponding h_1/D is linear for different values of crimp. This relationship is useful for engineering non-jammed structures for a range of values of crimp. The fabric parameters can be calculated from the above non-jammed linear relation between p_2/D and h_1/D for any desired value of warp crimp. Then h_2/D can be obtained from $(1-h_1/D)$ and for this value of h_2/D one can obtain the corresponding value of p_1/D for the desired values of weft crimp. Thus all fabric parameters can be obtained for desired value of p_2/D, picks per cm, warp and weft yarn tex, warp and weft crimp. One can choose any other four parameters to get all fabric parameters.

3.1.3 Straight cross threads

The intersection of horizontal line corresponding to $h_1/D=1$ gives all possible structures ranging from relatively open to jammed configurations. In this case $h_2= 0$, $h_1= D$; This gives interesting structures which have stretch in one direction only, enabling maximum fabric thickness and also being able to use brittle yarns. The fabric designer gets the options to choose from the several possible fabric constructions. These options include jamming and other non jammed constructions. Using the above logic it is also possible to get fabric parameters for:

1. fabric jammed in both directions.
2. fabric with maximum crimp in one direction and cross-threads being straight.
3. fabric which is neither jammed nor has zero crimp in the cross-threads.

3.2 Weavability limit

The maximum number of ends and picks per unit length that can be woven with a given yarn and weave defines weavability limit (Hearle et al., 1969). This information is helpful to the weavers in avoiding attempts to weave impossible constructions thus saving time and money. It also helps to anticipate difficulty of weaving and take necessary preparations. (Dickson, 1954) demonstrated the usefulness of theoretical weavability limit and found agreement with the loom performance. Most of the work in this area was done using empirical relationships. The geometrical model is very useful in predicting this limit for a given warp, weft diameter (tex) and any weave. Maximum weavability limit is calculated in the model by using jamming conditions for plain and non-plain weaves for circular and race track cross-sections.

3.2.1 Yarn diameter

Two important geometrical parameters are needed for calculating weavability for a general case. These are yarn diameter and weave factor.

Yarn diameter in terms of linear density in tex for a general case is given as:

$$d = \frac{\sqrt{T}}{280.2\sqrt{\varphi_{\rho_f}}} \tag{37}$$

Where d = yarn diameter (cm), T = yarn linear density (tex, i.e. g/km),

ρ_f = fiber density(g/cm^3), ρ_y = yarn density(g/cm^3), Φ is yarn packing factor.

This equation for the yarn diameter is applicable for any yarn type and fiber type. The packing factor depends on fiber variables such as fiber crimp, length, tex and cross-section shape.

Table 2 and 3 give the fiber density and yarn packing factor for different fiber and yarn type respectively.

Acetate	1.32
Cotton	1.52
Lycra	1.20
Nylon 6	1.14
Nylon 66	1.13-1.14
Polyester	1.38
Polypropylene	0.91
Rayon	1.52
Wool	1.32

Table 2. Fiber density, g/cm^3

Ring-spun	0.60
Open-end-spun	0.55
Worsted	0.60
Woolen	0.55
Continuous-filament	0.65

Table 3. Yarn packing factor

For blended yarns, average fiber density is given by the following

$$\frac{1}{\rho} = \sum_{1}^{n} \frac{p_i}{p_{ft}}$$ (38)

where $\bar{\rho}$ = average fiber density ,

p_i = weight fraction of the ith component,

p_{ft} = fiber density of the ith component and

n = number of components of the blend

3.2.2 Effect of variation in beta (d_2/d_1) on the relation between warp and weft cover factor for jammed fabrics

An increase in the value of beta from 0.5-2 increases the range of warp cover factors but raises the level for the weft cover factor. This means with an increase in beta higher weft cover factors are achievable and vice-versa. However it may be noted that for cotton fibers having higher fiber density the sensitivity range between the warp and weft cover factor is relatively large compared to polypropylene fiber as shown in figure 13a and 13b. This shows a very important role played by fiber density in deciding warp and weft cover factors for the jammed fabrics.

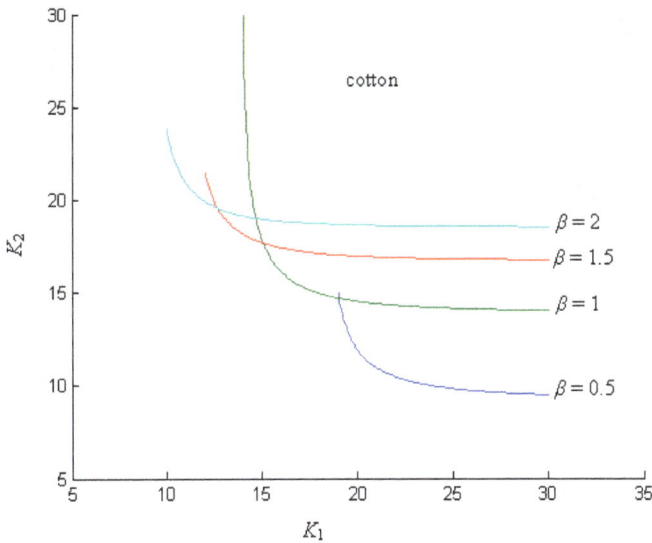

Fig. 13a. Effect of β on the relation between warp and weft cover factor

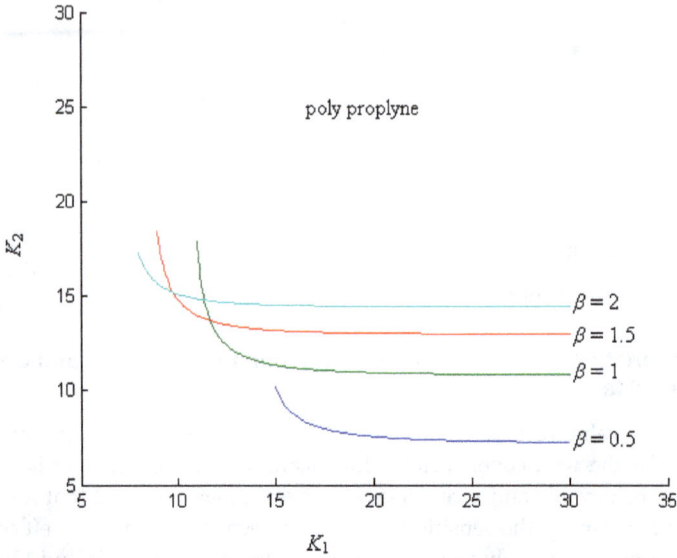

Fig. 13b. Effect of β on the relation between warp and weft cover factor

3.2.3 Equation for jammed structure for circular cross-section in terms of weave factor

Weave factor is useful in translating the effect of weave on the fabric properties. For circular cross-section the general equation for jammed cloth is desired.

Thread spacing P_{t1} for a non-plain weave per repeat is shown in figure 14 and is given as:

$$P_{t1} = I_2 p_1 + (E_1 - I_2) d_1 \qquad (39)$$

Average thread spacing $\overline{P_1} = \dfrac{I_2 p_1 + (E_1 - I_2) d_1}{E_1}$

That means, $\dfrac{E_1 \overline{P_1}}{I_2} = p_1 + \left(\dfrac{E_1}{I_2} - 1\right) d_1$

$$M_1 \overline{P_1} = p_1 + (M_1 - 1) d_1$$

$$\frac{p_1}{D} = M_1 \frac{\overline{P_1}}{D} - (M_1 - 1)\frac{d_1}{D}$$

$$\frac{p_1}{D} = M_1 \frac{\overline{P_1}}{D} - \frac{(M_1 - 1)}{1 + \beta} \qquad (40)$$

where $\beta = d_2 / d_1$

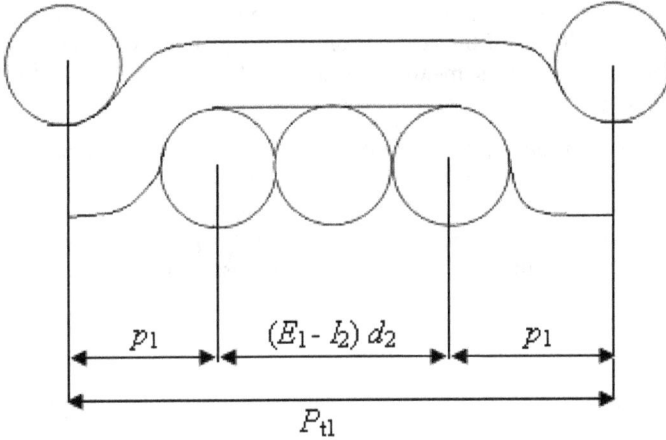

Fig. 14. Jammed structure for 1/3 weave (circular cross-section along warp)

Similarly, interchanging suffix 1, 2 we get

$$\frac{p_2}{D}= M_2\frac{\overline{p_2}}{D}-(M_2-1)\frac{d_2}{D}$$

$$\frac{p_2}{D}= M_2\frac{\overline{p_2}}{D}-(M_2-1)\frac{\beta}{1+\beta} \tag{41}$$

For a jammed fabric the following equation is valid:

$$\sqrt{1-\left(\frac{p_1}{D}\right)^2}+\sqrt{1-\left(\frac{p_2}{D}\right)^2}=1$$

$$\sqrt{1-\left(M_1\frac{\overline{p_1}}{D}-\frac{(M_1-1)}{1+\beta}\right)^2}+\sqrt{1-\left(M_2\frac{\overline{p_2}}{D}-\frac{(M_2-1)\beta}{1+\beta}\right)^2}=1$$

This equation can easily be transformed in terms of warp and weft cover factor (K_1 and K_2)

$$\sqrt{1-\left[\left(\frac{28.02\sqrt{\varphi\rho_f}M_1}{K_1}\right)-(M_1-1)\right)\frac{1}{1+\beta}\right]^2}+\sqrt{1-\left[\left(\frac{28.02\sqrt{\varphi\rho_f}M_2}{K_2}\right)-(M_2-1)\right)\frac{\beta}{1+\beta}\right]^2}=1 \tag{42}$$

3.2.4 Relation between fabric parameters for circular cross-section for different weaves

The effect of weave in the jammed structures is examined using the above equations for plain, twill, basket and satin weave. M, the weave factor value (average float length) for these weaves are 1, 1.5, 2 and 2.5 respectively for all the discussion which follows.

The relation between p_{1avg}/D and p_{2avg}/D is established and it is seen that with the increase in float length, the sensitivity of the curve decreases in general. Also the range of p_1/D and p_2/D values gets reduced. This means a weave with longer float length decreases the flexibility for making structures.

Figure 15 shows the relationship between the warp and weft cover factor for circular cross-section. It is interesting to note that the behavior is similar for different weaves. However with the increase in float, the curve shifts towards higher values of weft cover factor. It should be borne in mind that the behavior shown in this figure is for virtual fabrics. In real fabrics jammed structure is unlikely to retain circular cross-section.

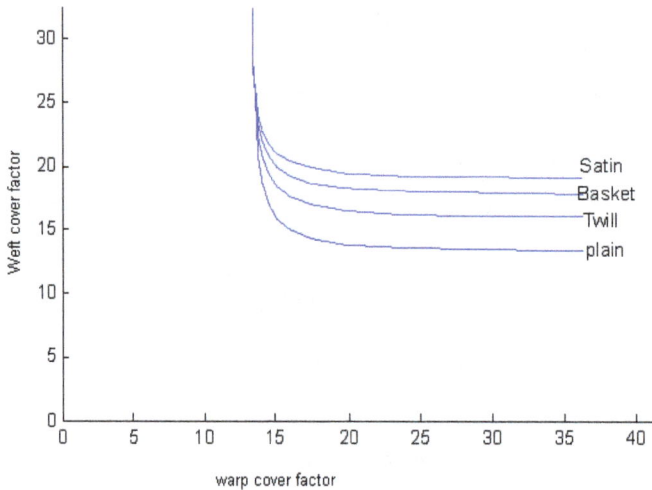

Fig. 15. Relation between warp and weft cover factor for jammed fabric (circular cross-section)

3.2.5 Equation for jammed structure for a race track cross-section in terms of weave factor

In jammed fabrics, the yarn cross-section cannot remain circular. The cross-section will change. It is easy to modify the geometry for circular cross-section by considering race track cross-section. Figure 16 shows the configuration of jammed structure for 1/3 weave for race track cross-section along weft direction of the fabric.

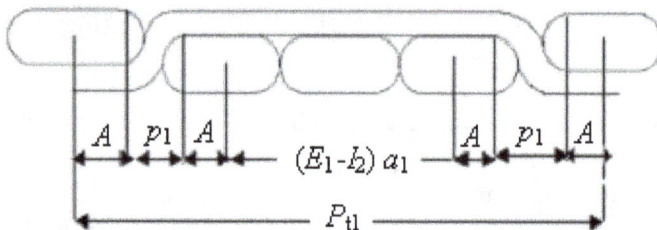

Fig. 16. Jammed structure for 1/3 weave (race track cross-section along warp)

Here,
$$A = \left(\frac{a_1 - b_1}{2} \right)$$

Thread spacing P_{t1} for a non-plain weave per repeat is given as:

$$P_{t1} = I_2 p_1 + (E_1 - I_2) a_1 + 4 \left(\frac{a_1 - b_1}{2} \right) \qquad (43)$$

Simillarly,

$$P_{t2} = I_1 p_2 + (E_2 - I_1) a_2 + 4 \left(\frac{a_2 - b_2}{2} \right) \qquad (44)$$

Where, p_1 and p_2 are horizontal spacing between the semi-circular threads in the intersection zone. Here, a and b are the major and minor diameters of race track cross-section.

The average thread spacing

$$\overline{P_1} = \frac{p_1}{M_1} + \left(1 - \frac{1}{M_1} \right) a_1 + \frac{4}{M_1 I_2} \left(\frac{a_1 - b_1}{2} \right) \qquad (45)$$

Similarly,

$$\overline{P_2} = \frac{p_2}{M_2} + \left(1 - \frac{1}{M_2} \right) a_2 + \frac{4}{M_2 I_1} \left(\frac{a_2 - b_2}{2} \right) \qquad (46)$$

As such analysis of circular thread geometry can be applied for the intersection zone of the race track cross-section.

$$L_1 = 4 \left(\frac{a_2 - b_2}{2} \right) + (E_2 - I_1) a_2 + I_1 \times l_1 \qquad (47)$$

Total warp crimp in the fabric is given by:

$$C_1 = \frac{L_1}{P_{t2}} - 1$$

p_1 and p_2 can be calculated from the jamming considerations of the circular thread geometry using:

$$\sqrt{1 - \left(\frac{p_1}{B} \right)^2} + \sqrt{1 - \left(\frac{p_2}{B} \right)^2} = 1$$

It should be remembered that p/B corresponds to the semi-circular region of the race track cross-section and is similar to p/D for circular cross-section. As such the values of p/D ratio can be used for p/B

$$\sqrt{1-\left(M_1\overline{\frac{P_1}{B}}\frac{4}{BI_2}\left(\frac{a_1-b_1}{2}\right)-(M_1-1)\frac{a_1}{B}\right)^2}+\sqrt{1-\left(M_2\overline{\frac{P_2}{B}}-\frac{4}{BI_1}\left(\frac{a_2-b_2}{2}\right)-(M_2-1)\frac{a_2}{B}\right)^2}=1$$

This equation can be simplified to the following usable forms.

$$\sqrt{1-\left(M_1\overline{\frac{P_1}{B}}-\frac{2(1-e)}{e(1+\beta)I_2}-\frac{(M_1-1)}{e(1+\beta)}\right)^2}+\sqrt{1-\left(M_2\overline{\frac{P_2}{B}}-\frac{2(1-e)\beta}{e(1+\beta)I_1}-\frac{(M_2-1)}{e(1+\beta)}\right)^2}=1$$

It is assumed that $e_1 = e_2 = e$

where $e = b/a$

The above equation can easily be transformed in terms of warp and weft cover factor as under:

$$\sqrt{1-\left(\frac{28.02\sqrt{\varphi\rho_f}M_1}{(1+\beta)K_1}\sqrt{1+\frac{4}{\pi}\left(\frac{1}{e}-1\right)}-\frac{2(1-e)}{e(1+\beta)I}-\frac{(M_1-1)}{e(1+\beta)}\right)^2}$$

$$+\sqrt{1-\left(\frac{28.02\sqrt{\varphi\rho_f}M_2\beta}{(1+\beta)K_2}\sqrt{1+\frac{4}{\pi}\left(\frac{1}{e}-1\right)}-\frac{2(1-e)\beta}{e(1+\beta)I}-\frac{(M_2-1)\beta}{e(1+\beta)}\right)^2}=1 \qquad (48)$$

3.2.6 Relationship between fabric parameters in race track cross-section

The relationship between fabric parameters such as p_2 and p_1, p_1 and c_2 for the race track cross-section in jammed condition is discussed below.

The parameters are similar to that for the circular cross-section but it shifts towards higher values of thread spacing.

Figure 17 shows the relationship between warp and weft cover factors for different weaves. As discussed above in real fabrics the weaves show distinct differences between them unlike in circular cross-section. Increase in float length decreases the scope of cover factors.

From these equations crimp and fabric cover can be evaluated using the above two equations along with:

$$d^2=b^2\left[\frac{\pi}{4}+\left(\frac{1}{e}-1\right)\right] \text{ and } \frac{b_2}{b_1}=\beta \qquad (49)$$

3.3 Square cloth

A truly square fabric has equal diameter, spacing and crimp.

$$p_1=p_2,\ c_1=c_2,\ d_1=d_2,\ h_1=h_2=D/2,\ \theta_1=\theta_2$$

From the basic equations of the geometrical model from the previous section we have:

$$\tan\theta/2 = \frac{2}{3}\left[\frac{p}{D} - \sqrt{\left(\frac{p}{D}\right)^2 - 0.75}\right] \tag{50}$$

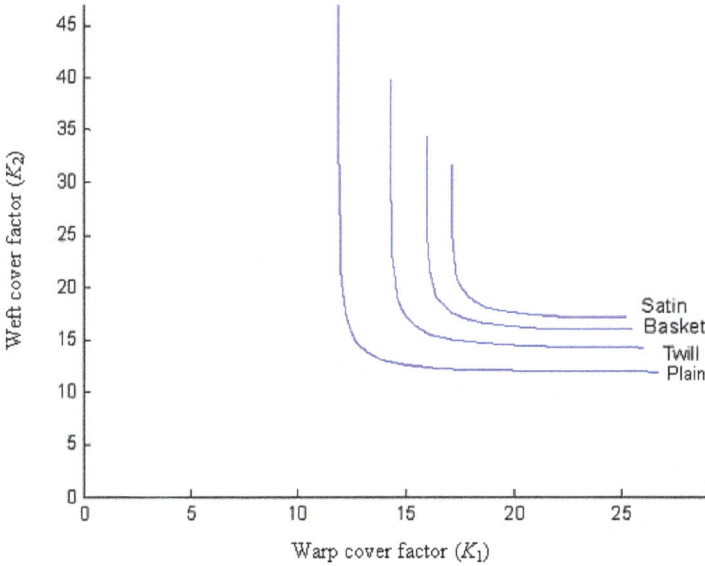

Fig. 17. Relation between warp and weft cover factor for jammed fabric (race track cross-section)

This is valid for all values of (p/D) $2 \geq 0.75$ or $p/D \geq 0.866$

$$p/d \geq 1.732;\ d/p \leq 0.5773$$

Also $D = 2d = h_1 + h_2 = 2 \times (4/3) p\sqrt{c}$

$$\sqrt{c} = \frac{3}{4}\frac{d}{p} = \frac{0.75}{p} \times \frac{\sqrt{Tex}}{280.2\sqrt{\varphi\rho_f}}$$

$$c = \left(\frac{0.02677K}{\sqrt{\varphi\rho_f}}\right)^2 \tag{51}$$

Crimp in % can be calculated from, $\%c = \left(\dfrac{K}{3.57}\right)^2$

For jammed square cloth

$$\cos\theta_1 + \cos\theta_2 = 1 \text{ will give}$$

$$\cos\theta = \tfrac{1}{2} \text{ and } \theta = 60^0$$

$$p = 2d\sin\theta, \; 1 = D\theta = 2d\frac{\theta}{3} \quad \therefore \frac{1}{d} = 2\frac{\theta}{3}$$

$$\frac{p}{d} = \sqrt{3} = 1.732$$

$$\Rightarrow \frac{p}{d} = 0.5773$$

$$\text{crimp} = \frac{1/d}{p/d} - 1 = 0.2092 = 20.9\%$$

Therefore complete cover is not possible with square cloth.

4. Crimp in the fabric

The crimp in fabric is the most important parameter which influences several fabric properties such as extensibility, thickness, compressibility and handle. It also decides quantity of yarn required to weave a fabric during manufacturing. Therefore control of crimp is vital for geometrical analysis of fabric structure.

4.1 Crimp interchange equation

Normally crimp interchange equation is used to predict the change in crimp in the fabric when it is extended in any direction by keeping the ratio of modular length to the sum of thread diameter (l_1/D and l_2/D)constant. An attempt is made by soft computing to exploit the crimp interchange equation in a different way instead of keeping the usual three invariants l_1, l_2 and D and the relationship between warp crimp (C_1) and weft crimp (C_2) is determined by varying l_1/D and l_2/D. Such a strategy enables bias of crimp in a preferred direction. This is a new concept and entirely a different use of crimp interchange equation.

Following equation gives a useful relationship between the two directions of the fabric.

$$D = h_1 + h_2 = h_1' + h_2'$$

Superscript represents changes in the fabric parameter after modification

$$D = h_1' + h_2' = \frac{4}{3}\left[p_2'\sqrt{c_1'} + p_1'\sqrt{c_2'} \right]$$

$$D = \frac{4}{3}\left[\frac{l_1\sqrt{c_1'}}{1+c_1'} + \frac{l_2\sqrt{c_2'}}{1+c_2'} \right]$$

$$\frac{l_1}{D}\frac{\sqrt{c_1'}}{(1+c_1')} + \frac{l_2}{D}\frac{\sqrt{c_2'}}{(1+c_2')} = \frac{3}{4} \tag{52}$$

The above equation is called crimp interchange equation. It gives the relationship between the warp and weft crimp for the new configuration after the application of stretch in warp/weft direction. It may be noted that the parameters l_1, l_2 and D are invariant; they have the same value in the original fabric and in the new configuration. This basically means it is assumed that the geometry in deformed fabric is same as in undeformed fabric.

In the crimp interchange equation one of the parameter c_1' or c_2' is determined based on the requirement of modification and the other parameter is calculated.

The most general manner of solving crimp interchange problems is getting relation between $\dfrac{\sqrt{C_1}}{1+C_1}$ and $\dfrac{\sqrt{C_2}}{1+C_2}$ for constant l_1/D and l_2/D.

4.2 Crimp balance equation

Textile yarns are not flexible as assumed in Peirce's geometrical model. They offer resistance to bending. The elastica model demonstrates the existence of inter yarn force at the crossover points during fabric formation. The crimp balance equation is an offshoot of this analysis. It shows the importance of bending rigidity of warp and weft yarns in influencing the ratio of crimp in both warp and weft directions.

The analysis using the rigid thread thread model [R] gives the value of inter yarn force

$$V = 16\,M\,\sin\theta/p^2$$

The balance of inter yarn force in two direction gives

$$V_1 = V_2$$

$$M_1 \sin\theta_1/\ p_2^2 = M_2 \sin\theta_2\ p_1^2$$

Since $\sin\theta\ ;\ \sqrt{C}$

$$\frac{\sqrt{C_1}}{\sqrt{C_2}} = \frac{M_2}{M_1}\left(\frac{p_2}{p_1}\right)^2 \tag{53}$$

$$\frac{C_1}{C_2} = \left(\frac{1+C_2}{1+C_1}\right)^4 \left(\frac{M_2}{M_1}\right)^2 \left(\frac{l_1/D}{l_2/D}\right)^4 \tag{54}$$

The solution for C_1 and C_2 for this equation is obtained in terms of M_2/M_1 and l_1/D and L_2/D using special algorithm in MATLAB.

4.3 Interaction of crimp interchange and crimp balance equations

The interaction of crimp interchange and crimp balance equations for given values of l_1/D, l_2/D and M_2/M_1 (ratio of bending moment of warp and weft) gives desired C_1 and C_2. It is impossible to solve these equations mathematically however soft computing facilitates solutions using iterations. It is the aim of this paper to facilitate fabric engineer in determining the fabric parameters for a given value of warp and weft crimp. This approach gives another alternative to engineer fabrics. The important variables of crimp balance equations are M_2/M_1, $l1/D$ and l_2/D. For a given crimp interchange equation in terms of $l1/D$ and l_2/D, the crimp balance equation gives intersections. The scales are also calibrated in terms of crimp.

Figures 18, 19 and 20 show the interaction of crimp interchange and crimp balance equation corresponding to $l_1/D = l_2/D$, $l1/D > l_2/D$ and $l_1/D < l_2/D$ respectively. It is interesting to note that in all these curves with the increase in M_2/M_1, warp crimp increases and weft decreases. Another interesting result can be seen from these figures when $l1/D$ not equal to l_2/D. $l_1/D > l_2/D$ or $l_1/D < l_2/D$ causes a reduction in a range and shift towards lower values for both C1 and C2

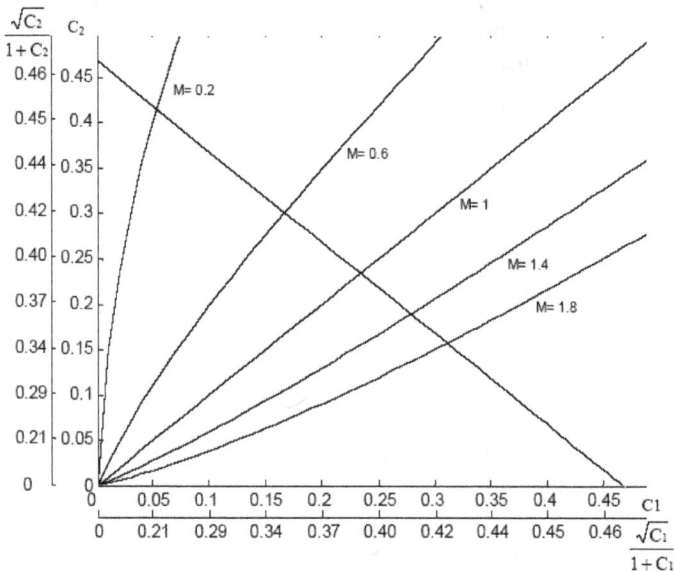

Fig. 18. Interaction of crimp interchange and crimp balance equations ($l_1/D = l_2/D$)

These three curves show very interesting ways in which the values of crimp in warp and weft can be varied in a wide range. Therefore the three parameters M_2/M_1, l_1/D and l_2/D can influence the crimp in warp and weft in a wide range and this is what gives maneuverability to the fabric designer.

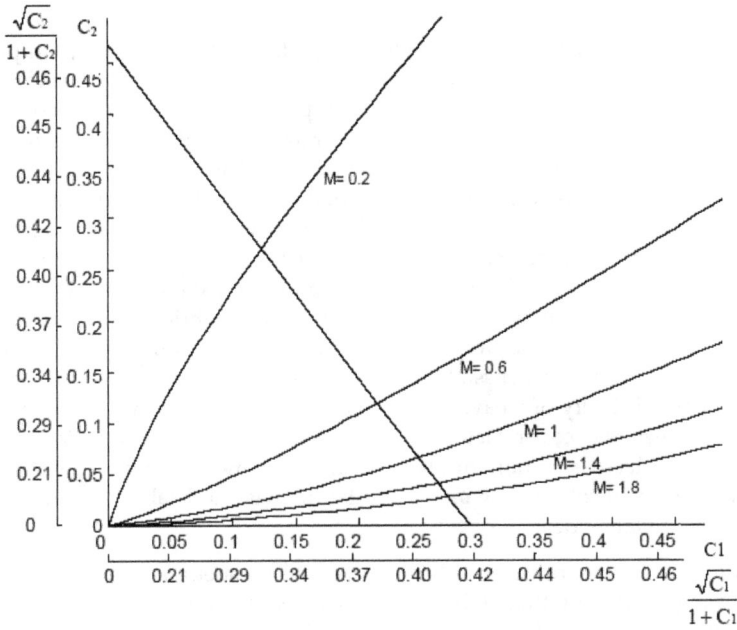

Fig. 19. Interaction of crimp interchange and crimp balance equations ($l_1/D > l_2/D$)

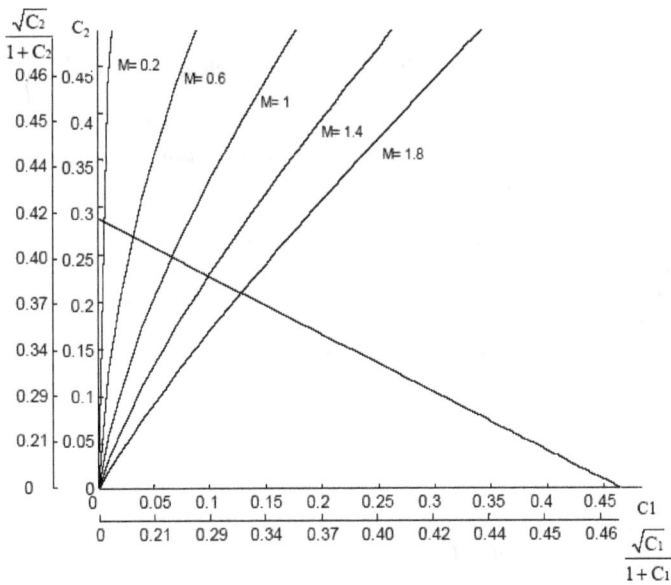

Fig. 20. Interaction of crimp interchange and crimp balance equations ($l_1/D < l_2/D$)

5. Conclusion

An attempt has been made to optimize engineering attributes of plain and non-plain weave fabrics as per requirement. Soft computing is used to solve fabric geometrical model equations and relationships between useful fabric parameters such as thread spacing and crimp, fabric cover and crimp, warp and weft cover are obtained. Such relationships help in guiding the direction for moderating fabric parameters. The full potential of Peirce fabric geometrical model for plain weave has been exploited by soft computing and the same is extended for non-plain constructions. The inter-relationships between different fabric parameters for jammed structures, non jammed structures and special case in which cross threads are straight are obtained using suitable computing techniques. It is hoped that the fabric designer will be benefited by the flexibility to choose fabric parameters for achieving any end use with desired fabric properties. This information is helpful to the weavers in avoiding attempts to weave impossible constructions thus saving time and money. It also helps to anticipate difficulty of weaving and take necessary steps in warp preparations. The relationship between the cover factors in warp and weft direction is demonstrated for circular and racetrack cross-section for plain, twill, basket and satin weave. Non plain weave fabric affords further flexibility for increasing fabric mass and fabric cover. As such they enlarge scope of the fabric designer.

Soft computing can successfully provide a platform to manoeuvre crimp in warp and weft over a wide range with only three fabric parameters; yarn tex, modular length of warp and modular length of weft yarn. This has enabled solutions by interaction of crimp interchange and crimp balance equations. This exercise offers several solutions for fabric engineering by varying the above three parameters.

6. References

Dickson, J. B. (1954). *Practical Loom Experience on Weavability Limits, Textile Research Journal*, Vol. 24, No. 12, 1083-1093.
Grosicki, Z. (1988). *Watson's Textile Design and Colour*, Newnes Butterworths.
Hearle, J. W. S.; Grosberg, P. and Backer, S. (1969). *Structural Mechanics of Fibers, Yarns and Fabrics*, Wiley Interscience.
Kemp, A. (1958). *Journal of the Textile Institute*, 49, T 44.
Love, L. (1954). *Graphical Relationships in Cloth Geometry for Plain, Twill, and Sateen Weaves, Textile Research Journal*, Vol. 24, No. 12, 1073-1083.
Newton, A. (1993). *Fabric Manufacture: A Hand book, Intermediate Technology Publications*, London.
Newton, A. (1991). *Tightness comparison of woven fabrics, Indian Textile Journal*, 101, 38-40.
Newton, A. (1995). The comparison of woven fabrics by reference to their tightness, J.Text. Inst., 86,232-240.
Nirwan, S. and Sachdev, S. (2001). B. Tech. Thesis, *I.I.T. Delhi*
Peirce, F. T. (1937). *Journal of the Textile Institute*, 28, T45-112.
Robinson, A. T. C. and Marks, R. (1973). *Woven Cloth Construction, The Textile Institute*.
Seyam, A. M. (2002). *Textile Progress, The Textile Institute*, Vol. 31, No. 3.
Seyam, A. M. (2003). *The Structural Design of Woven Fabrics: Theory and Practice, Textile Progress*, Vol.31, No. 3.
Singhal and Choudhury (2008). B. Tech. Thesis, I.I.T. Delhi.
Weiner, L. (1971). *Textile Fabric Design Tables, Technomic*, Stanford, USA.

2

3D Woven Fabrics

Pelin Gurkan Unal
Namık Kemal University Department of Textile Engineering
Turkey

1. Introduction

Composite material, also called composite, a solid material that results when two or more different substances, each with its own characteristics, are combined to create a new substance whose properties are superior to those of the original components in a specific application (Encyclopædia Britannica, 2011). Because the objective in manufacturing component is to produce a stiff and a strong material with a low density, these materials have found place in many application fields such as land transportation, marine, construction, aerospace and medical. There are two constituents in a composite material which are the reinforcement and matrix phases. From the matrix point of view, composites can be divided into three main categories; polymer, metal and ceramic matrix composites (Zhang, 2003). Based on the reinforcement mechanism, composites can be separated as particle reinforced (large particle, dispersion strengthened), fibre reinforced (continuous (aligned), and short fibres (aligned or random)) and structural composites (laminates, sandwich). Fibre reinforced composites have been first introduced in exterior parts of Corvette in 1953. Today, fibre reinforced composites are used in many application fields.

Textile structural composites usually consist of stacked layers known as 2D laminates, exhibit better in-plane strength and stiffness properties compared to those of metals and ceramics (Bilisik, 2010, 2011; Mohamed&Bogdanovich, 2009). However, the application of 2D laminates in some critical structures in aircraft and automobiles has also been restricted by their inferior impact damage resistance and low through thickness mechanical properties when compared against the traditional aerospace and automotive materials such as aluminium alloys and steel (Mouritz et al., 1999). These structures have low out-of plane properties because of the lack of third direction reinforcements which will result in low delamination resistances (Chou, 1992). In order to improve interlaminar properties of the 2D laminates, three dimensional (3D) textile preforms have been developed by using different manufacturing techniques like weaving, knitting, braiding, stitching, and non-woven manufacturing. Among these manufacturing techniques, sewing and 3D weaving are the promising technologies which address the shortcomings of the stack-reinforced composites (Padaki et al., 2010). Since manufacturing technology has a direct effect on the fibre orientation and fibre volume fraction of the preform, the properties of the end product will vary depending on the production and end-use requirements (Peters, 1998).

Although 3D woven preforms have been used for approximately forty years in different application fields, there is not a common understanding and definition of these fabrics which make them difficult to comprehend. Therefore, this chapter attempts to make a detailed overview of 3D woven fabrics, basic structure of 3D woven fabrics, definitions and classifications of 3D woven fabrics in comparison with 2D woven fabrics.

2. Definition, classification and weave structures of 3D woven fabrics

2.1 Definition of 3D woven fabrics

A basic common definition of 3D fabric is that these types of fabrics have a third dimension in the thickness layer. In 3D-fabric structures, the thickness or Z-direction dimension is considerable relative to X and Y dimensions. Fibres or yarns are intertwined, interlaced or intermeshed in the X (longitudinal), Y (cross), and Z (vertical) directions (Badawi, 2007).

3D fabrics can also be defined as "a single-fabric system, the constituent yarns of which are supposedly disposed in a three mutually perpendicular plane relationship" (Behera&Mishra, 2008). According to Chen, structures that have substantial dimension in the thickness direction formed by layers of fabrics or yarns, generally termed as the three-dimensional (3D) fabrics (Chen, 2011). Although all textiles have a 3D internal structure, macroscopically most can be regarded as thin 2D sheets. By 3D fabrics, (1) thick multilayer fabrics in a simple regular form or (2) made in more complicated 3D shapes, (3) hollow multilayer fabrics containing voids and (4) thin 3D shells in complex shapes are meant (Hearle&Chen, 2009). Khokar defined 3D woven fabrics as a fabric, the constituent yarns of which are supposed to be disposed in a three-mutually-perpendicular-planes relationship (Khokar, 2001).

2.2 Classification of 3D woven fabrics

When the classification of 3D woven fabrics are examined, it is observed that there are several classifications based on the shedding mechanisms, weaving process, geometries and configurations, interlacements and fibre axis according to the different researches (Khokar,1996 as cited in Soden&Hill, 1998; Chen, 2010; Soden&Hill, 1998;Bilisik, 2011).

Khokar classified the 3D fabrics as follows (Khokar, 1996 as cited in Soden&Hill, 1998);

1. The conventional 2D weaving process designed to interlace two orthogonal sets of threads (warp and weft). This produces *an interlaced 2D fabric* on a 2D weaving device.
2. The conventional 2D weaving process designed to interlace two orthogonal sets of yarns (warp and weft) with an additional set of yarns functioning as binder warps or interlacer yarns in the through-the-thickness or Z direction. This is referred to as multilayer weaving and produces *an interlaced 3D fabric* constituting two sets of yarns on a 2D weaving device.
3. The conventional 2D weaving process using three sets of yarns (ground warp, pile warp and pile weft) to produce pile fabrics, known as *2.5D fabrics*.
4. The conventional 2D weaving process using three sets of yarns to produce a non-interlaced fabric with yarns in the warp, weft and through-the-thickness directions. This produces *a non-interlaced 3D fabric* with three sets of yarns on a 2D weaving device.
5. The 3D weaving process designed to interlace three orthogonal sets of yarns. The weaving shed operates both row-wise and column-wise. This produces *a fully interlaced 3D fabric* where all three sets of orthogonal yarns interlace on a specifically designed 3D weaving machine.

6.   A non-woven, non-interlaced 3D fabric forming process designed to connect three orthogonal sets of yarns together with no interlacing (weaving), interloping (knitting), or intertwining (braiding). The fabric is held together by a special binding process.

However Soden and Hill added a new category as 4A to Khokar's classification for the fabrics that could be placed between categories 4 and 5 where the conventional 2D weaving process uses three sets of yarns to produce *an interlaced 3D fabric* with yarns in the warp, weft and through-the-thickness directions (Soden&Hill, 1998).

Regardless of the types of machines used, weaving technology is capable of constructing 3D fabrics with many different geometrical shapes. Chen studied the configurations and geometries of the 3D woven fabrics and classified 3D woven fabrics into four different categories, as listed in Table 1 (Chen, 2007 as cited in Chen et al., 2011).

Structure	Architecture	Shape
Solid	Multilayer Orthogonal Angle Interlock	Compound structure, with regular or tapered geometry
Hollow	Multilayer	Uneven surfaces, even surfaces, and tunnels on different level in multi-directions
Shell	Single layer Multilayer	Spherical shells and open box shells
Nodal	Multilayer Orthogonal Angle Interlock	Tubular nodes and solid nodes

Table 1. 3D textile structures and weave architectures (Chen, 2011).

2.3 Weave structures and properties

In 3D woven fabrics, generally multilayer, angle interlock and orthogonal weave architectures are the most widely used weave structures. While multilayer and angle interlock weave structures can be produced with conventional 2D weaving machines especially with shuttle looms, orthogonal weave architecture needs a special designed 3D weaving machine to be produced. Orthogonal weave structures consist of three sets of yarns that are perpendicular to each other (X, Y and Z coordinates). In this particular 3D woven fabric formation process, Z yarns interconnect all individual warp- and fill-directional yarns and thus solidify the fabric (Bogdanovich, 2007). Mechanical and structural properties of the composites having orthogonal weave architectures with various binding weaves and different numbers of layers were investigated (Chen&Zanini, 1997 as cited in Behera&Mishra, 2008). The results of the investigation are as follows:

•   Since straight yarns exist in the orthogonal structures, tensile stiffness and strength properties of these structures are well regardless of type of binding weave. However, tensile stiffness and strength values of these weave structures are directly proportional with the number of layers.

- Number of layers and binding weaves do not affect the breaking elongation of the orthogonal structures. This property is mostly dependent on the elongation properties of the yarns used.
- The shear rigidity and shearing hysteresis increases when more layers are involved.
- Tighter binding weaves and more layers of the orthogonal structure will produce higher bending stiffness and bending hysteresis.

Multilayer weave structures consist of multiple layers each of which have its own sets of warp and weft yarns. The connection of the layers is done by self-stitching (existing yarns) or central stitching (external sets of yarns). In angle interlock weave structures; there are at least two sets of yarns such as warp and weft. In some cases, in order to increase fibre volume fraction and in-plane strength, stuffer yarns can be added. Angle interlock weaves are divided into two groups; through thickness angle interlock and layer to layer angle interlock weaves. In through thickness angle interlock weaves, warp yarn travels from one surface of the 3D fabric to the other holding all the layers together, whereas in layer to layer angle interlock weaves warp yarn travels from one layer to the adjacent layer and back. A set of warp weaves together hold all the layers of the weave structure. Mechanical properties of 3D angle interlock woven composites are not as good as the mechanical properties of corresponding laminated composites. However, composites having angle interlock woven structure have advantages of enhanced delamination resistance, impact/fracture resistance, damage tolerance and dimensional stability (Naik et al., 2002). Higher through-the-thickness elastic and strength properties can be achieved by using 3D orthogonal interlock woven composites (Naik et al., 2001).

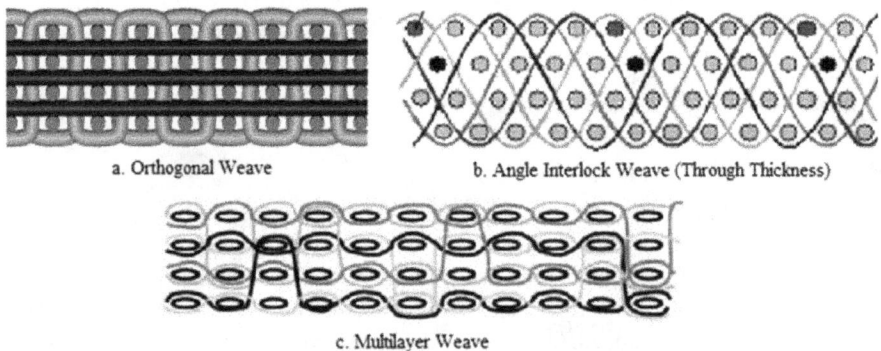

a. Orthogonal Weave   b. Angle Interlock Weave (Through Thickness)

c. Multilayer Weave

Fig. 1. Weave structures (Stig, 2009; Badawi, 2007;Chen, 2011).

Composites having different weave architectures will have different mechanical properties as well as structural stability. Chen et al investigated the mechanical and structural properties of the composites produced from 3D woven preforms having warp self-stitched multilayer weave and angle interlock weave (Chen et al., 1999). The results of the investigation are as follows:

- Increasing number of layers in multilayer weave structures results in a stronger structure. As the multilayer fabrics are warp self-stitched, the strength increase in weft direction is more significant than in the warp direction.

- The structural stability of multilayer structures increases when more layers are involved.
- For multilayer structures, the effect of weave combination is little on strength properties whereas it plays a significant role on structural stability.
- The increase in stitch density in multi-layer structures will generally reduce the strength of the structures, but its effect on the structural stability is not clear. The latter may be related to the distribution of the stitches and also to the weaving conditions.
- The number of layers of weft threads in angle-interlock structures mainly increases the tensile strength in the weft direction because of the construction.
- Angle-interlock structures permit more elongation in the warp direction than in the weft.
- The increase in the number of layers in angle-interlock structures makes the structures more difficult to bend; this is more significantly so in the weft direction than in the warp direction. However, the increase in the number of layers showed little influence on the shear rigidity.

3. Weaving

Weaving is an ancient tradition which dates back over seven millenniums. In traditional weaving, there are two sets of yarns, perpendicular to each other, interlace to form a woven fabric. While one set of the yarns that run lengthwise along the weaving machine direction are called warp, the other set of the yarns that run transversely from one side to the other side of the machine are called weft (a.k.a. filling).

There are three basic motions in order to produce a fabric by ensuring the interlacing between warp and weft yarns. These three essential motions are; shedding, weft insertion and beat-up. For the continuation of weaving process, warp yarns have to be let-off and the produced fabric has to be taken-up. These necessary two motions are auxiliary movements that are warp let-off and fabric take-up.

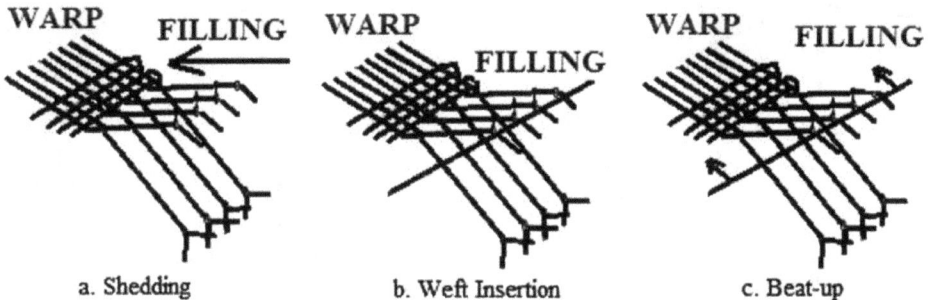

a. Shedding   b. Weft Insertion   c. Beat-up

Fig. 2. Basic motions on weaving machine (Lord&Mohamed, 1982).

In order to get different weave structures in traditional weaving, the movement of the warp yarns have to be controlled and changed before each weft insertion. To perform warp yarns movement on a loom, warps that follow the same interlacing pattern have to be grouped with the same frame called harness. In each harness, there are heddles that have an eye in the middle in which the warp yarns pass through. By lifting the harness up or down, the groups of warp yarns will move either upwards or downwards. Based on the pattern, there must be different harnesses for each group of warp yarns. In the case where each warp weaves a different pattern, a harness cord is provided for each heddle. There are four different shedding mechanisms to manipulate warp yarns; crank, cam, dobby and jacquard. Crank, cam or dobby mechanisms work together with harnesses. On the other hand in jacquard mechanism there are harness cords for each warp yarn, no harnesses. Thus patterning capability of jacquard mechanism is the highest among shedding mechanisms. By altering shedding motion, various weave structures such as plain, twill and satin can be produced.

When the shedding is opened based on the patterning, weft insertion takes place. Considering the weft insertion system of the weaving machine, weft insertion can be performed in different ways. In single-phase weaving machines, the weft yarn is carried from one side of the machine to the other side transversely by shuttle, projectile, rapier or jet systems. After the weft insertion, the reed beats up the last inserted weft and the produced fabric is taken up. For the next cycle, warp yarns have to be let off in order to open a shed. This cycle of operations are continued repeatedly to obtain the woven fabric in a sequence. This weaving process is called 2D weaving. Even this fabric production process is two dimensional (warp yarns are moved through fabric thickness and weft yarn is inserted at the open shed, consequently two orthogonal sets of yarns are interlaced), it is also possible to weave 3D woven fabrics known as multilayer structures. However, producing 3D woven fabrics with conventional weaving (2D weaving) does not mean that the process can be named 3D weaving. Because, the arrangement of the weaving motions unchanged whether a single warp sheet is used to produce sheet-like 2D fabrics or multiple warp sheets are used to produce multilayer 3D fabrics (Stig, 2009).

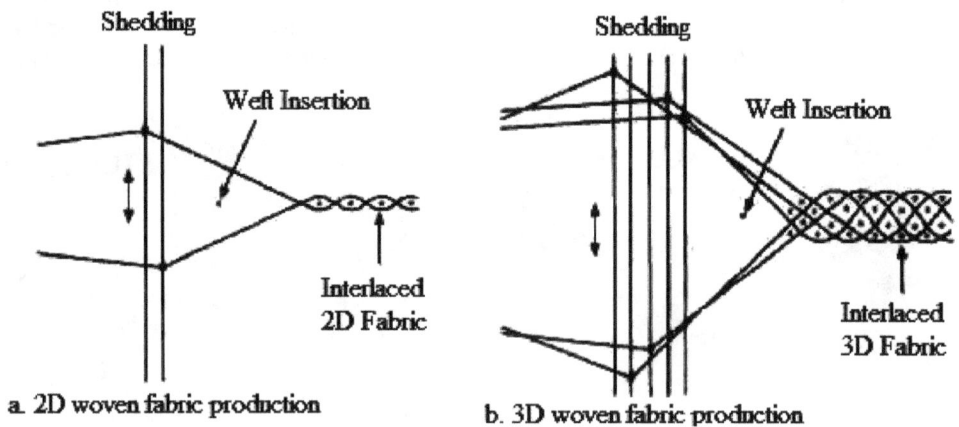

Fig. 3. Fabric production on conventional weaving machine (Khokar, 2001).

As previously mentioned, multilayer and angle interlock weave structures can be produced with 2D weaving. Unfortunately, composites made of 3D woven fabrics produced with 2D weaving have low in-plane stiffness and strength properties due to high crimp levels (Bogdanovich, 2007; Mohamed&Bogdanovich, 2009). A 3D woven fabric produced with 2D weaving is costly because only one weft insertion can take place during one cycle of weaving machine. Furthermore increasing the thickness of 3D woven fabrics while producing them with 2D weaving makes the costs more (Bogdanovich, 2007). Therefore, a new method of 3D woven fabric has been required and developed that is called 3D weaving. However, there is not only one method of 3D weaving. There are different methods of 3D weaving techniques according to the produced structure (angle interlock, orthogonal or fully interlaced 3D woven structures) and orientation of the yarn sets (uniaxial, multiaxial).

4. 3D woven fabric production

3D woven fabrics can be manufactured both with 2D and 3D weaving. The produced 3D fabrics are different from the properties point of view due to the differences in weaving methods. With 2D weaving, pleated or plissé fabric, terry fabrics, velvet fabrics and multilayer woven fabrics can be manufactured. Orthogonal 3D weave structures, fully interlaced 3D weave structures can be manufactured only by using special designed 3D weaving machines. In this section, manufacturing methods of different 3D woven fabrics will be mentioned.

4.1 Production of 3D woven fabrics with 2D weaving

In the case of 2D weaving, two sets of perpendicular yarns are interlaced, irrespective of whether it is woven as single- or multi-layer. Another set of yarns, known as pile or binder yarns, can be introduced in the direction of fabric thickness. Fabrics could be produced by 2D techniques, with different sets of warp yarns in the ways mentioned below (Gokarneshan & Alagirusamy; 2009):

1. By effective utilisation of warp and weft in single layer.
2. By the use of multi-layer warp and weft or multi-layer ground warp, binder warp and weft.
3. Conventional 2D process can also produce pile fabrics by utilising three sets of yarns, namely, single-layer ground warp, pile warp and weft.

4.1.1 Production of Plissé or pleated fabrics

Plissé or pleated material is a folded material, which can be achieved in different methods such as weaving, shrinking and finishing (Routte, 2002);

i.   Woven plissé: Produced by an additional device on a power loom; two warp systems of different tension achieve drape. Folding can also be achieved by suitable bindings. In knit goods folds are created by stitching.
ii.   Shrunk plissé is produced by the use of synthetic fibres with different shrinking properties.
iii. Finishing plissé: The material is laid in folds, which are thermally fixed in so-called pleating machines.

Plissé or pleated woven fabrics can be produced on weaving machines equipped with two warp beams in addition to a special pleated device or a variable beat up.

Pleated woven fabric is produced on weaving machines equipped with a special pleated device as follows:

a.   b.

Fig. 4. The appearances of a). a smooth pleated fabric b). a tough pleated fabric (Kienbaum, 1996 as cited in Badawi, 2007).

In this method, at the beginning of pleat formation, both back rest and breast-beam take the most far on the right lying position as illustrated at point A in Figure 5. When the intended pleat length is reached, they are farther on the left (B). The distance between the two limit points A and B is determined by the pleat length. After the last weft insertion within the pleat length and with beginning of the next inter-fabric part, the pleated length is formed by returning back of the back rest and breast-beam into the starting position (A), at that moment the back rest pull the tight warp to the back position (A) (Badawi, 2007).

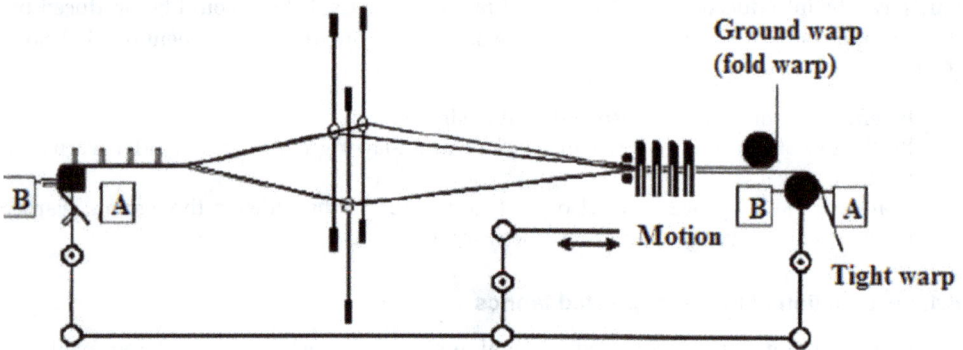

Fig. 5. Device for pleated fabrics weaving (Kienbaum, 1996 as cited in Badawi, 2007).

The coordination of pleated fabric and take-up mechanism are shown in Figure 6. The height of the formed pleat is equal to about half of the pleat length before the backward-movement of the tight warp yarns (Kienbaum, 1996 ; Hennig, 1968 as cited in Badawi, 2007).

Pleated woven fabric is also produced on weaving machines equipped with a variable sley beat up.

Manufacturing of pleated woven fabric on weaving machines equipped with a variable sley beat up can be briefly explained as follows;

The mechanism allows the beating-up point of the sley to be shifted by small but precise steps from the normal beating-up point. While the pleat is being woven, the fabric take-up remains idle so that the weft density is achieved by shifting the beating-up point of the sley.

With a weft density of 40 picks/cm, this means that the beating-up point needs to be shifted only 0.25 mm weft by weft. At the end of the woven pleat, which may be up to 20 mm long, the device must return to its normal beating-up position.

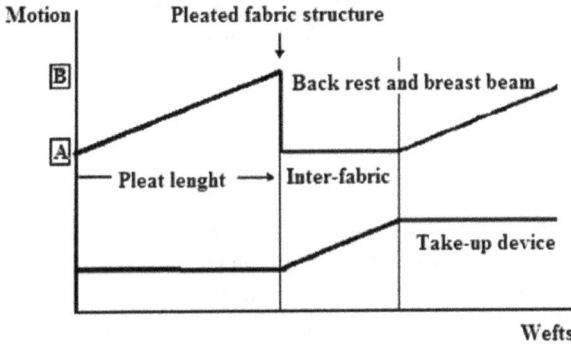

Fig. 6. Movement coordination of pleated fabric and take-up device (Kienbaum, 1996 as cited in Badawi, 2007).

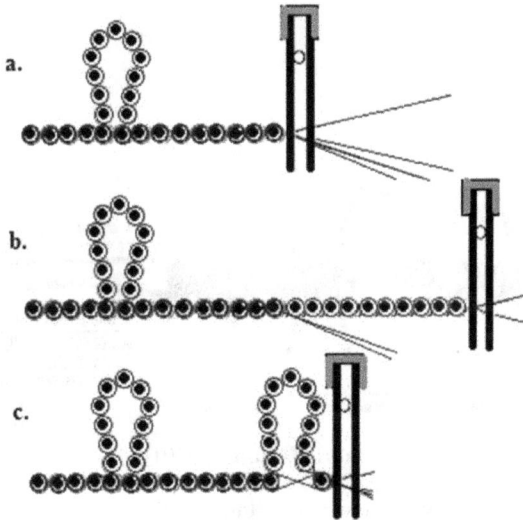

Fig. 7. Beat-up motion during pleat formation (redrawn from Marfurt, 1998).

Weaving a pleat uses only some of the warp threads which are wound onto a separate warp beam. The small proportion of warp threads in the pleat area is compensated for by a greater weft density. The remaining warp threads are left lying underneath the fabric while the pleat is being woven (a). The fabric is no longer taken off, and the sley stays back a given distance in each weft (b). Once the pleat has reached the desired length, all the warp threads are again used in the weaving of the fabric. The sley executes its complete movement, known as full beating. The threads wound onto the separate warp beam yield, and the pleat falls into line (c) (Marfurt, 1998).

4.1.2 Production of terry fabrics

The production of terry fabrics is a complex process and can only be realized on special designed weaving machines. Two warps are processed simultaneously in the production of terry fabrics which one of them is the ground warp that are highly tensioned and the other is the pile warp that is lightly tensioned. A special weaving method enables loops to be formed with lightly tensioned ends on either one surface or two surfaces (Adanur, 2001). Two specialized mechanisms are used in terry weaving machines such as reed control and fabric control.

Two picks are inserted at a variable distance –the loose pick distance- from the cloth fell. The loose pick distance is varied according to the desired loop height. When the third pick is beaten up, the reed pushes the pick group, on the tightly tensioned ground warps, towards the fell and loose pile warp ends woven into the pick group are uprighted and form loops. Depending on the weave, loops are thus formed on one or both sides of the fabric. In general, the reed has two beat up positions which do not impose alternative movements to the warp, fabric and various components of the weaving machine. The sley has a special mechanism built in which allows different beat-up positions for pile formation (Adanur, 2001).

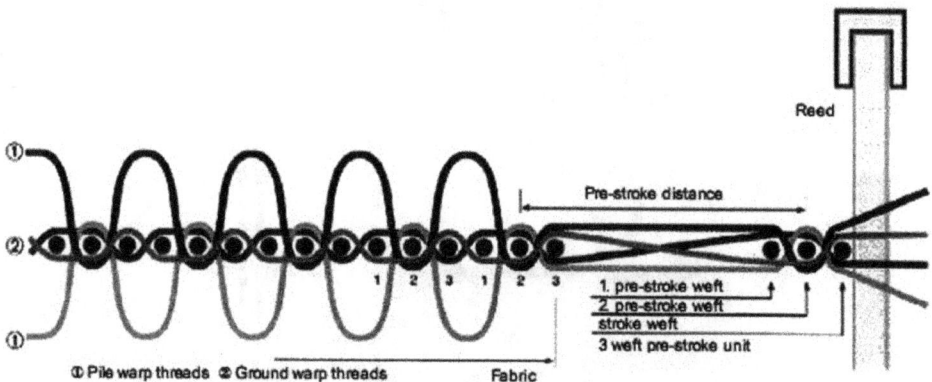

Fig. 8. Structure of a three-pick terry fabric (Adanur, 2001).

In the second system, the sley motion is constant on the other hand cloth fell is moving. Using this principle the fabric is shifted towards the reed by means of a positively controlled movement of the whip roll 6 and a terry bar together with the temples on the beat-up of the fast pick. The sturdy reed drive is free of play. It provides the necessary precision for the beat-up of the group of picks. A compact, simplified whip roll system 6 with the warp stop motions arranged on two separate levels improves handling and has a decisive influence on reducing broken ends. With the help of electronics the precision of measuring the length of pile yarn is improved. The tensions of the ground and pile warps 1 and 2 are detected by force sensors 3 and 9 and electronically regulated. In this way warp tension is kept uniform from full to the empty warp beam. To prevent starting marks or pulling back of the pile loops the pile warp tension can be reduced during machine standstill (Badawi, 2007).

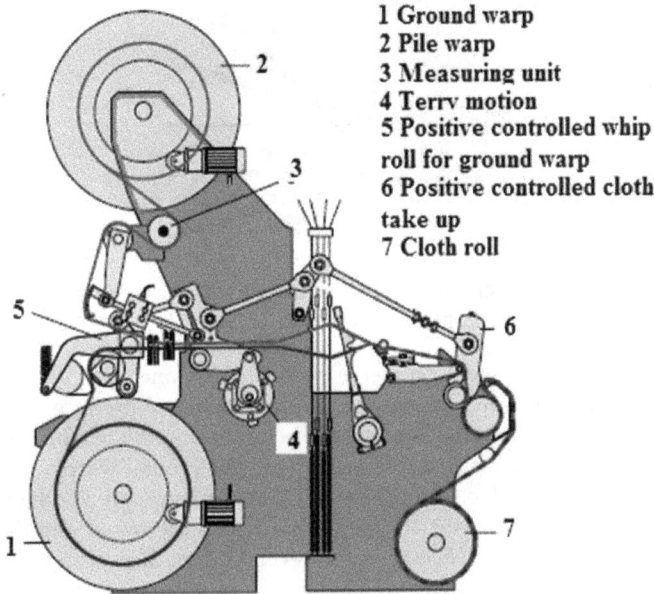

1 Ground warp
2 Pile warp
3 Measuring unit
4 Terry motion
5 Positive controlled whip
roll for ground warp
6 Positive controlled cloth
take up
7 Cloth roll

Fig. 9. Fabric control mechanism (Dornier, 2007).

4.1.3 Production of velvet fabrics

Velvet fabrics are a class of pile fabrics which are divided into two as warp pile fabrics and weft pile fabrics known as velveteen according to the pile direction. Warp pile fabrics, also known as velvet, can be produced with two weaving methods; wire weaving technique and face to face (a.k.a. double plush) weaving technique. The advantage of face to face weaving technology is two fabrics are effectively woven at the same time one above the other joined together by the pile warp ends which cross from top cloth to bottom cloth according to the design and during this weaving process a knife situated between the two cloths continuously traverses the width of the fabric cutting the pile warp threads to create two cloths each with a cut warp pile surface. It is important to appreciate that a surface pile tuft is formed only when a pile end crosses from the top cloth into the bottom cloth and is cut on the loom by the traversing knife, and it is in this way that the surface pile design and colour are created. When it is not required on the surface of the fabric the pile is woven or 'incorporated' into the ground structure either in the top or bottom cloth (Fung&Hardcastle, 2001). Therefore, two different pile structures such as cut pile or loop pile can be obtained based on the pile cut or not.

Weaving machines based on face to face weaving technique are equipped with a 3 position shedding device (dobby or Jacquard machine), so as to form two overlapped and properly spaced out sheds and to permit to the pile warp to tie up the two fabrics together. Into each of the two shed a weft is inserted, usually by means of a pair of superimposed rods driven by the same gear (Castelli et al., 2000). With this technique, 3D woven spacer fabrics can also be easily woven.

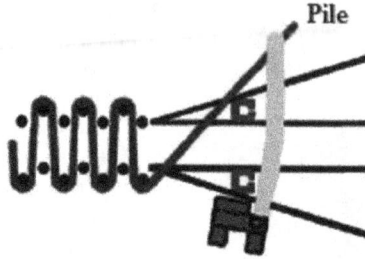

Fig. 10. Face to face weaving technique of producing velvet fabrics
(Van De Wiele as cited in Chen, 2011).

In wire weaving technique, there is one set of ground warp, one set of ground weft and an extra set of warp yarns to form piles on the fabric. In order to produce a fabric with this technique, firstly a group of ground warps is raised according to the fabric pattern and the weft yarn is inserted to make its first interlacing with the ground warps. Then, pile warps are raised and a rod is inserted to the opened shed through the entire width of the fabric. To complete the weaving cycle, the remaining ground warps are raised and then again weft yarn is inserted. This weaving cycle is repeated several times; then the rods are slipped out by forming a loop pile. In order to produce cut pile velvet, rods equipped with knives can be used. In some special types of weaving machines with wire weaving techniques, for the production of velvet fabrics, weft yarn is inserted in the bottom shed of a double shed opening while steel rods or wires are inserted in the top shed to obtain piles. Pile yarns are supplied from a creel that all the ends come from a separate package utilising a negative system of yarn feed controlled by friction tension devices. Again in this method, during extraction of the wires, the piles can be cut or uncut or a combination of both. The weft insertion is performed with a rapier and weft insertion rates can be up to 200 rpm. However, the wire insertion reduces the speed of the weaving machine. Through the combination of ground warps and the weft the base fabric is obtained. The pile ends are woven over the wires and fixed into the base fabric in such a way that loops are being formed over the wires. A certain number of wires is woven into the fabric (10, 12, 16, 20 or 24 wires in total). Each wire is inserted into the shed between the weaving reed and the fabric border. The wires that are woven into the fabric are being extracted one by one from the fabric. For each insertion of a wire another wire is extracted. The wire that has been extracted is reinserted into the shed. Pile wires are specially made very fine steel rods rolled in several passages into the final dimension as requested for the specific pile height that one wants to obtain.

4.1.4 Production of spacer fabrics

Spacer fabrics can be produced both on conventional weaving and special 3D designed weaving machines. These fabrics are classified as even and uneven surfaces according to Chen (2011).

Spacer woven fabrics with even surfaces can be produced on conventional weaving machines with the weft insertion system of shuttle. The weaving loom has a conventional heddle harness (50) system comprising individual heddles 51 to 56 controlling warp yarn groups 11, 12, 21, 22, 31 and 32. As in all weaving looms, shedding and movement of the harnesses in timed sequence with shuttle and reed movement is controlled with the shedding mechanism. Figure 12 represents the opened shedding for 21 and 22 warp yarns for weft insertion, shuttle (57). On the loom, totally three shuttles are used, one for each fabric ply. With arranging the fabric take-up motion in time sequenced with shedding mechanism, this kind of fabric can be easily woven (Koppelman&Edward, 1963).

Fig. 11. Different views of spacer fabrics (Chen, 2011).

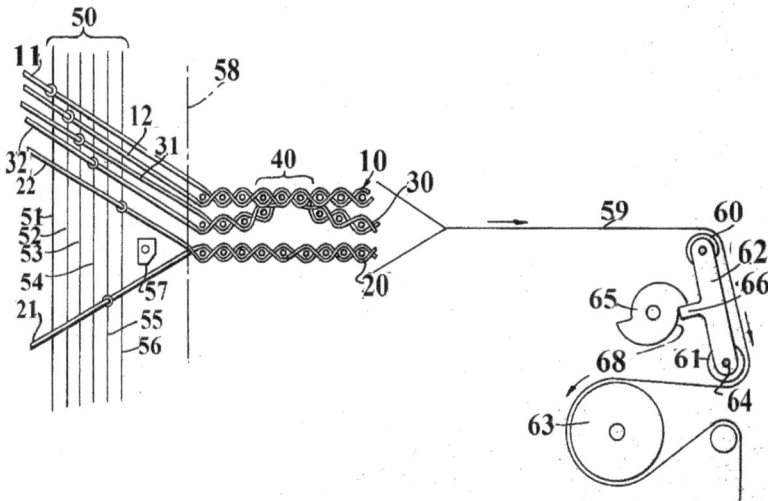

Fig. 12. Method of weaving a hollow 3D woven fabric on a conventional loom (Koppelman&Edward, 1963).

Another design of 3D spacer fabric consisting of double ribs between the bottom and top layers or a 3D spacer fabric with an I shape can be produced on a conventional weaving machine (Rheaume, 1976). For the production of this type of fabric, the weaving loom has to have four separate shuttles and to include eight separate harnesses, each of which control different groups of warp yarns. Both of these two fabrics have the properties of foldability while being produced on the loom. When the fabrics are taken off the loom and get rid of the stresses, they open up and have the cross sections of V-shaped and I-shaped (Figure 13).

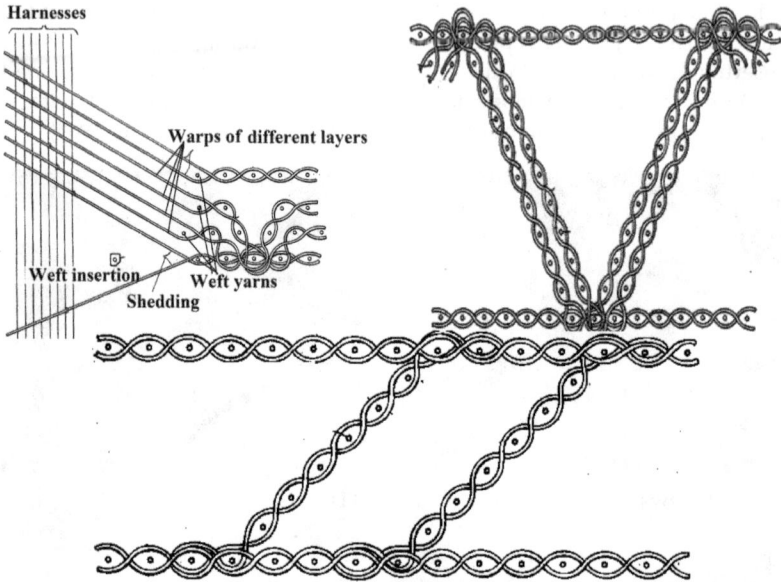

Fig. 13. Method of weaving a hollow 3D woven fabric consisting of two ribs on a conventional loom and appearance of fabric cross section and I-shaped hollow fabric (Rheaume, 1976&1970).

Another type of 3D woven spacer fabric with double ribs connecting the upper and lower layers which is designed to be used in lightweight composite materials is given in Figure 14. However, the structure of this fabric is different compared to the others mentioned up to now. The double ribs connecting the top and bottom layers also constitute the upper and lower layers interchangeably. The weaving of this fabric is possible with warp-let off and fabric take up modifications of the narrow weaving machine. The weaving is performed in three stages; upper and lower ground fabrics weaving, wall-fabrics weaving, and backward movement of the floated tight yarns (formation of wall-fabric) (Badawi, 2007).

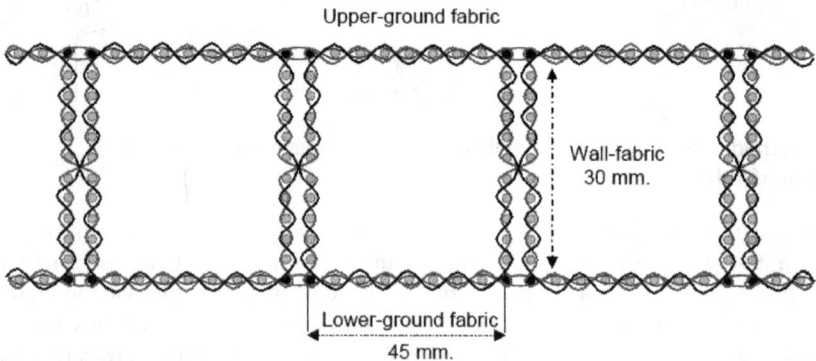

Fig. 14. The structure of 3D woven spacer fabric designed for lightweight composites (Badawi, 2007)

4.1.5 Production of shell fabrics

Shell fabrics are a special class of 3D woven fabrics since the structure of these fabrics may have only one layer or multiple layers; however the end product is always three dimensional. The importance of these types of fabrics is increasing since these types of woven fabrics are widely used in helmets, bra cups in fashion and clothing, female body armour and car door lining material (Chen&Tayyar, 2003).

Shell (a.k.a. doomed) fabrics can be produced with weaving, or cut and sew. Cut and sew technique has been the most commonly method used to produce shell fabrics but seams are a big disadvantage in technical applications, where the continuity of fibres is important. Seams definitely reduce the level of reinforcement and protection. Furthermore, cut and sew creates extra waste of materials and labour (Chen&Tayyar, 2003).

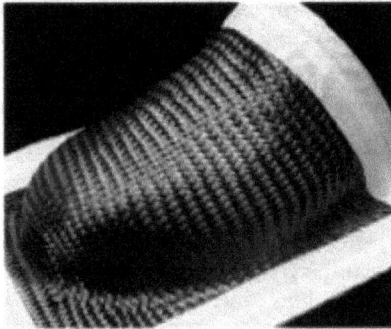

Fig. 15. An example of a shell fabric woven with conventional weaving (Chen, 2011)

In conventional weaving, shell fabrics can be produced with using discrete take up and combination of different weaves. In order to produce a shell woven fabric, one can use a mixture of weaves with long and short floats. For instance, the plain weave, the tightest, is arranged in the middle, where a 2/2 twill is used in middle ring, and a five-end satin, with the longest average float length which is the loosest weave of the three types, is used for the outer ring. In a fabric with constant sett (the same warp and weft densities), the areas woven with plain weave tends to occupy a larger area and therefore will grow out of the fabric plane; the part of the fabric with the five-end satin tends to be squeezed, thus enhancing the domed effect. Consequently, the height difference between the lower and higher planes forms a dome. This method is a quick, easy, and economical way to produce fabrics that require relatively small domed effects. However, it appears that for fabrics requiring larger domed effects, the weave combination method is not sufficient (Chen&Tayyar, 2003).

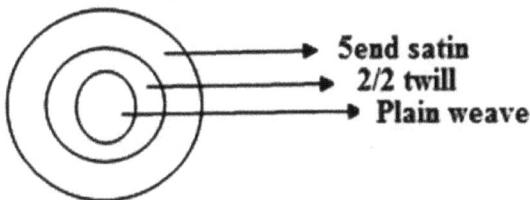

Fig. 16. Weave combination to produce shell fabric (redrawn from Chen&Tayyar, 2003)

There is a balance between warp let-off and fabric take up of the weaving machine in normal weaving. Otherwise, there will be variations in the weft density which is an undesired situation. However, in weaving of shell fabrics, the variations of the weft density are required. But these variations have to be under control. In order to achieve a controlled variation in the weft density, the width of the loom is needed to be divided into several sections and each section is required to be maintained in its own balanced fabric take up and warp let off. This can be achieved by a method in which the fabric is taken forward at different rates across the sections (Chen&Tayyar, 2003). To produce the shell fabric given in Figure 16 as an example, the loom has to be divided into three parts and three warp beams are required controlled individually. By this way, the ratio of fabric take up and warp let off will be constant in each section.

A special fabric take up system which consists of many discs electronically controlled to perform individual take-up movement, besides individually controlled warp let off system on conventional weaving machines allow producing 3D shell woven fabrics (Busgen, 1999).

4.1.6 Production of non-interlaced uniaxial orthogonal 3D fabrics

Khokar (2002) named this kind of fabrics as NOOBed fabrics which is an acronym for Non-interlacing, Orthogonally Orientating and Binding. These fabrics are different in structure when compared to the classical weaving structures (Figure 17.a). There are three different yarns that are positioned in three coordinates (x, y, and z). However, the yarns are not interlaced with each other as in the conventional weaving. Orthogonal fabrics are divided into two groups as uniaxial and multiaxial.

A conventional weaving machine is modified to produce a fabric which has three dimensions of yarns (Figure 17). In this fabric, Ground warps GW are arranged in rows and columns and are positioned in the x direction. In y direction, weft yarns are positioned and these yarns are used to bind ground warps in the row direction. In z direction, extra binder warp threads are used which are supposed to bind ground warps in the column direction (Figure 17.a). Ground warp yarns pass over a roller RL and through two sets or columns of horizontal spacing or separating bars SBR (1-5) and SBL (1-5) arranged in pairs. They pass then between heald wires. The horizontal bars SBR, SBL are designed to open a warp gap between adjacent horizontal rows or layers of ground warps to facilitate the weft insertion. for each such horizontal row, two bars are required; one of which is under appropriate row of threads and serves to raise it SBR, and the other SBL above which serves to lower it. Binder warp threads are controlled by heald frame by raising or lowering the frame. The weaving is performed as follows; when the heald frame HF is in the lowered position, all the bars SBR and SBL are also in the lowered position except SBR1 and SBL1 which are in the raised position in order to form a warp gap for the weft insertion. Thus, the first weft is inserted in the fabric. After the first weft insertion, second bars of SBR and SBL join SBR1 and SBL1 in the raised position are raised and the second weft is inserted into the warp gap. This process is continued until the last weft thread of the first vertical row of weft threads has been inserted. Then reed beats up the inserted wefts by moving forward. The heald frame HF is moved up to insert the vertical binder warp. The lowest SBR5 and SBL5 bars are lowered to form a gap for the weft insertion. Pair by pair all the bars are lowered and after each lowering, the corresponding weft is inserted until the second vertical row of weft threads is completed. Then the reed again beats up the second row and once again the heald frame is lowered to insert the warp binder vertically (Greenwood, 1974).

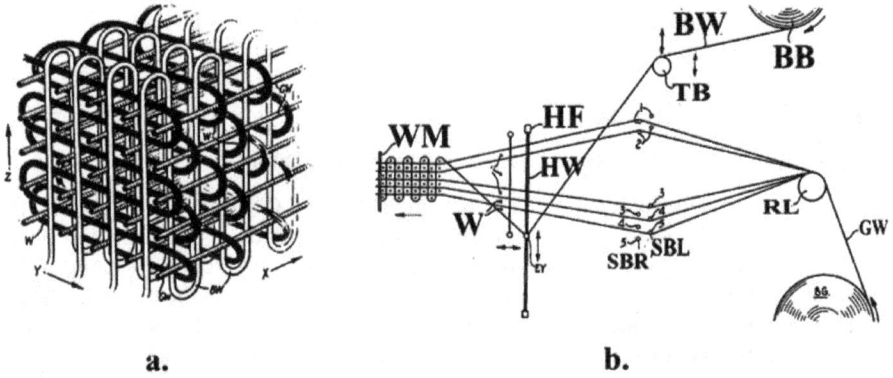

a.   **b.**

Fig. 17. An example of a uniaxial 3-D fabric and the modified conventional weaving loom (Greenwood, 1974)

Uniaxial 3D variable shaped fabric can also be produced on conventional weaving machines (Mohamed&Zhang, 1992). In this system, two groups of weft yarns, Y1 and Y2 are used for weft insertion with one weft group (Y1) being inserted from one side for the flange and other weft yarn group (Y2) being inserted from the other side for the web portion of the inverted T cross-shape (Figure 18.b). Two selvage yarns, Sa and Sb, are required to hold the

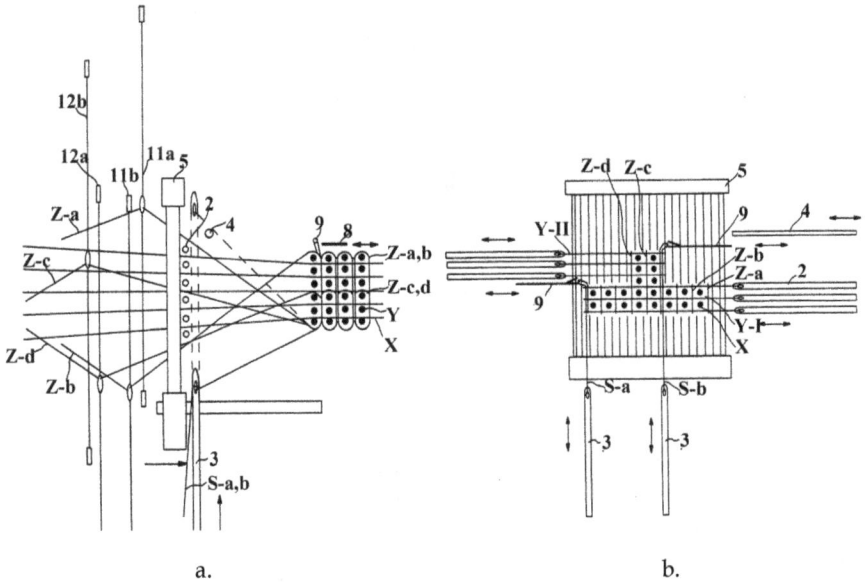

a.   b.

Fig. 18. Production of a uniaxial 3D variable shaped fabric on modified weaving machine (Mohamed&Zhang, 1992)

fore end loops formed by the two groups of filling yarns, Y1 and Y2, respectively. Preferably, four harnesses, 11a, 11b, 12a, 12b, are used to control two sets of vertical Z yarns, Za-Zd. One of set of Z yarns, Za, Zb, is inserted for the flange portion of the inverted T shape fabric, and the other set of Z yarns, Zc, Zd, is inserted for the web portion of the inverted T cross-sectional shape fabric (Mohamed&Zhang, 1992).

4.2 Production of 3D woven fabrics with 3D weaving

4.2.1 Production of non-interlaced uniaxial orthogonal 3-D fabrics

This method of producing an integrated nonwoven 3D fabric F (Figure 19) comprises disposal of axial yarns Z in a grid form and in accordance with the required cross sectional profile, and traversing horizontal and vertical sets of binding yarns X and Y about the corresponding rows and columns of axial yarns in a closed-loop path to bind the fabric directly. The device is essentially composed of a plate (P) having two sets of profiled tracks (D and C) existing in a mutually perpendicular configuration and in the same plane on the front face of the plate (P); two sets of binder yarn spool carriers (K and L); two pairs of tracking arrangement such that each pair is situated at the terminal sides to contain between it all the tracks of sets D and C respectively for guiding the binder yarn carriers in a closed-loop path; and openings (B) in plate (P), arranged in rows and columns, to allow the axial yarns Z to pass through, a creel (J) to supply axial yarns Z, and a fabric take-up unit (H) (Khokar, 2002).

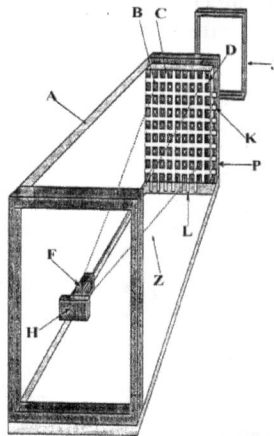

Fig. 19. Production of a uniaxial 3D fabric on a special designed 3D weaving machine (Khokar& Domeij, 1999)

The warp yarns Y are arranged in multiple layers each of which has a number of yarns which run in one horizontal plane in parallel relation with or at an equal space from adjacent yarns that are passed through a reed 1 through number of holes formed therein at uniform intervals in both horizontal and vertical directions. The warp yarns of the respective layers are in vertical alignment, forming regularly spaced vertical warp rows. Weft inserting device 6 comprises a number of elongated picking plates 7 which are spaced from each other at the same distance for secure insertion of wefts into the spaces between the respective layers of the tensioned warp yarns.

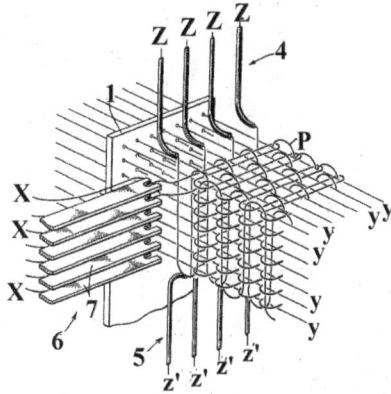

Fig. 20. Special designed 3D weaving machine that produces a uniaxial 3-D fabric (Fukuta et al., 1974).

In order to pick in weft yarns X and vertical yarns Z into the horizontally and vertically aligned warp yarns, the weft inserting device 6 is first picked transversely or perpendicularly to the warp yarns while maintaining the upper and lower vertical yarn inserting devices 4 and 5 in the upper and lower retracted positions as shown in Figure 20. Each of the weft yarns X being inserted between the warp layers in double fold forming a loop at the fore end thereof. The weft inserting device 6 is temporarily stopped when the looped fore ends of the weft yarns are projected out of the warp yarns on the opposite side for threading a binder yarn P (Fukuta et al., 1974).

As shown in Figure 21, base 10 supports movable upper and lower frames 12 and 13 with holes for supporting a plurality of filaments 15 that extends in the vertical (Z-axis) orientation. Identically working filament feed units 20 and 20' alternately insert yarns in the X- and Y-axes directions, respectively. First, filaments 21 from supply bobbins are woven through the spaced rows between filaments 15 along the X-axis by advancing the needles 22 by pushing rods 25. A pin 30 is inserted in the Y-axis direction to lie across the top of filaments 21 outside the last row of filaments 15 to tamp filaments 21 down. Needles 22 are then retracted from filaments 15, forming a tightly looped first course of X-axis filaments that is restrained by pin 30. Similarly, the course of Y-axis filaments is woven next by advancing threaded needles 22', inserting pin 30' on top of filament 21' in the X-axis direction and retracting needles 22'. As the filament layers build up, pins 30 and 30' are removed. To increase the fabric's density, all the filament layers are compressed. The fabric integrity results primarily from inter-yarn friction (King, 1976 as cited Khokar, 2002).

In Weinberg's special designed 3D weaving machine, it is possible to form sheds between layers of planar warp yarns, so that the orthogonal weft yarns can easily be inserted in any predetermined directions. Planar warp yarns are threaded through two parallel and perforated plates. The distance between these two plates is enough to accommodate the shedding and weft insertion. The top plate can slide on the warp yarns. The base plate is used to anchor the ends of the warp yarns (Weinberg, 1995).

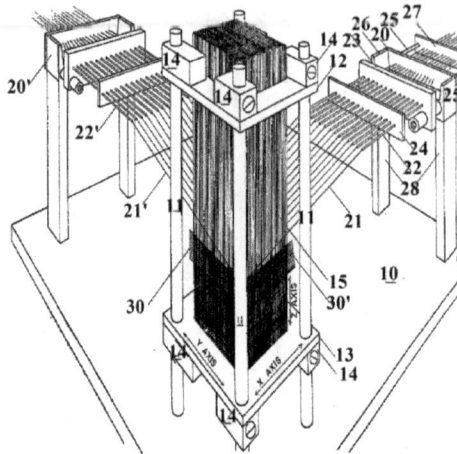

Fig. 21. King's special designed 3D weaving machine (King, 1976).

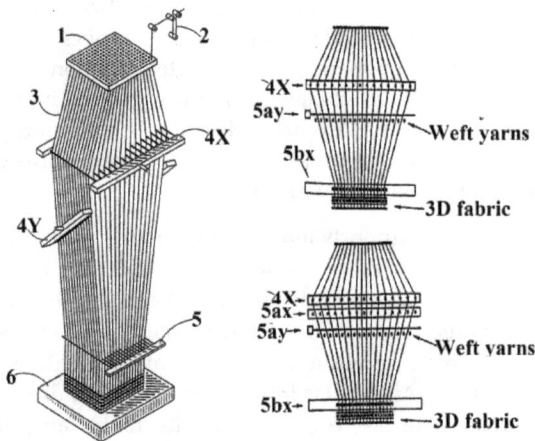

Fig. 22. Weinberg's special designed 3D weaving machine (Weinberg, 1995).

4.2.2 Production of non-interlaced multiaxial orthogonal 3D fabrics

One of the main problems using multilayer woven fabrics in preforms is insufficient in-plane and off-axis properties of composites. Conventional weaving machines which are capable of producing multilayer fabrics cannot produce fabrics that contain fibres or yarns orientated at ±45° in the plane of the preform. With conventional machines, it is only possible to manufacture fabrics with fibres or yarns oriented at angles of 0° and 90°. It is also possible to orient the fibres or yarns at angles of ±45° in through the thickness. However, these oriented yarns at the angle of ±45° in through the thickness will not affect the in-plane and off-axis properties of composites in a positive way. The more recent machinery developments have therefore tended to concentrate upon the formation of preforms with multiaxial yarns (Tong et al., 2002).

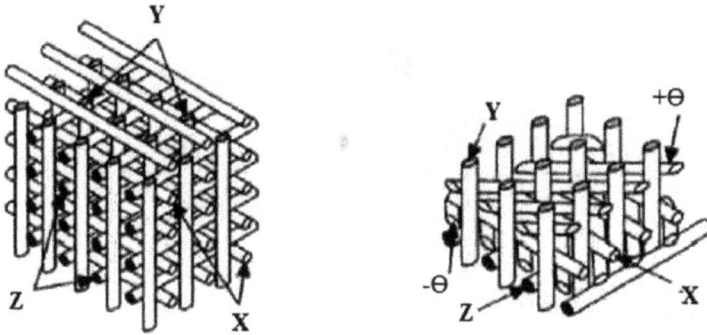

Fig. 23. Uniaxial (on the right) and multiaxial (on the left) orthogonal fabrics (Khokar, 2002).

A set of linear yarns Z, X, ±θ, arrayed in multiaxial orientation in the directions of the fabric's length, width, and two bias angles respectively, is bound using a set of binding yarns Y in the fabric-thickness direction. The yarns Y could be of either single or double type. The corresponding bindings occur above and under the set of Z, X, ± θ yarns and they form two surfaces of the fabric. The resulting 3D fabric has the three sets of linear yarns X, Y and Z in a mutually perpendicular configuration and, additionally, the linear yarns ± θ in bias directions (Khokar, 2002).

Anahara (1993) et al, invented a special weaving machine to manufacture a multiaxial orthogonal 3D fabric. In the aforementioned fabric, there are five axes of yarns used to construct the structure. First of all, there are warp yarns used in the length wise direction of the fabric (z). Similarly, there are weft yarns used in the width direction of the fabric (x). The first and second bias yarns B1 and B2 are arranged at an angular relationship of ±45°. In other words, the 3D fabric F has a five axis structure in which fabrics have four axes in one plane (Figure 24) and are interconnected by the lines of the vertical yarn y.

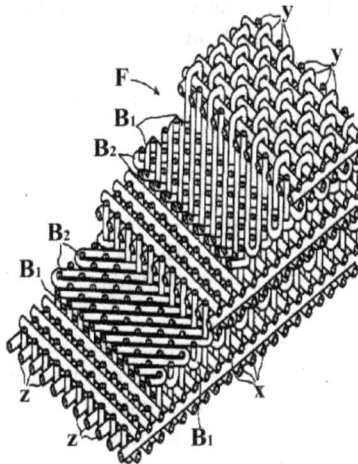

Fig. 24. Multiaxial 3D woven fabric structure (Anahara et al., 1993).

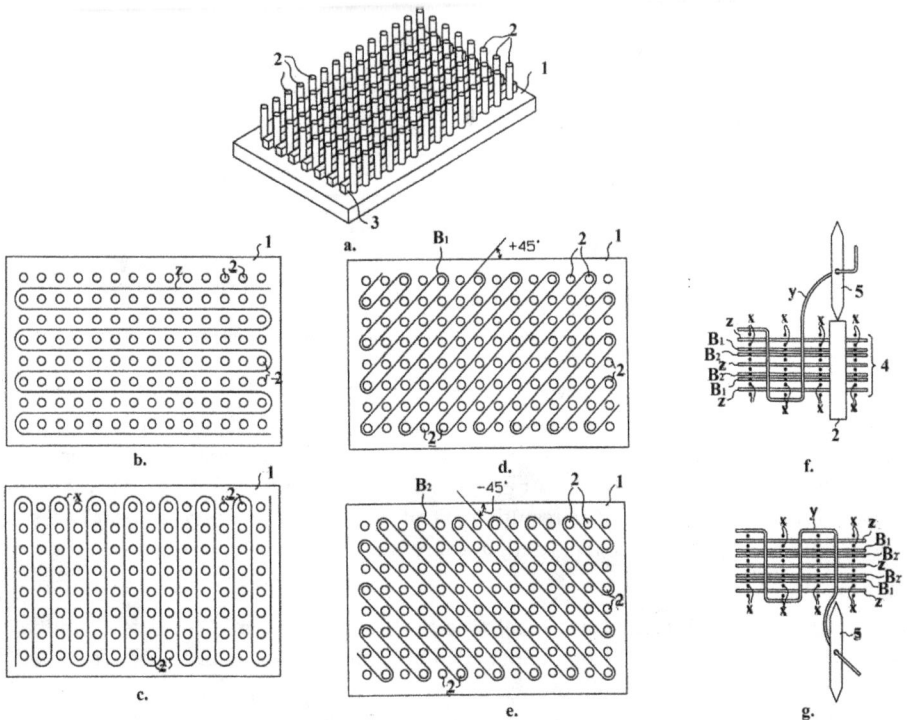

Fig. 25. Multiaxial 3D woven fabric manufacturing method (Anahara et al., 1993).

In the production of such a fabric F, a flat base 1 is used as shown in Figure 25 (a). There are a number of pins 2 that can be unfastened which allows the yarns to be arranged in different axes. The support bar 3 can be disposed between the pins 2 on the base 1. The lines of the weft x, warp z and first and second bias B1 and B2 are arranged in a way that these yarns run between the pins 2 and to be looped back, in engagement with those pins 2 which are located along the peripheral portion of the base 1. The weft layer, warp layer and bias yarn layers are inter-laminated in order. Firstly, the lines of warp yarns z are arranged in parallel in the length wise direction of the fabric in a way that they are being repeatedly looped back and forth around the pins 2 as shown in Figure 25 (b). Similarly, the lines of weft yarns x are arranged in parallel in the width direction of the fabric in such a way that they are being looped back and forth around pins 2 located at the right and left sides of the base 1 shown in Figure 25 (c). as shown in Figure 25(d), the lines of bias yarns B1 are inserted at an angle of +45° with respect to the lengthwise direction of the fabric while being repeatedly looped back and forth around the pins 2. Similarly, the lines of bias yarns B2 are inserted again in the length wise direction of the fabric that are being repeatedly looped back and forth around pins 2 however at the angle of -45° as shown in Figure 25(e). After the individual layers are completed one on another in a predetermined order, the pins 2 are removed from the base and are replaced by vertical warp yarns y through a needle Figure 25(f and g) (Anahara et al., 1993).

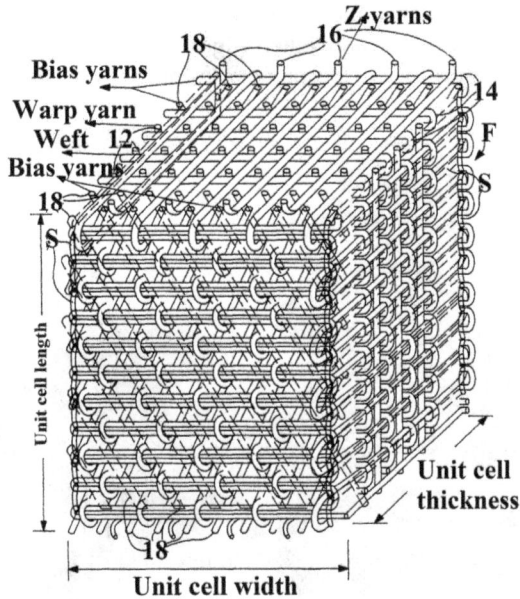

Fig. 26. Multiaxial 3D orthogonal woven fabric (Mohamed et al., 1995).

In the invention of Mohamed et al., the woven preform consists of multiple warp layers 12, multiple weft yarns 14, multiple z yarns 16 that are positioned in the fabric thickness and ±bias yarns as shown in Figure 26. The ±bias yarns 18 are located at the back and front of the fabric which are connected with the other sets of z yarns. In the manufacturing of this preform, warp yarns 12 are arranged in a matrix of columns and rows based on the required cross-sectional shape. After bias yarns oriented at ±45° to each other on the surface of the preform, weft yarns 14 are inserted between the rows of warp yarns and loops of weft yarns are locked with the help of two selvages at both edges of the fabric. Z yarns 16 are then inserted and passed across each other between the columns of warp yarns 12 to cross weft yarns 14 in place. The weft insertion takes place again as mentioned before and the yarns are returned to their initial positions. Z yarns 16 are now returned to their starting positions passing between the columns of warp yarn 12 by locking ±45° bias yarns 18 and weft yarns in their place. The inserted yarns are beaten up against the fabric formation line and a take up system removes the fabric frim the weaving zone. This is only a one cycle of the machine. By repeating this cycle, 3D multiaxial orthogonal woven fabric can be produced within the desired fabric length (Mohamed et al., 1995).

A three-dimensional multiaxial cylindrical woven fabric (Figure 27) having a core, comprises five sets of yarns: axial (14), circumferential (16), radial (18) and two sets of bias yarns (12) that are orientated ±45° with reference to the longitudinal axis of the cylindrical fabric. The bias yarns (12) occur at the outer and inner surfaces. The fabric is produced using a multiaxial circular weaving apparatus (100) that comprises mainly four units: feeding unit (110), machine bed (130), beat-up unit (180) and take-up unit (190). The steps in the operation of the weaving machine are: rotation of positive and negative bias yarn carriers by one carrier distance; rotation of circumferential yarn carriers by one carrier distance; moving radial yarn carriers between outer and inner edges of the machine bed; beating-up the inserted yarns; and taking-up the woven preform from the weaving zone (Bilisik, 2000 as cited in Khokar, 2002).

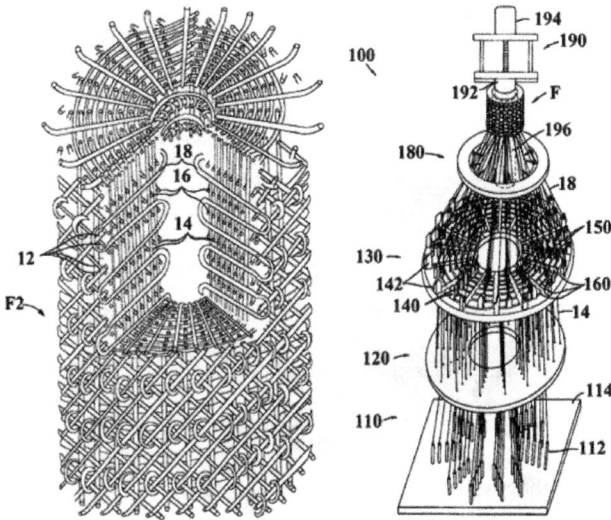

Fig. 27. Multiaxial 3D circular woven fabric structure and apparatus (Bilisik, 2000).

4.2.3 Production of interlaced 3D fabrics

The aim of producing a full interlaced 3D woven fabric is to provide a flexible-structure composite which exhibits high mechanical strength against repeatedly exerted loads and, at the same time, enjoys the advantage of light weight. This process is first developed by Fukuta (Fukuta et al., 1982).

In this manufacturing method, X and Y referred as horizontal and vertical weft yarns respectively, are interlaced with the rows and columns of Z multi-layer warp yarns respectively. In this method, shedding of multi-warp Z yarns is not performed only in the fabric thickness direction like in orthogonal 3D fabric formation but it is performed also across the fabric width. To do this, a dual shedding is needed.

In addition to dual shedding that enables column-wise and row-wise sheds to be formed, in order to produce a fully interlaced 3D woven fabric; a grid-like multiple-layer warp (Z), and two orthogonal sets of wefts (X—set of horizontal wefts and Y—set of vertical wefts) are required (Khokar, 2001).

The dual shedding is performed as shown in Figure 29 (a-i). In Figure 29 (a), the grid-like multiple layer warp yarns Z are in their initial position. Multiple synchronized column-wise sheds are formed (Figure 29 (b)) in which vertical wefts Y are to be inserted (Figure 29(c)) and after the insertion of vertical weft yarns Y all the sheds are closed. The produced fabric structure up to now is given in Figure 29 (d) which is a result of interlacement of vertical weft yarns Y and grid-like warp yarns Z. Then the warp yarns Z are subjected to form a shed in the row-wise direction (Figure 29 (f)) into which horizontal weft yarns X are to be inserted (Figure 29(g)). The result of interlacing horizontal weft yarns X and grid-like warp yarns Z are shown in Figure 29(h). When the operations of column-wise and row-wise shedding are performed sequentially, and the corresponding wefts are inserted backward and forward in the aforementioned sheddings, the structure of plain-weave 3D fabric is formed that is shown in Figure 29(i).

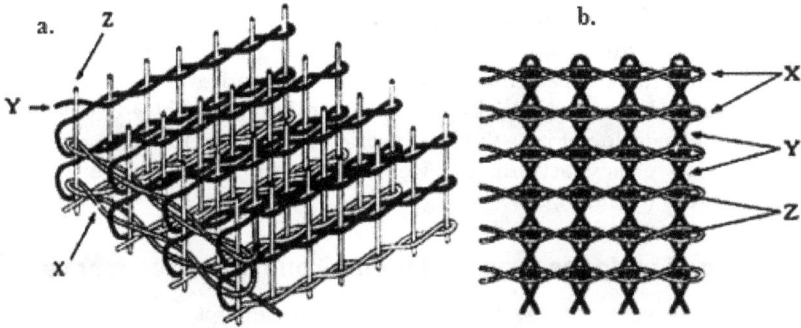

Fig. 28. Fully interlaced 3D woven fabric structure isometric view (a) and orthogonal view (b) (Fukuta et al., 1982).

However, these fabrics suffer from the crimp and fibre damage problems (Mohamed & Bogdanovich, 2009). As the shedding operation alternately displaces the grid-like arranged warp yarns Z in the thickness and width directions, two mutually perpendicular sets of corresponding vertical wefts Y and horizontal wefts X are inserted into the created sheds. The warps Z, therefore, interlace with the sets of vertical Y and horizontal X wefts, thus creating a fully interlaced 3D woven fabric. Due to the interlacing, the resulting structure has crimped fibres in all three directions, which would be detrimental for potential applications of this type of fabric as a composite reinforcement

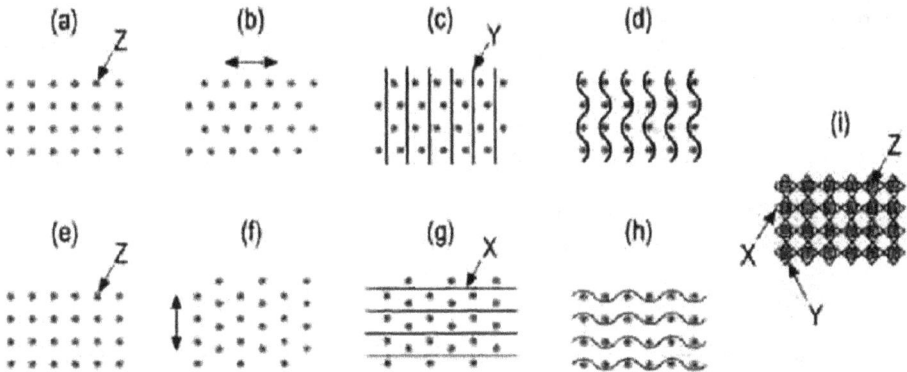

Fig. 29. Dual-directional shedding and corresponding picking for weaving fully interlaced 3D fabric (Khokar, 2001).

5. Advantages of 3D weaving process

The production of 3D woven fabrics on conventional weaving machines is an inefficient process since conventional 2D weaving machine inserts weft yarns one at a time. Contrary to the aforementioned situation, 3D weaving looms that are special designed to produce 3D woven fabrics allow simultaneously insert multiple layers of warp yarns and weft yarns. The simultaneous insertion of an entire column of weft yarns makes the linear productivity of the 3D weaving process independent of layers (Lienhart, 2009).

Another important distinction is that in 2-D weaving process, yarns in the warp direction are passed through heddles, and must be pulled past neighbouring warp yarns, above or below the filling insertion. Repeated motions through the heddles and through other planes of warp tend to abrade fibres, especially brittle technical fibres. In the 3-D Weaving process individual warp planes do not pass through heddles, and are not forced to repeatedly cross neighbouring warp planes; accordingly, weaving-induced fibre damage in this case is significantly reduced (Lienhart, 2009).

6. Comparison of preforms produced from 2D and 3D woven fabrics

3D woven preforms were first developed in 1970's in an attempt to replace expensive high temperature metal alloys in aircraft brakes (Mouritz et al., 1999). In order to produce the preform of the brake component that was produced with 3D weaving process, Avco Corporation developed a specialised 3D weaving loom that performs weaving of hollow cylindrical preforms in which carbon fibres were aligned in radial, circumferential and axial directions. Research and development of 3D woven preforms remained at a low level until the mid-1980s since problems of using traditional 2D laminates in aircraft structures were encountered. One of the main problems that was faced at that time by the aircraft manufacturers was preforms produced from traditional 2D laminates were expensive to produce complex structures. The second problem was the low impact resistance of the traditional 2D laminates since aircraft maintenance engineers complaint of damage impacts from dropped tools during maintenance. These problems led more research and development in the field of composites produced from 3D woven fabrics. Nearly thirty years of know-how in the field of 3D weaving makes the preforms produced from 3D woven fabrics have superior properties compared to those of the preforms produced from traditional 2D laminates.

While preforms produced from traditional 2D laminates can only be processed into relatively simple and slightly curved shapes, preform for a composite component with a complicated shape can be made to the near-net-shape with 3D weaving. This ability of 3D weaving producing near-net-shape preforms can reduce the production costs thanks to the reduction in material wastage, need for machining and joining, and the amount of material handled during lay-up.

The second advantage of 3D weaving is preforms can be produced on conventional weaving looms only by making minor modifications to the machinery. This minimises the investment cost of producing preforms made of 3D woven fabrics. However, a range of specialised looms have been developed that have higher weaving speeds and are capable of weaving more complex shapes than traditional looms which have been modified (Mouritz et al., 1999).

With the use of 3D weaving, fabrics having different through-thickness properties can be produced. Amounts and types of binder yarns such as carbon, glass, Kevlar and ceramic fibres in through-thickness can be used to tailor the properties of a composite for a specific application (Mouritz et al., 1999).

Composites produced from 3D woven fabrics have higher delamination resistance, ballistic damage resistance and impact damage tolerance. These aforementioned properties have been a major problem in composites produced with traditional 2D weaving used in military aircraft structures (Mouritz et al., 1999).

7. Conclusion

Today, textile structural composites are widely used in many application fields that usually consist of stacked layers known as 2D laminates, exhibit better in-plane strength and stiffness properties compared to those of metals and ceramics. However, the application of 2D laminates in some critical structures in aircraft and automobiles has also been restricted by their inferior impact damage resistance and low through thickness mechanical properties when compared against the traditional aerospace and automotive materials such as aluminium alloys and steel. In order to improve interlaminar properties of the 2D laminates, three dimensional (3D) textile preforms have been developed by using different manufacturing techniques like weaving, knitting, braiding, and stitching. Among these manufacturing techniques, sewing and 3D weaving are the promising technologies which address the shortcomings of the stack-reinforced composites.

In order to comprehend 3D weaving technology and its products, the production techniques and their principles have been reviewed in detail within this chapter.

8. References

Adanur S. (2001). *Handbook of Weaving*, Technomic, ISBN 1-58716-013-7, Pennsylvania, USA.

Anahara, M., Yasui, Y., Sudoh, M. and Nishitani, M. (1993). *Three Dimensional Fabric with Symmetrically Arranged Warp and Bias Yarn Layers*, Patent No. USP 5 270 094.

Badawi S.S. (2007). *Development of the Weaving Machine and 3D Woven Spacer Fabric Structures for Lightweight Composites Materials*, PhD Thesis, Technical University of Dresden, Dresden, Germany.

Behera B.K., Mishra R. (2008). 3-Dimensional Weaving, *Indian Journal of Fibre&Textile Research*, Vol.33, pp.274-287.

Bilisik K. (2011). Multiaxis Three Dimensional (3D) Woven Fabric, In: *Advances in Modern Woven Fabrics Technology*, Vassiliadis S., pp.79-106, InTech, Retrieved from: http://www.intechopen.com/books/show/title/advances-in-modern-woven-fabrics-technology

Bilisik K. (2010). Multiaxis 3D Weaving: Comparison of Developed Two Weaving Methods Tube-Rapier Weaving versus Tube-Carrier Weaving) and Effects of Bias Yarn Path to the Preform Properties. *Fibers and Polymers*, Vol.11, No.1, pp.104-114

Bilisik, A.K. (2000). *Multiaxial Three-dimensional (3-D) Circular Woven Fabric*, Patent No. USP 6 129 122.

Bogdanovich A.E. (2007). Advancements in Manufacturing and Applications of 3-D Woven Preforms and Composites, *Sixteenth International Conference on Composite Materials (ICCM-16)*, Kyoto, Japan, July 8-13 2007

Busgen A. (1999). *Woven fabric having a bulging zone and method and apparatus of forming same*, Patent No. USP 6000442.

Castelli G., Maietta S., Sigrisi G., Slaviero I.M., (2000). *Reference Books of Textile Technology: Weaving*, Italian Association of Textile Machinery Producers Moral Body, Milano, Italy.

Chen X., Taylor L.W., Tsai L.J. (2011). An Overview on Fabrication of Three-Dimensional Woven Textile Preforms for Composites, *Textile Research Journal*, Vol. 81, No.9, pp. 932-944

Chen X., Tayyar A.E. (2003). Engineering, Manufacturing, and Measuring 3D Domed Woven Fabrics. *Textile Research Journal*, Vol.73, No.5, pp.375-380.

Chen X., Spola M., Paya J. G., Sellabona P. M. (1999). Experimental Studies on the Structure and Mechanical Properties of Multi-layer and Angle-interlock Woven Structures, *Journal of the Textile Institute*, Vol.90:1, pp.91-99.

Chou T. W. (1992). *"Microstructural Design of Fibre Composites"*, Cambridge University Press, ISBN 978-0-521-35482-0, New York, USA

Dornier, (2007). ServoTerry, In: *Lindauer Dornier*, 03.10.2011, Available from: http://www.lindauerdornier.com/weaving-machine/servoterryae/air-jet-terry-weaving-machine?set_language=en&redirect=1

Encyclopædia Britannica. (2011). Composite Material, In: *Encyclopædia Britannica Online*, 25.09.2011, Available from: http://www.britannica.com/EBchecked/topic/130093/composite-material

Fukuta K., Nagatsuka Y., Tsuburaya S., Miyashita R., Sekiguti J., Aoki E., Sasahara M., (1974). Three dimensional fabric and method and loom construction for the production thereof. Patent No. USP 3834424.

Fukuta, K., Onooka, R., Aoki, E., Tsuburaya, S. (1982), A three-dimensional latticed flexible-structure composite, Patent No. USP 4336296.

Fung W., Hardcastle M. (2001). *Textiles in Automotive Engineering*, Woodhead Publishing Ltd., ISBN 1-85573-493-1, Cambridge, England.

Gokarneshan N., Alagirusamy R.(2009). Weaving of 3D fabrics: A critical appreciation of the Developments. *Textile Progress*, Vol. 41, No. 1, pp. 1–58

Greenwood, K. (1974). *Loom*, US Patent No. 3818951.

Hearle J. W.S., Chen X. (2009). 3D Woven Preforms and Properties for Textile Composites, *Seventeenth International Conference on Composite Materials*, Edinburgh, UK, July 27-31 2009

Khokar N. (2002). Noobing: A Nonwoven 3D Fabric forming Process Explained, *Journal of the Textile Institute*, Vol.93, No.1, pp. 52-74.

Khokar, N. (2001). 3D-Weaving: Theory and Practice, *Journal of the Textile Institute*, Vol.92 No.2, pp.193-207.

Khokar N., Domeij, T. (1999). Device for Producing Integrated Nonwoven Three-dimensional Fabric. Sweden. Patent No. SE 509 944.

King, R.W. (1976). Apparatus for Fabricating Three-Dimensional Fabric Material. US Patent No. 3955602.

Koppelman, E., Edward A.R. (1963). *Woven Panel and Method of Making Same*, US Patent No. 3090406.

Lienhart B. (2009). 3Tex Preforms - The Metal Alternative, 3Tex Company Brochure.

Lord, P.R., Mohamed, M.H. (1982). *Weaving: Conversion of Yarn to Fabric* (Second Edition), Merrow Publishing Co., ISBN 0-900-54178-4, Watford, England.

Marfurt, P. (1998), Decorative Pleats, In: *Sulzer Technical Review 1/98*, 02.10.2011, Available from:http://www.sulzerpumps.com/PortalData/7/Resources/03_NewsMedia/S TR/1998/1998_01_marfurt_e.pdf

Mohamed M. H., Bogdanovich A. E. (2009) Comparative Analysis of Different 3D Weaving Processes, Machines and Products, *Proceedings of 17th International Conference on Composite Materials (ICCM-17)*, July 27-31, 2009, Edinburgh, UK

Mohamed M. H., Zhang Z.H. (1992). *Method of Forming Variable Cross Sectional Shaped Three Dimensional Fabric*, US Patent No. 5085252.

Mohamed, M.H. and Bilisik, A.K. (1995). *Multi-layer Three-dimensional Fabric and Method for Producing*, US Patent No. 5 465 760.

Mouritz A.P., Bannisterb M.K., Falzonb P.J., Leongb K.H. (1999). Review of Applications for Advanced Three-Dimensional Fibre Textile Composites. *Composites: Part A*, Vol.30 pp.1445–1461

Naik N.K., Azad SK. N.M., Durga Prasad P., Thuruthimattam B. J. (2001). Stress and Failure Analysis of 3D Orthogonal Interlock Woven Composites, *Journal of Reinforced Plastics and Composites*, Vol. 20, No. 17, pp.1485-1523

Naik N.K., Azad SK. N.M., Durga Prasad P. (2002). Stress and Failure Analysis of 3D Angle Interlock Woven Composites, *Journal of Composite Materials*, Vol. 36, No. 1, pp.93-123

Padaki N.V., Alagirusamy R., Deopura B. L., Fangueiro R., (2010). Studies on Preform Properties of Multilayer Interlocked Woven Structures Using Fabric Geometrical Factors. *Journal of Industrial Textiles*, Vol. 39(4) pp.327-345

Peters, S.T. (1998). Introduction, Composite Basics and Road Map, In: *Handbook of Composites*, Peters, S.T., pp.1-21. Chapman&Hall, ISBN 0-412-54020-7, London, UK

Rheaume J.A. (1970). Three-dimensional woven fabric. Patent No. US3538957.

Rheaume J.A. (1976). Multi-ply woven article having double ribs. Patent No. US3943980.

Rouette H.K. (2002). *Encyclopedia of Textile Finishing*, Springer, ISBN 3-540-65490-9.

Soden J.A., Hill J. (1998). Conventional Weaving of Shaped Preforms for Engineering Composites, *Composites Part A*, Vol.29A pp.757–762

Stig F. (2009). *An Introduction to the Mechanics of 3D-Woven Fibre Reinforced Composites*, Licentiate Thesis, Stockholm, Sweden.

Tong L., Mouritz A.P. and Bannister M.K. (2002). *3D Fibre Reinforced Polymer Composites*, Elsevier, ISBN 0-08-043938-1, Netherlands.

Weinberg, A. (1995). Method of Shed Opening of Planar Warp for High Density Three Dimensional Weaving. US Patent No. 5449025.

Zhang, C. (2003). *Characterization and Modelling of 3D Woven Composites*, PhD Thesis, North Carolina State University, Raleigh, USA.

3

The Physical Properties of Woven Fabrics for Emotional Garment According to the Weaving Loom Characteristics

Seung Jin Kim and Hyun Ah Kim
School of Textiles, Yeungnam University, Gyeongsan,
Korea Institute for Knit Industry, Iksan,
Korea

1. Introduction

Many efforts for making good quality fabrics for emotional garment have been performed by SME weavers and finishers. And weaving machinery companies are researching about loom mechanism applied by low warp and filling tensions and loom mechanism for good quality fabrics. The fabric defects complained by garment manufacturers are stop marks, streaky phenomena on the warp direction, thickness variation and color differences between edges on the right and left sides of the fabrics, which are partly due to the tension variation of warp and filling directions. Therefore, many researches(Basu, 1987; Islam & Bandara, 1996, 1999) related to the fabric defects and weaving loom mechanism were carried out and many patents related to the loom were presented by loom makers. Many researches related to the warp and filling tensions during weaving were performed with relation to the fabric defects. Fabric physical property is largely affected by various factors such as constituent yarn physical property and fabric structural parameters. But, the fabric physical property for emotional garment is also affected by weaving loom characteristics. Among weaving loom characteristics, warp and weft yarn tensions during weaving are the most important parameters which affects fabric physical properties and quality. And warp and weft yarn tensions are different according to loom characteristics i.e. according to air-jet, rapier and projectile. Even though same rapier loom, these tensions are slightly different according to the mechanism of rapier loom. Many researches(Islam & Bandara, 1996, 1999) related to the warp and weft yarn tensions during weaving were carried out with relation to the stop marks and other fabric quality. On the other hand, air-jet insertion in air-jet loom and its mechanical mechanism were also performed with variation of the air flow and weft yarn tension.(Natarajan et al, 1993; Adanur & Mohamed, 1988, 1991, 1992)

Recently, many simulation studies(Belforte et al, 2009; Simon et al, 2005, 2009) related to the air-jet nozzle on the air-jet loom were investigated. And new concept and recent innovations in loom were also studied.(Bilisik & Mohamed, 2009; Gokarneshan et al, 2010; Kopias, 2008) The warp yarn tension and weavability related to the end break during weaving were studied with relation of yarn physical property and weave limit.(Lappage, 2005; He et al, 2004; Bilisik & Demiryurek, 2011; Seyam, 2003) But, the fabric property related to the yarn tension on the air-jet loom was investigated using yarn tension meter(DEFAT) by Sabit Adanur and Jing Qi. Weft yarn tension was measured with yarn physical parameters such as yarn count, twist multiplier, yarn hairiness and yarn elongation. Fabric physical properties such as weight and thickness, air permeability, dimensional stability and abrasion resistance were analysed with average weft yarn tension of air-jet loom. Fabric stiffness, drape coefficient and wrinkle recovery were also measured and discussed with average weft yarn tension. Many weavers are using foreign looms made by Japan, and European countries such as Italy, Germany and Belzium. Especially, polyester fabrics woven by rapier looms show many defects such as thickness differences and color differences between edges on the right and left sides of the fabrics. Many weavers are thinking that the physical properties of fabrics including these defects are also different between fabrics woven by these various kinds of looms. And they are wondering how is the tension difference among various looms and how is the difference of the fabric mechanical properties according to the looms and the fabric positions with relation to the warp and weft weaving tensions on the various looms, respectively. But, any investigations about fabric physical properties according to the loom characteristics and about warp and weft tension variations according to the warp position among looms were not found yet. Therefore, this topic surveys the fabric physical properties according to the weaving looms, for this purpose, warp and filling yarn tensions during weaving were measured on the various looms and the fabric mechanical properties due to warp and weft tension differences were analysed using KES-FB system. In addition, weavability was also analysed by measuring warp tension variation according to the looms and the warp position. And the relationship between shed amount and warp tension on one fixed heald frame was surveyed according to the various looms and also fabric thickness according to the fabric width was measured for analysing fabric thickness variation with weaving loom characteristics.

2. The importance of the fabric mechanical properties for emotional garment's formability

The fabric formability of the worsted and wool/polyester blend fabrics widely used for suit garment for men and women is very important physical property. Formability is defined as ability of the fabric to be re-shaped from a plane fabric to the 3D form of clothing(Pavlinic, 2006). Fabric formability was predicted by many researchers(Lindeberg, 1960; Niwa et al, 1998; Shishoo, 1989; Postle & Dhingra, 1989; Ly et al, 1991). And fabric mechanical properties were used in the predicting fabric formability by Lindberg et al(Lindeberg, 1960), Niwa et al(Niwa, 1998), Yokura et al(Yokura, 1990) and Morooka et al(Morooka & Niwa, 1978). Lindberg formerly proposed formability by fabric bending and compression properties.

But, many equations related to the garment formability were suggested after developing KES-FB and FAST systems which are measuring devices of fabric mechanical properties.

Postle et al and Ly et al proposed formability equations using fabric mechanical properties measured by FAST system. Shishoo et al also suggested formability equation using KES-FB System. But, Niwa et al have published many papers related to the garment formability as a TAV(total appearance value) using fabric mechanical properties measured by KES-FB System.

On the other hand, many researches about mechanical property of the woven fabric according to the yarn and fabric parameters were carried out using KE-FB and FAST systems (Oh & Kim,1993). Among them, the PET synthetic fabric mechanical properties according to weft filament yarn twists, yarn denier and fabric density were analysed and discussed with these yarn and fabric structural parameters. On the other hand, the worsted fabric mechanical properties according to the looms such as rapier and air jet were also analysed and discussed with weaving machine characteristics (Kim & Kang, 2004; Kim & Jung, 2005). Similar studies were also performed using the PET and PET/Tencel woven fabrics (Kim et al., 2004). The researches related to the fabric mechanical property according to the dyeing and finishing processes were also carried out (Kim et al., 1995; Oh et al., 1993). According to the these studies, many factors such as the fabric structural parameters and processing parameters on the weaving and dyeing and finishing processes affects on the fabric mechanical properties which are governing garment's physical properties. Among these process parameters, weaving process is one important process which affects the fabric mechanical properties due to warp and weft tensions during weaving.

On the other hand, the large companies for production of worsted fabric have sequential production line such as spinning, weaving, dyeing and finishing processes, but some small companies have only one production line such as weaving, dyeing or finishing. So, large fabric lot processed in large companies is divided and delivered to the small companies by small fabric lot. Therefore, large quantity of fabrics are woven by various looms such as projectile, rapier and air-jet in various small weaving companies, and then, they are finished by various small finishing companies. It is known that these production system makes fabric physical properties such as hand, fabric thickness and shrinkage non-homogeneous. It is investigated that these non-homogeneity of the fabric physical properties may be originated from the difference of loom even though the loom setting is same.

Many researches related to the warp and weft yarn tensions during weaving were performed with relation to the stop marks on the fabrics. Among them, Helmut Weinsdörfer investigated that the distribution of the warp end tension over the warp width and how it is influenced by the weaving machine setting. This analysis carried out on the poplin fabric using Sulzer projectile loom and a comparative investigations performed on a downproof fabric using a flexible rapier loom with rod type temples and a projectile loom with needle temples. In addition, he studied warp yarn tension variation according to the shed geometry, warp brake setting and loom speed using narrow fabric loom(Jacob Muller). But these researches are only contributed to the weavability related to the mechanism of weaving machine, and there were no investigations about fabric physical properties according to the warp and weft tension differences on the positions of the fabrics such as center and edges and according to the different looms itself. Many weavers are using various kinds of looms made by Japan, and European countries such as Italy, Germany and Belzium. Especially, polyester fabrics woven by rapier looms show many defects such as thickness differences and color differences between edges on the right and left sides of the fabrics. Many weavers are thinking that the physical properties of fabrics including these defects are also different between fabrics woven by various types of looms.

3. Experimental

3.1 Worsted fabrics

3.1.1 Weaving of worsted fabrics on the air-jet and rapier looms

Worsted fabrics specimens were woven using Picanol rapier loom(model GTX-4-R) and air-jet loom(model PAT-4-R-A), respectively. Table 1 shows loom characteristics used for making fabric specimens and Table 2 shows fabric design related to the yarn and fabric structural parameters.

Loom characteristics	Rapier GTX-4-R	Air jet PAT-4-R-A
Harness motion	Electronic dobby	Electronic dobby
Weft insertion	Rapier	Main nozzle & relay nozzle
Let off motion	Let off continuously with electronic control	Let off continuously with electronic control
Winding grey fabrics system	Max. diameter : 600mm Range of density : 4.5~340ppi	Max. diameter : 600mm Range of density : 5.8~183ppi
Micro processor	Pick finding, let off tension	Pick finding, let off tension

Table 1. The Characteristics of loom used for making the specimen.

Fiber composition	Yarn count		Yarn twist (tpm)		Fabric structure	Density (per 10 Cm)		
							Grey fabric	Finished fabric
Wool 100%	Wp	Nm	Wp	Z770/	5 harness,	Wp	338	376.9
	Wf	2/72	Wf	S830	satin	Wf	220	224.4

Table 2. Specification of weave design

3.1.2 Finishing process of the worsted fabrics

The grey fabrics woven by rapier and air-jet looms were overlocked for processing in the finishing process simultaneously. Table 3 shows the finishing process and its condition.

Processes	Conditions
Gas singeing	100 m/min., gas 9 bar, both side singed
Sewing for making sack	12 mm/stitch
Scouring	Soaping for 20 min., rinsing for 30 min., Soaping for 45 min., rinsing for 50 min.
Continuous crabbing	80°C, 90°C, 95°C, 95°C, 95°C, 20°C
Shearing	20 m/min., 2 times for surface, once for back
Continuous decatizing	20 m/min.
Kier decatizing	19 m/min., pressure 30 kg/cm²

Table 3. Finishing processes and conditions

The specimens for measuring fabric mechanical properties were prepared by grey and finished fabrics woven by rapier and air-jet looms, respectively. Table 4 shows preparation of specimens and Fig. 1 shows the position on the fabrics related to the specimen number shown in the Table 4. 5 kinds of specimens were selected as one center position and 4 sides positions.

Fabric	Loom	Sample No.	Remark	Fabric	Loom	Sample No.	Remark
Ingray fabric (A)	Rapier (a)	1	Center	Finished fabric (B)	Rapier (a)	1	Center
		2	Side			2	Side
		3	Side			3	Side
		4	Side			4	Side
		5	Side			5	Side
	Air-jet (b)	6	Center		Air-jet (b)	6	Center
		7	Side			7	Side
		8	Side			8	Side
		9	Side			9	Side
		10	Side			10	Side

Table 4. Preparation of specimen.

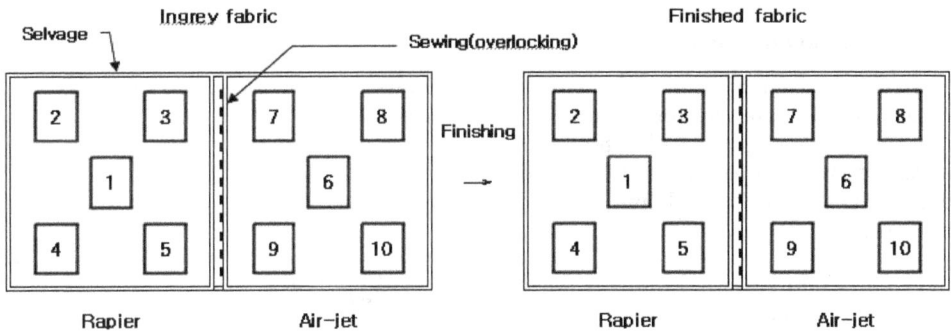

Fig. 1. Sampling position of specimen.

3.1.3 Weaving of the worsted fabrics on the projectile and air-jet looms

For surveying the warp and weft tension differences between projectile and air-jet looms and analysing the mechanical properties of the worsted fabrics for emotional garment with relation of these two looms characteristics, worsted fabric specimens were woven using projectile(Sulzer) and air-jet looms(Picanol PAT and OMNI), respectively. Table 5 shows looms characteristics used in this experiment. Table 6 shows specification of weave pattern.

	Projectile	Air jet	
	Sulzer-pu	Picanol	
		Pat	Omni
R P M	360	630	700
Reed width(mm)	2200	1830	
Harness motion	Mechanical dobby	Electronic dobby	
Weft insertion	projectile	Nozzle & sub nozzle	
Let-off motion	Electronic let-off	Electronic let-off	
Range of picking(mm)	9.1 - 230	5.8 - 183	
Microprocessor	Let off motion	Pick find, let off m/o	

Table 5. The characteristics of looms used for the test

Fiber Composition		Yarn count & TPM(Nm/tpm)	Fabric structure	Density (per 10cm)		Remark
				Grey fabric	Finished	
WP	Wool 93% Nylon 7%	1/40, 770 sirofil	5 harness satin	378	421	WP: 18ᴰ×5=90 WF:68 pick/in Width: 70.3″66.0″59.0″ Length: 97.5m 96.5y 91.0y
WF	Wool 100%	1/30, 1020		268	283	

Table 6. Specification of weave pattern

3.1.4 Finishing process of the worsted fabrics

Grey fabrics woven by Sulzer and two air-jet looms were cut by 3 yards, respectively and these were overlocked for processing in the finishing processes. Table 7 shows the finishing process for making finished fabrics. The specimens for measuring fabric mechanical properties were prepared by grey and finished fabrics woven by Sulzer and two air-jet looms, respectively. Fig. 2 shows the sampling positions on the fabrics for measuring fabric mechanical properties.

Process	Conditions
Gas singeing	100 m/min, gas pressure: 9 bar, both side singed
Solvent scouring	25 m/min,
Scouring	50°C, Soaping for 20min, rinsing for 30min Soaping for 45min, rinsing for 50min
Dry	110°C, Over feeding ratio 5%
Fabric dyeing	100°C,
Dry	110°C, Over feeding ratio 5%
Shearing	20 m/min, 2 times for surface, ones for back
Continuous decatizing	15 m/min,
Kier decatizing	19 m/min, pressure: 30kg/cm²

Table 7. Finishing process and conditions

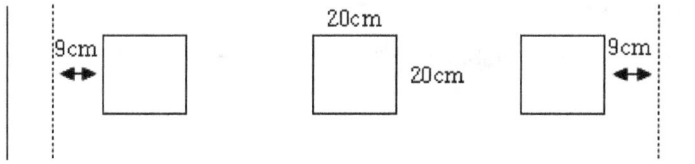

Fig. 2. Preparation of specimens for KES-FB System test.

3.1.5 Weaving of the worsted fabrics on the three kinds of rapier looms

For surveying the warp and weft tension differences among 3 types of rapier looms and analyzing the mechanical properties of these worsted fabrics for emotional garment with relation of these 3 kinds of rapier looms characteristics, 5 harness satin weave worsted fabrics were woven using FAST-R, THEMA-11-E and Picanol-GTX looms, respectively. Table 8 shows the specification of weave pattern of these worsted fabrics. Table 9 shows the characteristics of these 3 kinds of looms. Finishing process was same as shown in Table 7.

Fiber Composition		Yarn count (Nm, tpm)	Fabric structure	Density (per 10cm)		Remark
				grey fabric	finished	
WP	wool 93% nylon 7%	1/40, 770 sirofil	5 harness satin	378	421	WP: 18^D×5=90ends WF: 68picks Width: 70.3"66.0"59.0" Length: 97.5m 96.5y 91.0y
WF	Wool 100%	1/30, 1020		268	283	

Table 8. Specification of weave pattern

Division	Loom		
	FAST-R	THEMA-11-E	PICANOL-GTX
R P M	520	550	580
Reed width (mm)	2200	2100	1900
Harness motion	electronic dobby	electronic dobby	electronic dobby
Weft insertion	rapier	rapier	rapier
Let-off motion	Electronic let-off	Electronic let-off	Electronic let-off
Range of picking (mm)	4.8 - 282	7.6 - 198	4.5 - 340
Microprocessor	pick find let off m/o	pick find let off m/o	pick find let off m/o

Table 9. The characteristics of loom used for the test

The sampling position on the fabrics for measuring fabric mechanical properties was same as shown in Fig. 2.

3.2 Polyester filament fabrics

3.2.1 Weaving of PET filament fabrics by 2 kinds of rapier looms (Omega® and Picanol®)

PET fabrics were woven using 2 kinds of rapier looms (Omega® and Picanol®) for analysing the tension differences and fabric mechanical properties due to warp and weft tension differences. Table 10 shows weave design of woven fabrics. Table 11 shows the characteristics of two rapier looms.

Fiber composition		Yarn count	Fabric structure	Density/inch		Remark
				Grey	Finished	
Warp	Polyester 100 %	75D / 36F	5 Harness	168	261	42D ×4 =168end/in Pick: 86pick/in
Weft	Polyester 93.5 % Polyurethane 6.5 %	100D/192F + 30D spandex covering		86	98	

Table 10. Specification of weave design

Division Loom	OMEGA (Textec, Korea)	PICANOL-GTX (Belgium)
Maximum RPM	520	580
Maximum reed width	2100 (mm)	1900 (mm)
Harness motion	Electronic dobby	Electronic dobby
Let off motion	Electronic let off	Electronic let off
Microprocessor	Pick find motion Let off motion	Pick find motion Let off motion

note: running speed : 470rpm

Table 11. The characteristics of loom used in this study

3.2.2 Finishing process of the PET fabrics

These grey fabrics woven by two rapier looms were processed on the finishing process which is shown in Table 12.

Process	Condition
Cylinder dryer	130 °C × 60 m/min
Scouring	Speed : 35 m/min , Temperature : 60-95-60°C
Pre-setting	210 °C × 27 m/min
Dyeing	130 °C × 40 min
Final-setting	210 °C × 30 m/min

Table 12. Finishing processes and conditions

3.2.3 Weaving of PET filament fabrics on the 2 kinds of rapier looms (Omega® and Vamatex®)

For surveying the tension differences between Omega and Vamatex looms and analysing fabric mechanical properties using KES-FB system according to warp and weft tension differences, fabric was designed as 5 harness satin weave using 150d/48f warp and 200d/384f weft polyester filaments, and was woven by Omega®-Panter rapier loom by Textec Co. Ltd and P1001es rapier loom by Vamatex Co. Ltd., respectively.

These grey fabrics were processed on the same dyeing and finishing processes which was shown in Table 12. Weavability was also analysed by measuring warp tension variation according to the warp position. The relationship between shed amount and the warp tension on one fixed heald frame was surveyed, and the relationship between end breaks and warp and weft tensions was also discussed.

Table 13 shows specification of weave pattern. Table 14 and Fig. 3 show the characteristics of rapier looms used in this study.

	Fiber Composition	Yarn Count	Fabric Structure	Density/inch	
				Grey	Finished
Warp	Polyester 100%	150D / 48F	5 Harness satin	102.5	158
Weft	Polyester 93.5% Polyurethane 6.5%	200D/384F + 40D spandex		72	83

Table 13. Specification of weave pattern

Loom Division	Omega-Panter	Vamatex-P1001es
R.P.M (Max.)	466 (520)	423 (580)
Maximum reed width	2100	1900
Harness motion	Electronic dobby	
Let-off motion	Electronic let-off	
Microprocessor	Pick find motion Let-off motion	

Table 14. The characteristics of rapier looms used in this study

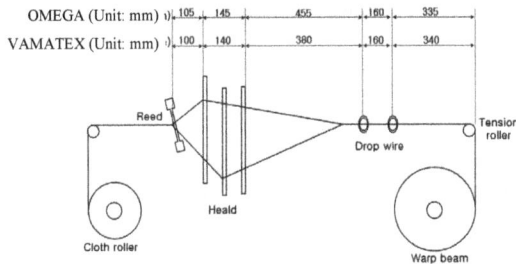

Fig. 3. Specification of test looms.

4. Measurement and assessment

4.1 Assessment in the weaving process

4.1.1 Warp and weft tensions

Weaving tensions on 7 kinds of weaving machines were measured using Dafat tension meter which is shown in Fig. 4. Measured position was between tension roller and drop wire on the loom. Various yarn tensions on the each heald frame from 1st to 5th were measured at the vicinity of the center of loom.

Yarn tension along full width of each loom was also measured on the 5th heald frame from left side to right on the back of the loom. The weaving efficiency in each loom was measured by number of end breaks both warp and weft per 100,000 picks.

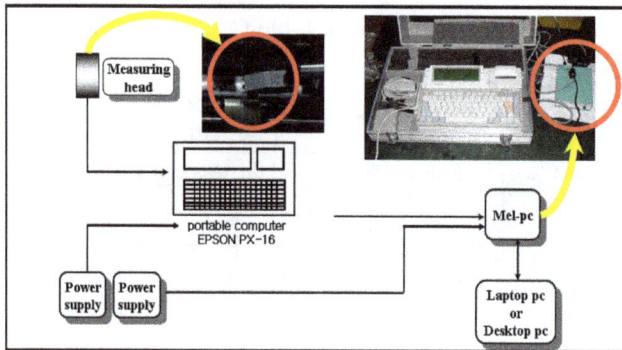

Fig. 4. DEFAT tension meter

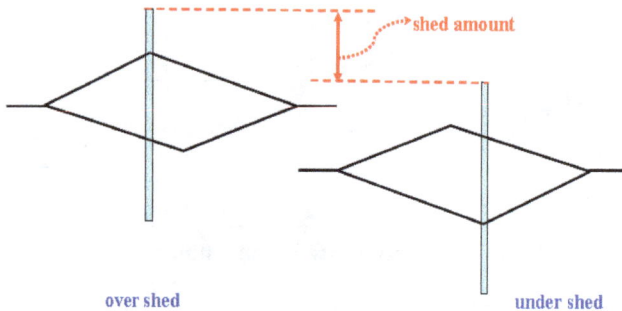

over shed   under shed

Fig. 5. Diagram of shed amount

4.1.2 Measurement of shed amount

At the upper state of the heald frame, the distance from the fixed guide of heald frame to the upper line of frame is the amount of upper shedding, the lowest state of the heald frame is the amount of lower shedding.

The warp movement is calculated by the difference between the amount of upper shedding and lower shedding, which is shown in Fig. 5.

4.1.3 Process shrinkages on the warp and weft directions

Fabric shrinkages of PET on the warp and weft directions in the each step on the dyeing and finishing processes were calculated using warp and weft fabric densities as equation (1) and (2) shown in bellows.

Warp shrinkage (%)

$$= \frac{\text{Weft density before process step} - \text{Weft density after process step}}{\text{Weft density before process step}} \times 100 \quad (1)$$

Weft shrinkage (%)

$$= \frac{\text{Fabric width before process step} - \text{Fabric width after process step}}{\text{Fabric width before process step}} \times 100 \quad (2)$$

4.2 Measurement of fabric physical properties

4.2.1 Fabric thickness

The positions for measuring fabric thickness to the direction of the fabric width and to the longitudinal direction on the right and left sides of the fabric were selected as shown in Fig. 6 and Fig. 7. The specimens for measuring fabric mechanical properties were prepared as shown in Fig. 8.

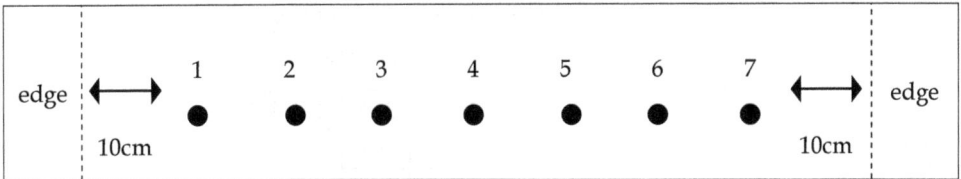

Fig. 6. Measured points of thickness to the direction of the fabric width.

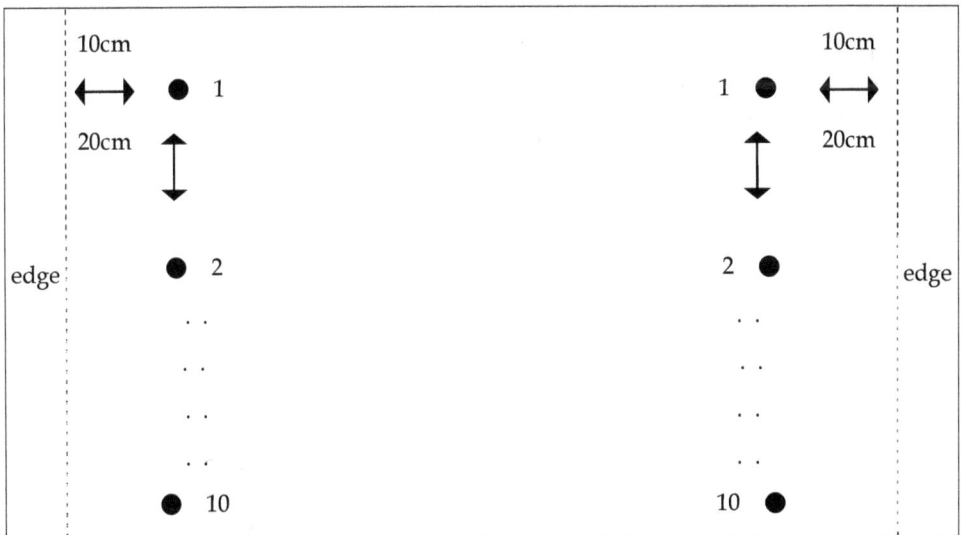

Fig. 7. Measured points of thickness on the right & left sides of Fabric

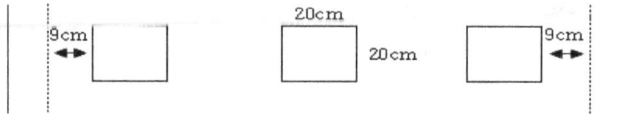

Fig. 8. Preparation of specimens for the test using KES-FB System

4.2.2 Fabric mechanical properties

16 mechanical properties of gray and finished fabrics such as tensile, bending, shear, compression and surface woven from 7 kinds of looms were measured by KES-FB system.

5. Results and discussions

5.1 Loom efficiency of worsted fabrics

5.1.1 Projectile and air-jet loom

Table 15 shows loom efficiency and stop number of each loom. Fig. 9 and 10 show rpm, efficiency and percentage of end break of Table 15.

Division	Loom	RPM	Effi.	Keeping looms /person	Pick	Loom stop number				Stop %	
					Hour	Total	Warp	Weft	Other	Warp	Weft
Proje-ctile	SULZER-PU	300	85.5	8	100,000	34.1	21.0	4.9	8.2	61.6	14.4
					hour	5.3	3.2	0.8	1.3		
Air - Jet	PICANOL-OMNI	500	67.0	10	100,000	46.6	9.9	35.8	0.9	21.2	76.8
					hour	9.4	2.0	7.2	0.2		
	PICANOL-PAT	520	65.0	10	100,000	46.2	22.2	21.6	2.4	48.1	46.8
					hour	9.4	4.4	4.4	0.5		

Table 15. Efficiency and stop number of weaving loom for the test

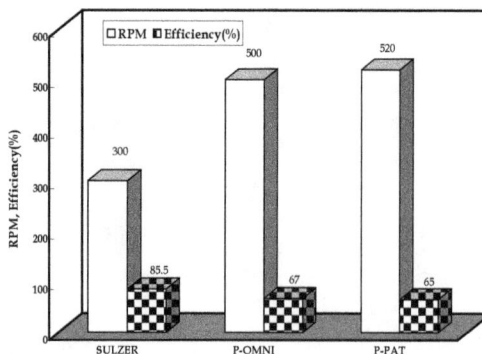

Fig. 9. Diagram between rpm and efficiency of various looms

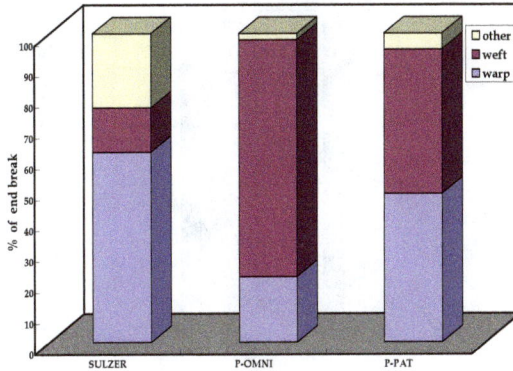

Fig. 10. Percentage of end break of warp and weft to the three looms

As shown in Fig. 9, Sulzer shows high efficiency, on the one hand, air-jet looms, both P-Omni and P-PAT show low efficiency. As shown in Fig. 10, Sulzer loom has high warp-break but air-jet loom has high weft-break.

5.1.2 Rapier looms

Table 16 shows loom efficiency and stop number of each rapier looms. Fig. 11 and 12 show rpm, efficiency and percentage of end break of Table 16.

Loom	RPM	Effi.	Keeping loom/person	Pick	Loom stop number				Stop %	
				hour	total	warp	weft	other	warp	weft
FAST - R	382	82.1	7	100,000	30.4	20.9	8.2	1.3	68.8	27.0
				hour	5.8	4.0	1.5	0.3		
THEMA-11-E	399	75.6	8	100,000	45.1	28.4	11.1	5.6	63.0	24.6
				hour	8.1	5.1	2.0	1.0		
PICANOL-GTX	402	67.8	10	100,000	41.7	31.7	6.2	3.8	76.0	14.9
				hour	6.8	5.2	1.0	0.6		

Table 16. Efficiency and stop number of weaving loom for the test

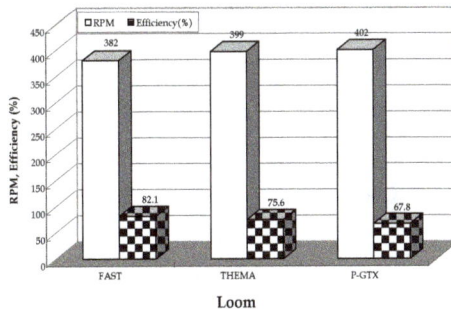

Fig. 11. Diagram between rpm and efficiency of test looms

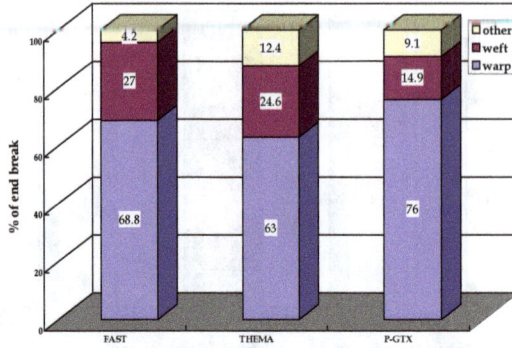

Fig. 12. Percentage of end break of warp & weft to the three looms

As shown in Table 16, warp breaks are much higher than weft breaks. It is shown that the loom efficiency of FAST is the highest. The reason why seems to be low rpm and low keeping loom per person of FAST loom. It is also seen in Fig. 11 and 12 that Picanol has high warp break and FAST has high weft break compared to other looms.

5.2 Warp and weft tensions of the worsted fabrics according to the loom characteristics

5.2.1 Projectile and air-jet looms

Fig. 13 shows warp tension variation according to the warp position on the Sulzer loom. As shown in Fig. 13, warp yarn tension variation on the vicinity of the center part of the fabric is higher than those of left and right parts of the fabrics and the tension of the right side of fabric is a little lower than that of left side of fabric. The warp yarn tension variation between right and left sides of fabric makes color difference between right and left of fabric, this phenomena deteriorates garment quality in clothing factory.

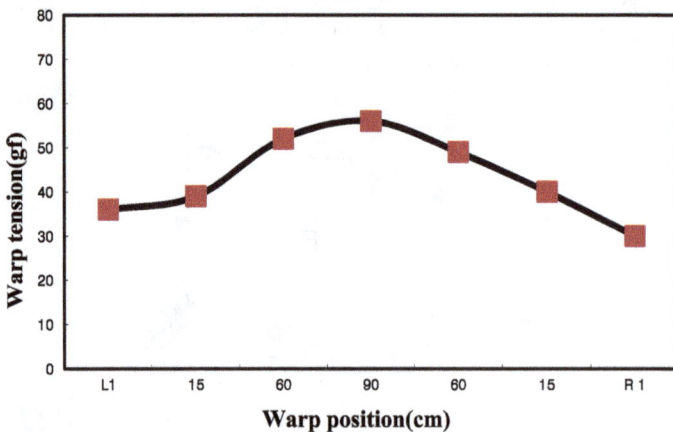

Fig. 13. Warp tension according to the warp position (SULZER Loom)

Fig. 14. shows one cycle warp tension variation which was measured on the 15cm position to the left of the loom.

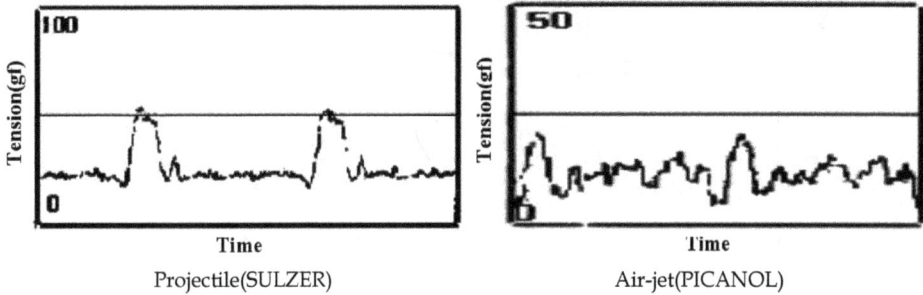

| Projectile(SULZER) | Air-jet(PICANOL) |

Fig. 14. The graphs of warp yarn tension of two looms

The warp tension variation during one cycle of the Sulzer loom was ranged between 22gf and 52gf, but, Picanol loom was ranged from 6gf to 23gf and shows 4 kinds of successive peaks, on the other hand, Sulzer has one large peak.

Table 17 shows shedding amount and warp tension of the each loom with bar and ring temple respectively. It was shown that warp tensions were slightly increased with increasing shedding amount in both Picanol Omni and PAT, But in Sulzer loom, is preferably decreased with increasing shedding amount.

Loom		Heald position	1	2	3	4	5	Average	Percentage	Warp breakage	Percentage
Bar temple	P-OMNI	Shedding amount(mm)	73	82	93	102	118	93.6	108.1	9.9	47.4
		Tension(gf)	36	36	39	40	43	38.0	115.2		
Ring temple	SUL-PU	Shedding amount(mm)	85	86	86	90	88	89.0	102.8	21.0	100.5
		Tension(gf)	66	65	58	59	56	60.8	184.2		
	P-PAT	Shedding amount(mm)	87	92	104	111	17	102.2	118.0	22.2	106.2
		Tension(gf)	40	45	54	50	56	49.0	148.5		

Table 17. The shedding amount and warp tension of the test weaving looms

Fig. 15. shows relationship between warp yarn tension and shed amount of each looms.

As shown in Fig. 15, the shed amount of P-Omni loom with bar temple is larger than that of Sulzer loom with ring temple, but warp tension of P-Omni has lower value by 37% compared to Sulzer, and also has lower value by 23% compared to P-PAT loom. This phenomena demsnstrates that air-jet loom with bar temple can contribute to the increase of weavability.

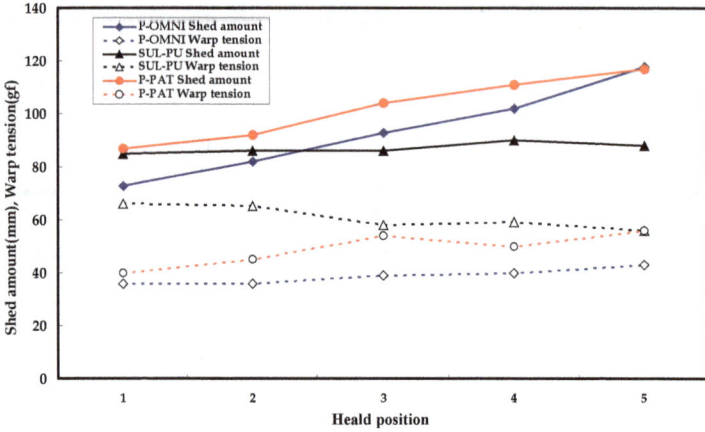

Fig. 15. Relation between warp yarn tension and amount of shed

Fig. 16 shows one cycle weft tension variation of Sulzer loom. As shown in Fig. 16, one peak is revealed by an instant tension during flying of projectile. The weft tension variations of ari-jet looms (P-PAT and P-Omni) could not measure because of much movement of nozzle.

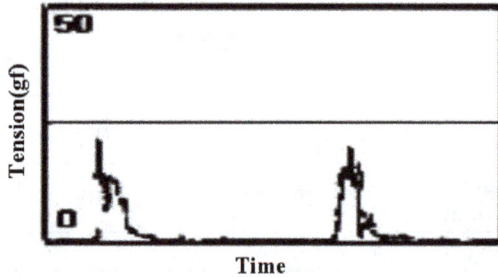

Fig. 16. The graph of weft yarn tension of sulzer loom

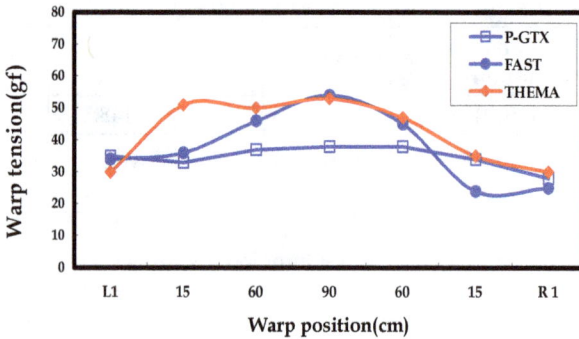

Fig. 17. Warp tension according to the warp position

5.2.2 Rapier loom

Fig. 17. shows warp tension variations at the full widths of the 3 types of looms (P-GTX, FAST, THEMA looms).

As shown in Fig.17, warp tension variation of Picanol loom(P-GTX) attached with bar temple according to the position of loom width direction is smaller than those of FAST and THEMA looms attached with ring temples. It was shown that the warp tensions on central part of the looms is much higher than those on left and right sides of the looms both FAST and THEMA looms used by ring temples. The tension on the right side of the loom is about 10% lower than that on left side of the loom. Fig.18 shows one-cycle warp tension variation of the 3 types of rapier looms on the 15cm position from left side of the loom.

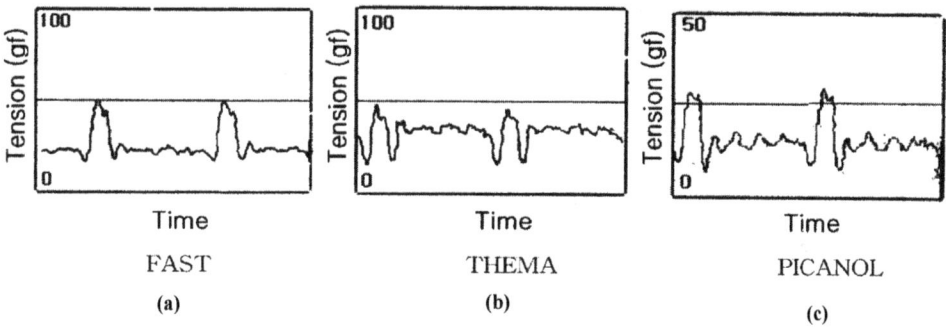

Fig. 18. The graph of warp yarn tension on the three rapier looms.

As shown in Fig.18, tension on FAST loom was distributed ranged between 25gf and 50gf, and ranged between 35gf and 49gf for THEMA loom, and ranged between 13gf and 28gf for Picanol. Especially, the tension variation on the Picanol loom revealed 4 types continuous small peaks and one large peak, of which weave design was 5 harness satin. For comparing warp tension variation according to the weave pattern, warp tension variation according to the loom position and loom types was measured on the 7th heald frame with 8 harness satin weave pattern and 2/60Nm warp yarn count.

Fig.19 shows warp tension variations on the THEMA and P-GTX looms. As shown in Fig.19 warp yarn tension of the 8 harness satin was much higher than those of 5 harness satin(Fig.13). It was shown that yarn tension change according to the warp position on the Picanol loom with bar temple was much less than that of the THEMA loom with ring temple.

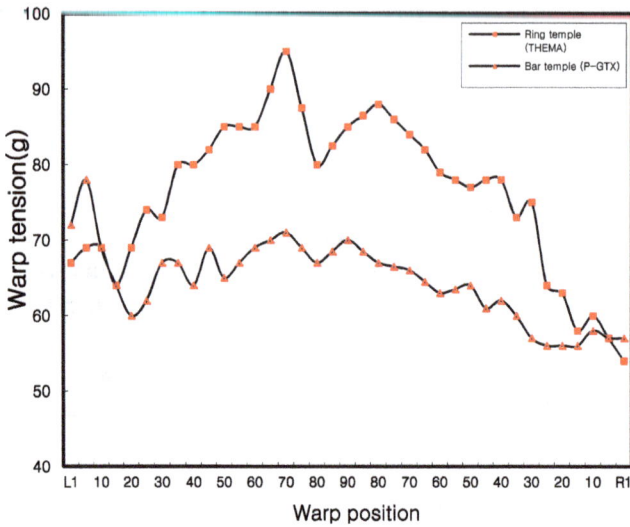

Fig. 19. Warp tension variation of warp position according to the loom with temple.

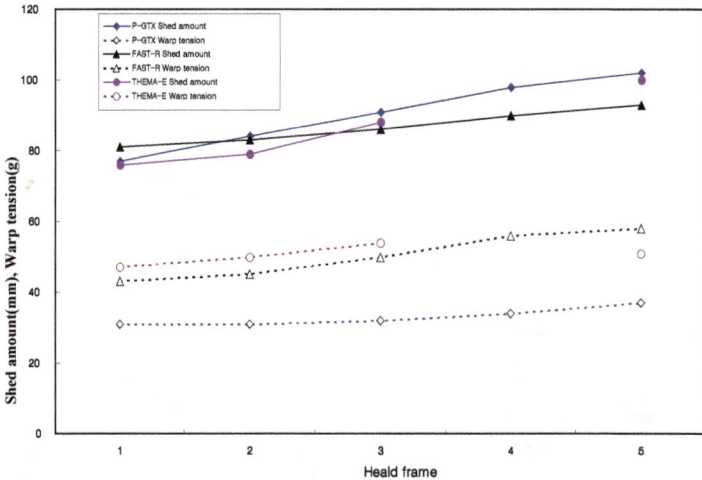

Fig. 20. Relation between warp yarn tension and amount of shed.

Table 18 shows shedding amount and warp tension of the 3 types of rapier looms attached with bar and ring temples. Fig.20 shows these variations according to the heald frame.

As shown in Fig.20, shedding amount was increased from 1st heald frame to 6th one, and warp tension was also increased from 1st heald frame to 6th one, which means that warp tension is proportional to shedding amount. Table19 shows weft tension and end break of weft on the 3 kinds of rapier looms.

Loom			1	2	3	4	5	6	mean	end break of warp
					Heald number					
Bar temple	P-GTX	Shedding amount(mm)	77	84	91	98	102	-	90.4	31.7
		warp tension(gf)	31	31	32	34	37	-	33.0	
Ring temple	FAST-R	Shedding amount(mm)	81	83	86	90	93	-	86.6	20.9
		warp tension(gf)	43	45	50	56	58	-	50.4	
	THEMA-E	Shedding amount (mm)	76	79	88	-	100	105	89.6	28.4
		warp tension(gf)	47	50	54	-	51	53	51.0	

Table 18. The shedding amount and warp tension of the test weaving looms

Loom		FAST-R	THEMA-E	PICANOL-GTX
Weft tension	Max.	84	84	99
	Min.	0	0	0
RPM		382	399	402
Break number of weft		8.2	11.1	6.2

Table 19. Weft tension and end break of weft on the 3 looms.

Fig.21 shows histogram of these data.

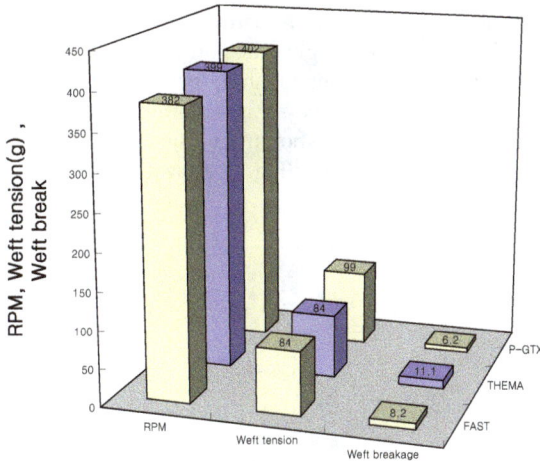

Fig. 21. Diagram between end break and yarn tension of weft to the three test looms.

It was shown that loom rpm and weft tension are less correlated and end break of weft on the Picanol was less than those of FAST and THEMA even though weft yarn tension of Picanol was highest.

Fig.22 shows one cycle weft tension variation on the 3 kinds of rapier looms.

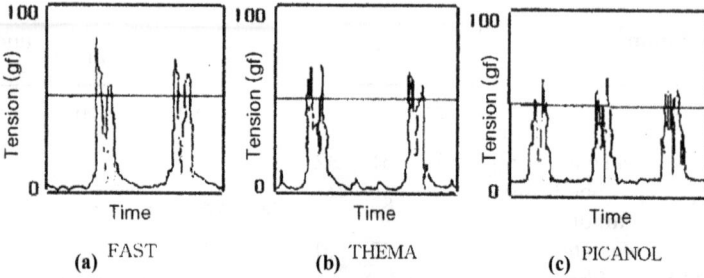

(a) FAST   (b) THEMA   (c) PICANOL

Fig. 22. The graph of weft yarn tension variation of three rapier looms.

As shown in Fig.22, 2 kinds of high peaks revealed on the rapier looms, weft is gripped on the gripper, at this moment, tension is highly loaded and 1st rapier handed the weft yarn to 2nd rapier, 2nd high yarn tension peak is at this moment highly loaded, so 2 kinds of peaks are shown on this Fig.22.

And it was shown that maximum peak tension was ranged from 65gf to 70gf, but in FAST loom, ranged from 85gf to 90gf and Pocanol shows the lowest tension value.

5.3 The physical properties of the worsted fabrics according to the loom characteristics

5.3.1 Fabric extensibility

For surveying the effects of the looms and finishing process to the fabric extensibility, tensile properties of gray and finished fabrics were measured using KES-FB system. For five kinds of looms, gray fabrics of left, center and right sides on the fabric were used as a specimens and then gray fabrics were processed on the finishing process. The processing method in the finishing was adopted by two ways. One way was continuous processing with five kinds of gray fabrics by sewing(overlocking) as shown in Fig. 1, the other way was discrete processing with five kinds of gray fabrics. Fig.23 shows extensibility of these gray and finished fabrics with various looms.

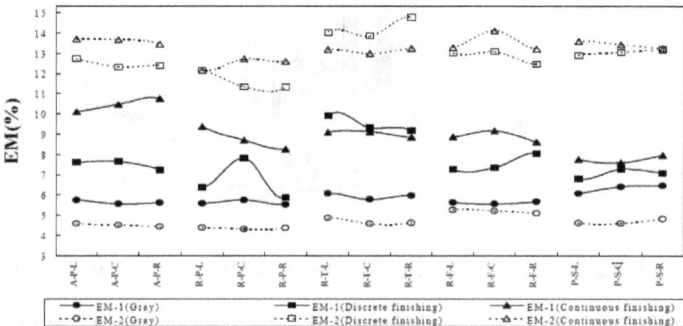

Note : A-P-L : Air jet Picanol (OMNI) Left   R-P-C : Rapier Picanol (GTX) Center
R-T-R : Rapier Thema (11E) Right   R-F-L : Rapier Fast (11E) Left
P-S-C : Projectile Sulzer (PU) Center   1: warp, 2: weft

Fig. 23. Fabric extensibility with various looms. (EM-1 : Warp, EM-2 : Weft)

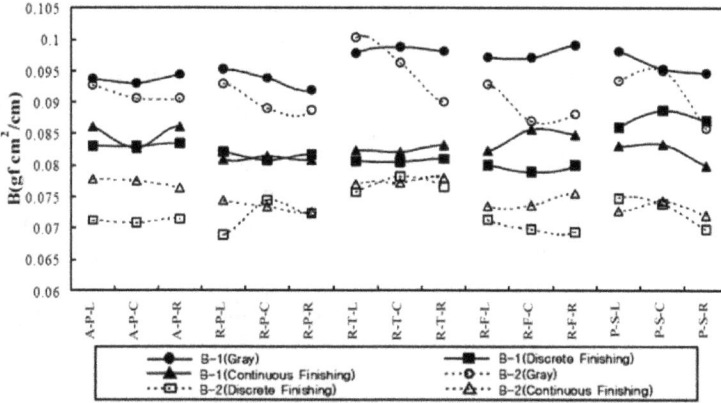

Fig. 24. Bending property of gray and finished fabrics woven by various looms.

As shown in Fig.23, for the warp extensibility of gray fabric, projectile(Sulzer) and rapier(THEMA) showed high values, then these looms showed high warp yarn tension and low shed amount for weaving as shown in Fig.15 and 20.

This means that the higher warp yarn tension and the lower shed amount, the more extensible of gray fabric. And the variation of extensibility on the right, center and left sides of gray fabric woven by Sulzer and THEMA weaving looms is also larger than those of other looms.

But, it is shown that these variations of gray fabric among various looms are less than those due to the method of finishing process. As shown in Fig.23, the warp extensibility of finished fabric for the continuous (—▲—) and discrete (—■—) finishing shows quite difference compared to gray fabric. And comparing between continuous and discrete finishing, the variation of warp extensibility among various looms by continuous finishing (—▲—) is smaller than that of discrete finishing (—■—). That results means that discrete finishing makes fabric extensibility deviating each other.

Especially, the variation on the right, center and left sides of fabric of warp extensibility of finished fabric (—▲—) woven by air-jet(Picanol A-P-L,C,R) and rapier(Picanol-GTX R-P-L,C,R) looms is larger than that of other looms. And comparing with weft extensibility of finished fabric between continuous and discrete finishing processes, continuous finishing is more even than that of discrete finishing. Among five looms, the variation of fabric extensibility of air-jet(Picanol-OMNI, A-P-L,C,R) and projectile(Sulzer, P-S-L,C,R) looms is the smallest both warp and weft directions, gray and finished fabrics, continuous and discrete finishing, respectively.

5.3.2 Fabric bending property

Fig.24 shows bending property of gray and finished fabrics woven by various looms.

First, at the state of gray fabric, warp bending rigidity of gray fabric woven by Picanol looms(air-jet and rapier), which showed low warp yarn tension as shown in Fig.15 and 20, shows low values compared with other rapier looms(THEMA, FAST) and projectile(Sulzer). After gray fabric were finished, the effects of high tension during weaving were remained for the case of continuous finishing, i.e. the variation of bending rigidity among right, center and left sides on the fabrics woven by Rapier looms(Picanol and THEMA) was not shown on the finished fabrics, which showed lower warp tension variation during weaving as shown in Fig.13 and 17. For the rapier(FAST) and projectile(Sulzer) looms, the bending rigidities on the center of the finished fabrics showed the highest values comparing to the right and left sides on the fabrics, which is originated from high warp tension during weaving. And it is shown that there was no variation of the bending rigidity according to the finishing method for the fabrics subjected under low warp yarn tension during weaving, on the other hand, for the fabrics subjected under the high warp yarn tension during weaving, the variation was high as shown in Fig.24.

5.3.3 Fabric shear property

Fig.25 shows shear rigidities of gray and finished fabrics with various looms.

Shear modulus of gray fabrics like bending rigidity showed the variation according to the weaving looms as shown in Fig.25, i.e. High weaving tension makes shear modulus of gray fabric high, i. e. shear modulus of gray fabrics (—●—) woven by Picanol (air-jet and rapier), which showed low warp tension during weaving, were lower than those of gray fabrics woven by other rapier looms (Thema, FAST) and projectile (Sulzer) as shown in Fig.25. But these variations disappeared after finishing, then shear modulus of finished fabric between continuous (—▲—) and discrete (—■—) finishing showed large difference. These phenomena demonstrate the importance of finishing process to the fabric shear property, which can be compared to the importance of weaving process to the fabric bending property.

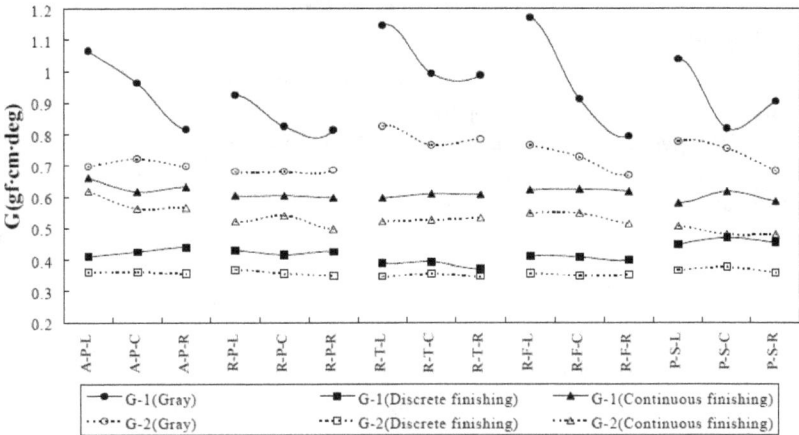

Fig. 25. Shear rigidities of gray and finished fabrics with various looms.

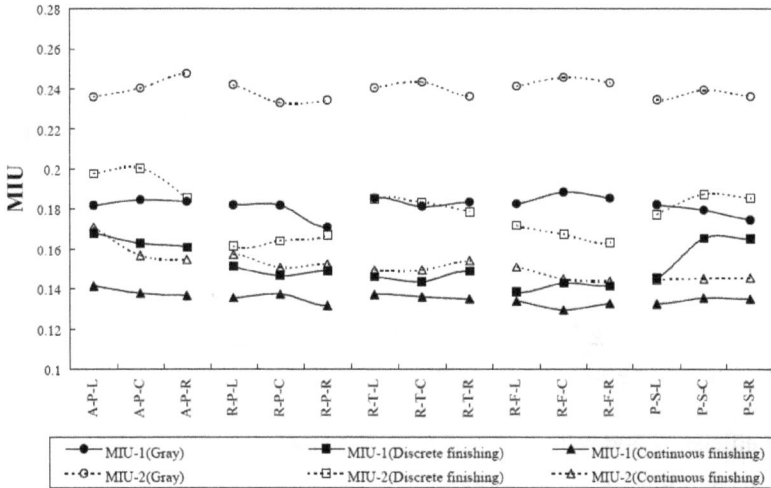

Fig. 26. Coefficient of friction of gray and finished fabrics with various looms.

5.3.4 Fabric surface property

Fig.26 shows coefficient of friction of gray and finished fabrics woven by various looms.

As shown in Fig.26, the variation of coefficient of friction (MIU) of gray fabrics (—●—) according to the various looms was less than that between right and left sides on the fabric, and the variation of the MIU of finished fabrics by continuous method (—▲—) according to the various looms was also much less than that between right and left sides on the fabrics. But the variation of finished fabric by discrete finishing method (—■—) showed big differences according to various looms and right, center and left sides on the fabric. This result means that discrete finishing makes fabric surface property deviating.

These phenomena also demonstrate the importance of finishing process to the fabric surface property.

5.3.5 Fabric thickness

Fig.27 shows fabric thickness of gray and finished fabrics with various looms.

As shown in Fig.27, the variation of the fabric thickness after continuous finishing (—▲—) did not show anymore among various looms and right, center and left sides on the same fabric. But for the discrete finishing (—■—), these variation was shown among looms and according to the position on the fabric. This shows that finishing process is still important like weaving tension for the control of even fabric thickness.

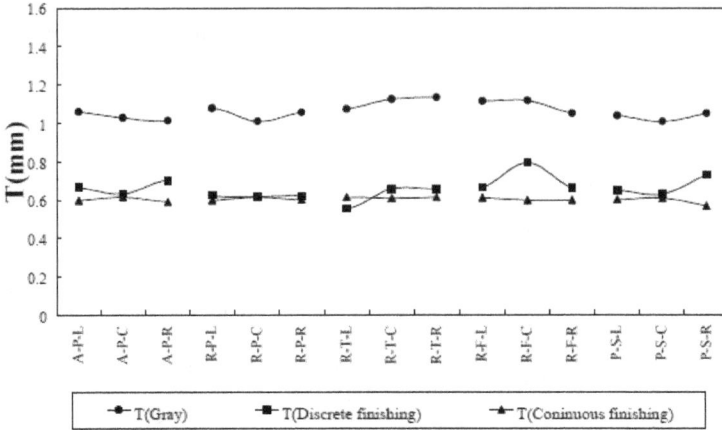

Fig. 27. The fabric thickness according to the various looms.

5.4 Loom efficiency and warp and weft tension variations of PET fabrics

5.4.1 Omega and Picanol rapier looms

Table 20 shows the efficiency and stop number of 2 rapier weaving looms (Omega and Picanol) during weaving PET fabric. Fig.28 shows warp tension according to the warp position on the 2 kinds of rapier looms.

Division Loom	RPM	EFF (%)	Stop number of loom				Stop (%)		Remark
			Warp	Weft	Other	Total	Warp	Weft	
OMEGA	470	95.62	2.8	2.8	2.8	8.4	33.3	33.3	WP: 75D
	465	73.74	39	4	27	70	55.7	5.7	
	461	98.57	0	0	2	2	0	0	
	465	97.18	3	0	3	6	50.0	0	
	461	85.26	0	24	14	38	0	63.2	
Average	464.4	92.07	8.8	6	9.6	24.4	36.1	24.6	
PICANOL-GTX	472	96.85	9.7	2.8	-	12.5	77.6	22.4	WP: 75D
	466	95.85	3	6	1	10	30.0	60.0	
	470	94.62	3	2	1	6	50.0	33.3	
	469	89.44	2	22	2	26	7.7	84.6	
	461	83.64	2	33	1	36	5.6	91.7	
Average	467.6	92.08	3.4	13	1	17.4	19.5	74.7	

Table 20. Efficiency and stop number of weaving loom during the test

As seen in Table 20, efficiency showed same value as 92% both Picanol and Omega looms, respectively. It is shown that warp breakage of Omega loom was much higher than that of Picanol loom. On the other hand, weft breakage of Omega was less than that of Picanol. The reason why is due to high warp tension of Omega loom and high weft tension of Picanol loom as shown in Fig.28 and 31. As shown in Fig.28, the warp tension according to the full width of Omega loom was much higher than that of Picanol. And the warp tensions of center parts of the loom were higher than those of both edges of fabric. It is explained that the filling yarn in the middle was held firm and tightly stretched by both sides, as it has to be beaten into the warp ends, in the edge zones, the filling can relax a little from the selvedge, the extent of relax is dependent on the filling insertion system, temples and the selvedge clap, as a result, the filling is woven in a little less firmly in the middle than at the selvedges, this means that they must bind more firmly in the center than at the edges, however the length of all the warp ends coming from the warp beam are practically the same length, those in the middle must elongated more. And it is shown that the average warp tension on center area of the Omega loom is 40~45gf and 35~40gf for Picanol loom, on edge part of the Omega loom is 30~35gf, 25~30gf for Picanol loom, so Omega loom shows 15~20gf higher tension than that of Picanol loom. In addition, high tension variation on edge part of Omega loom is shown, on the other hand tension variation on center part of Picanol loom is also shown. Fig. 29 shows real warp tension variations of Omega and Picanol looms, respectively. It is shown that Omega's warp yarn tension is much higher than that of Picanol. And 4 successive peaks and one high peak are shown both Omega and Picanol looms.

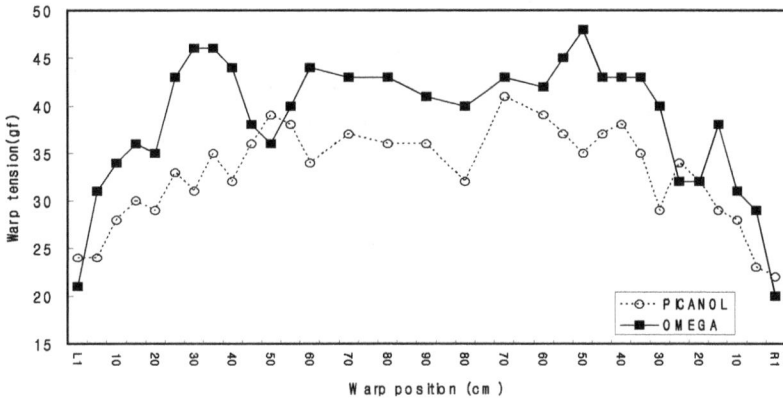

Fig. 28. Warp tension according to the warp position.

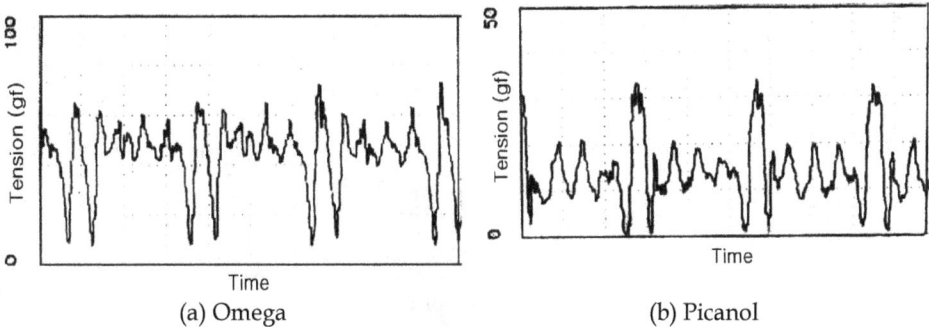

(a) Omega   (b) Picanol

Fig. 29. The graph of warp yarn tension of OMEGA and Picanol rapier looms.

Fig. 30 shows the shed amount and warp tension of each 10 heald frames at the center part of the loom both Picanol and Omega looms, respectively. As shown in Fig.30, shed amounts both Picanol and Omega looms were increased from front heald to back one even though warp tension of each heald is almost same, but very slowly increased. That means that there is some relationship between shed movement and warp yarn tension on the each heald frame, and it is needed for clean shedding. Fig. 31 shows weft yarn tension variation of Omega and Picanol, respectively. As shown in Fig. 31, weft yarn tension of Picanol was higher than that of Omega. And it was shown that Picanol has distinct 2 peaks but Omega has unstable and subtle 4 or 5 peaks.

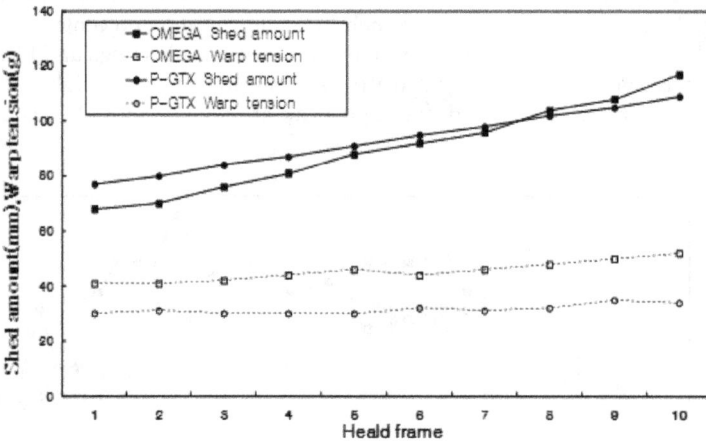

Fig. 30. Relation between warp yarn tension and amount of shed.

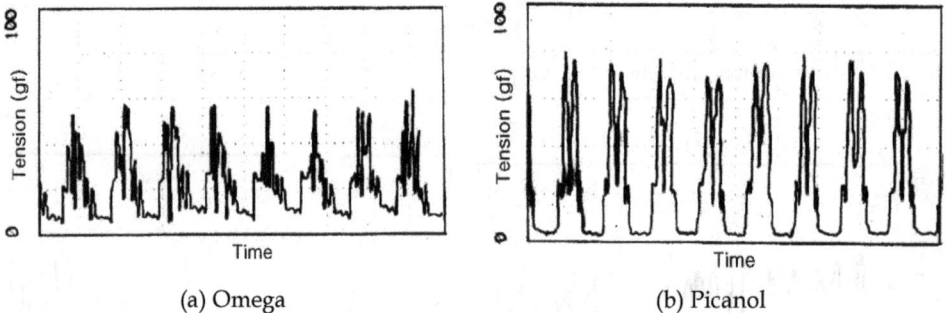

(a) Omega   (b) Picanol

Fig. 31. The graph of weft yarn tension of Omega and Picanol rapier looms.

5.4.2 Omega and Vamatex rapier looms

Table 21 shows the efficiency and stop number of two rapier looms (Omega and Vamatex). Fig. 32 shows warp tension according to the warp position on the Omega and Vamatex looms.

Loom	RPM	Efficiency (%)	Stop number of loom				Stop (%)	
			Warp	Weft	Other	Total	Warp	Weft
Omega	466	97.45	13	3	0	16	81.2	18.8
Vamatex	423	99.57	3	0	0	3	100	0

Table 21. Efficiency and stop number of weaving loom for the test

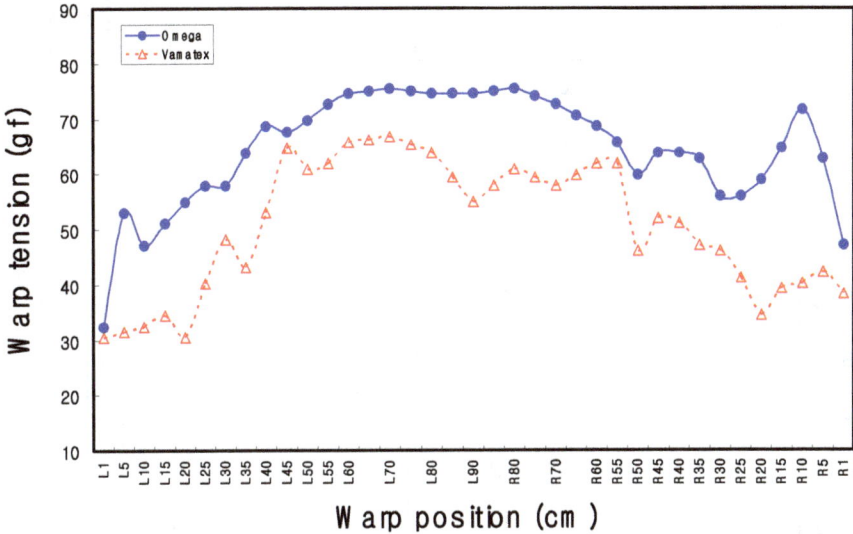

Fig. 32. Warp tension according to warp position

As shown in Table 21 and Fig. 32, efficiency shows 97.45% and 99.57%, respectively. The warp tensions according to the full width of Omega loom were higher than those of Vamatex. And it was shown that the warp tension of center part of the loom was higher than those of both edges parts of the loom. Fig. 33 shows one cycle warp yarn tension variation of 2 types of rapier looms. As shown in Fig. 33, warp yarn tension distribution in Omega was ranged from 50gf to 60gf, but ranged from 40gf to 60gf for the Vamatex. And four or five successive peaks and one high peak were shown in the Fig. 33.

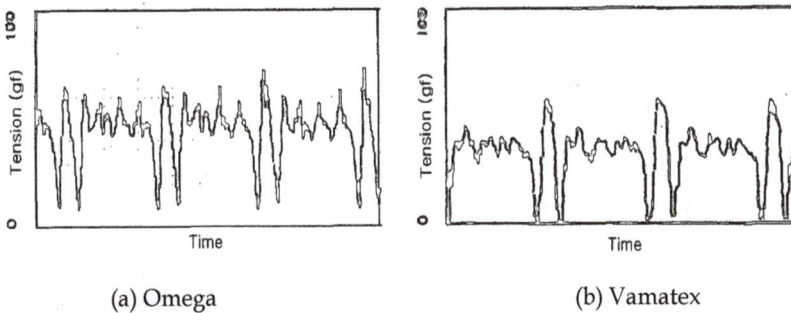

(a) Omega   (b) Vamatex

Fig. 33. The variation of warp yarn tension of rapier looms.

Table 22 and Fig.34 show the shed amount and warp tension of each 10 heald frames at the center of the loom both Omega and Vamatex.

Loom	Heald	1	2	3	4	5	6	7	8	9	10
OMEGA -Panter	shed amount (mm)	70	74	78	82	84	87	93	98	97	98
	warp tension (gf)	72	71	76	79	80	71	71	78	80	85
VAMATEX	Shed amount (mm)	63	68	74	79	83	88	93	98	103	108
	Warp tension (gf)	48	51	49	57	57	65	61	66	64	69

Table 22. The shedding amount and warp tension of the test weaving looms

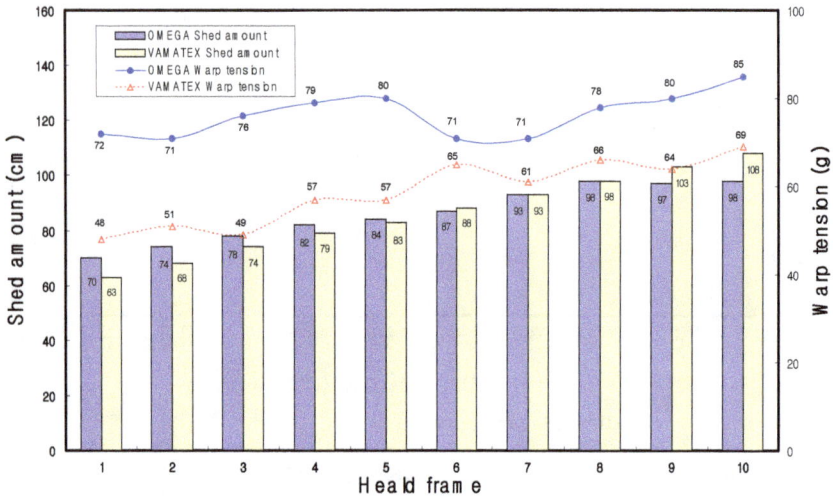

Fig. 34. Relation between warp yarn tension and shed amount.

It is shown that shed amount was increased from 1st heald to 10th one, and warp yarn tension was proportional to the shed amount.

Table 23 shows weft yarn tension and end break of weft yarn on the 2 kinds of looms.

Fig. 35 shows one cycle weft yarn tension variation of the Omega and Vamatex looms.

Loom	RPM	Max. weft tension	Number of end break
Omega-Panter	466	81.0	3
Vamatex	423	81.3	0

Table 23. Weft tension and end break of weft on the two looms.

As shown in Table 23, maximum weft yarn tension was 81gf in Omega and 81.3gf in Vamatex, but end break of Omega was 3 and zero for Vamatex. The reason why seems to be due to high weft yarn tension fluctuation in Omega loom which is shown in Fig. 35. As shown in Fig.35, weft yarn tension variation of Vamatex was much more stable and lower compared to Omega loom.

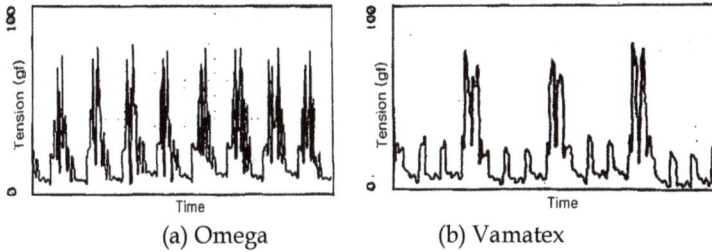

(a) Omega   (b) Vamatex

Fig. 35. The variation of weft yarn tension of 2 rapier looms.

5.5 Fabric physical properties according to the rapier looms

5.5.1 Comparison between Picanol and Omega

Fig. 36 shows the diagram of relative fabric mechanical properties between Picanol and Omega looms. It is shown that the tensile properties in the warp direction of the fabrics woven by Omega loom were higher than those of woven by Picanol loom, the same phenomena in bending properties were shown, which seems to be due to the higher warp tension of the Omega than Picanol loom. But that tendency was not shown in the weft direction. That phenomena shows that warp yarn tension during weaving on Omega loom affects fabric tensile and bending properties, on the other hand, weft yarn tension on Picanol loom does not affect so much. Contrary to the tensile and bending properties, the shear properties of the fabrics in the warp direction woven by Picanol loom was higher than that of woven by Omega loom, but, in the weft direction, the fabric shear properties woven by Omega loom was much higher than that of Picanol loom. And there was no difference of the compression properties between fabrics woven by Picanol and Omega looms. This result demonstrates that shear deformation of fabrics was combined with deformation of warp and weft yarns, high warp yarn tension during weaving on the Omega loom makes low shear rigidity and shear friction of the fabrics in the warp direction, and high weft yarn tension during weaving on the Picanol loom makes low shear rigidity and shear friction of the fabrics in the weft direction.

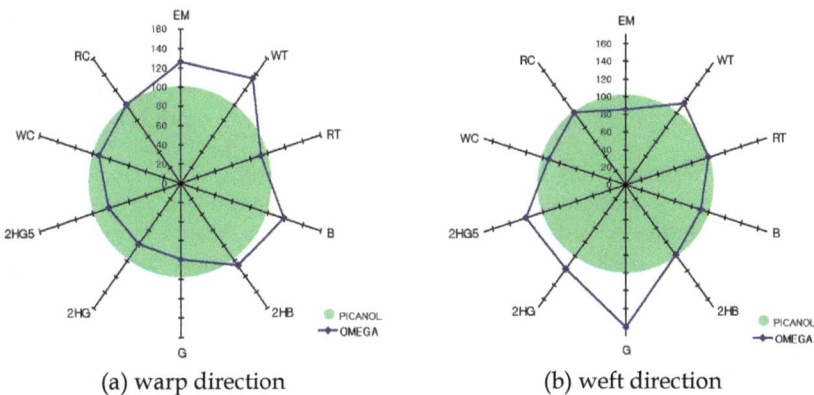

(a) warp direction   (b) weft direction

Fig. 36. The diagram of relative fabric mechanical properties between Picanol and Omera looms.(Picanol: shadow(100%), — : Omega)

Fig. 37 shows the diagram of the fabric mechanical properties between Picanol and Omega according to the fabric positions.

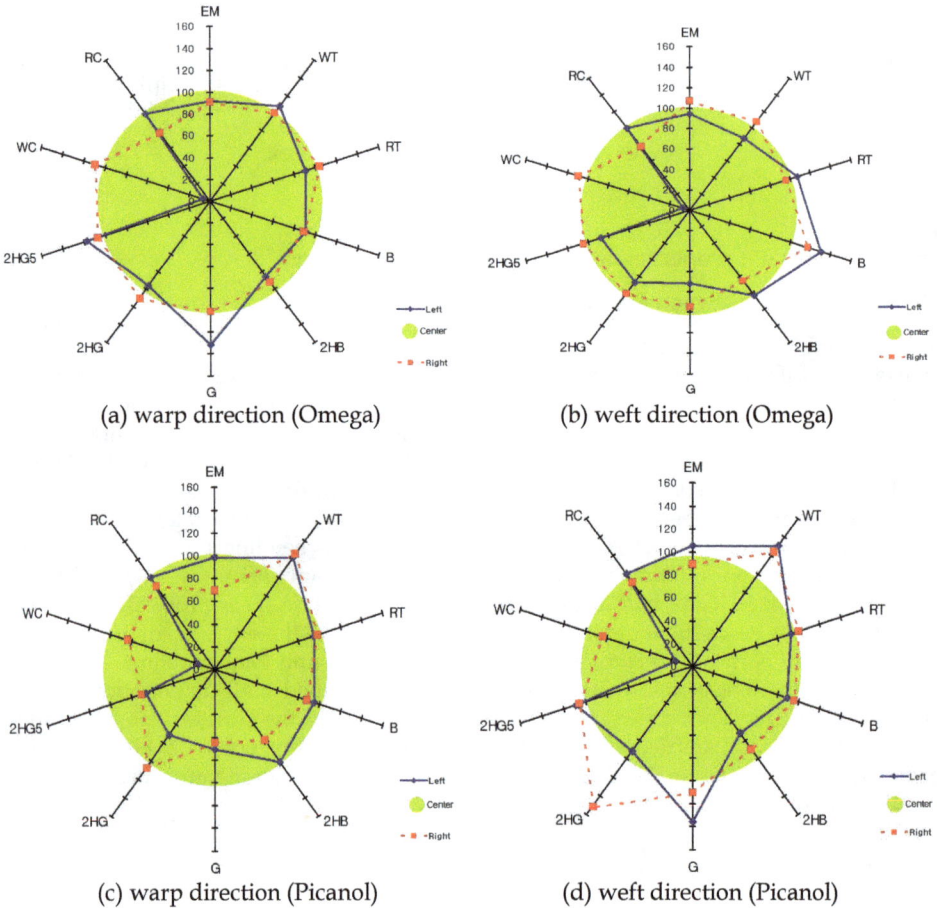

(a) warp direction (Omega)

(b) weft direction (Omega)

(c) warp direction (Picanol)

(d) weft direction (Picanol)

Fig. 37. The diagram of relative fabric mechanical properties between Picanol and Omega according to the fabric positions.(center : shadow(100%), — : left, ⋯ : right)

As shown in Fig. 37, concerning the tensile properties according to the fabric position such as right, center and left sides, the fabric woven by Omega loom doesn't show big difference of mechanical properties according to the fabric position, but Picanol shows big difference according to the fabric position compared with Omega. In addition, comparing Fig. 37 (a) and (c) in warp direction and Fig. 37 (b) and (d) in weft direction, the differences of the fabric mechanical properties according to the position of the fabrics woven by Picanol were higher than those of the fabrics woven by Omega loom.

It seems to be originated from high fluctuation of warp and weft yarn tensions of Picanol loom during weaving as shown in Fig. 28 and 29. Especially, the shear properties variation according to the position of the fabric woven by Picanol was larger than that of

Omega, which makes homogeneity of the fabric hand and tailorability of garment deteriorating.

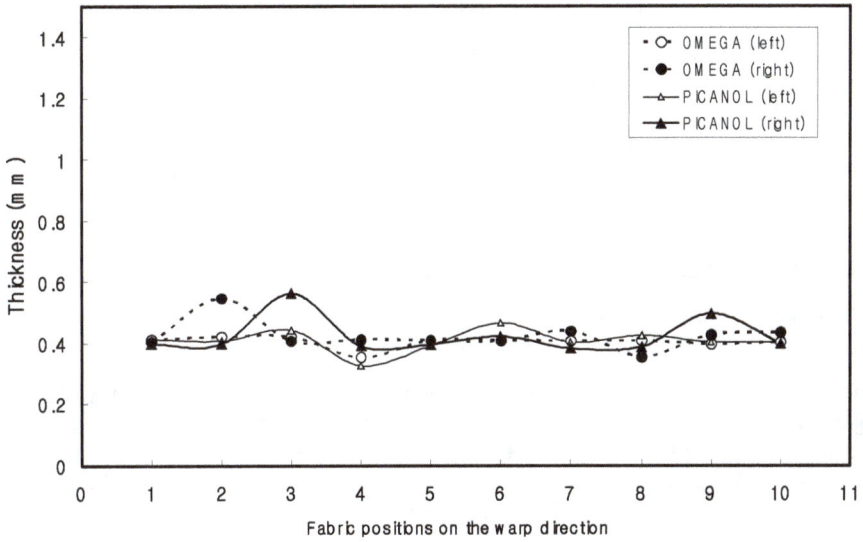

Fig. 38. The thickness variation on the right and left and sides of the finished fabrics.

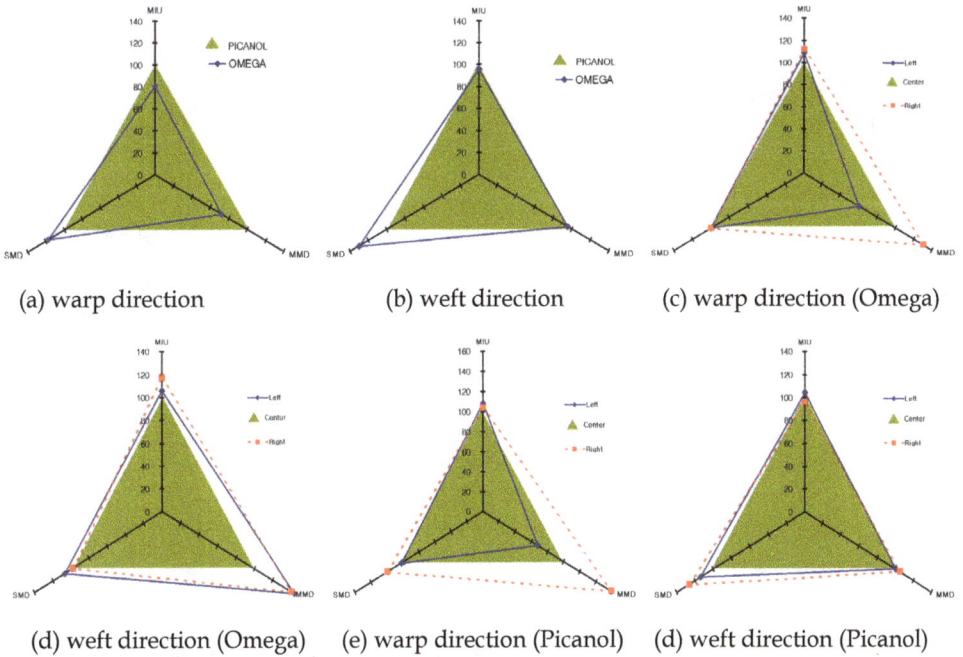

(a) warp direction　　(b) weft direction　　(c) warp direction (Omega)

(d) weft direction (Omega)　(e) warp direction (Picanol)　(d) weft direction (Picanol)

Fig. 39. Fabric surface properties according to the looms and fabric positions.

Fig. 38 shows the thickness variation on the right and left sides of the finished fabrics on the 10 positions of the fabric warp direction. As shown in Fig.38, one or two positions along fabric warp direction on the right and left sides of the fabric showed a little thick positions comparing to the other positions of the fabrics woven by both Picanol and Omega looms.

Fig. 39 shows fabric surface properties between Picanol and Omega looms and according to the fabric positions such as right, left and center. As shown in Fig. 39, MIU(coefficient of friction) and MMD(deviation of MIU) of the fabrics woven by Omega was lower than those by Picanol loom but, SMD(surface roughness) showed higher value than Picanol. But especially the differences of these values according to the fabric positions were much higher than those of looms.

5.5.2 Comparison between Vamatex and Omega

Fig. 40 shows diagram of process shrinkage of the warp and weft directions according to loom on the each dyeing and finishing processes.

As shown in Fig. 40, any differences of each process shrinkages between Vamatex and Omega could not find, which means that two grey fabrics woven by Vamatex and Omega were proceeded at the same process conditions on the dyeing and finishing processes. It can be seen that 20% of weave contraction was occurred and 30% thermal shrinkage after scouring and drying was occurred, and 12% relaxing expansion on the pre-set, dyeing and final set was occurred.

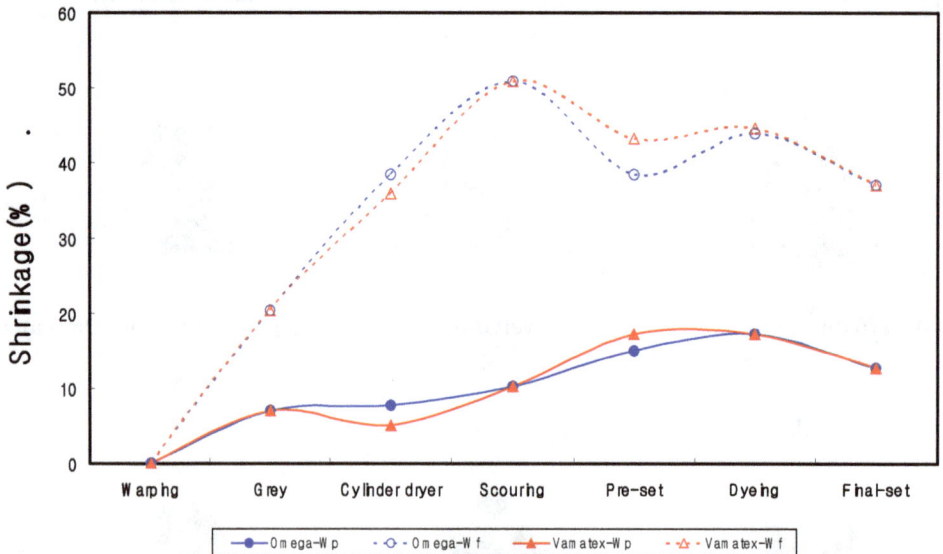

Fig. 40. Shrinkage of the fabrics according to the weaving machine.

Fig. 41 shows comparison diagram of fabric mechanical properties between Vamatex and Omega, which shows relative values of fabric woven by Omega to the mechanical properties of the fabric woven by Vamatex.

(a) warp direction   (b) weft direction

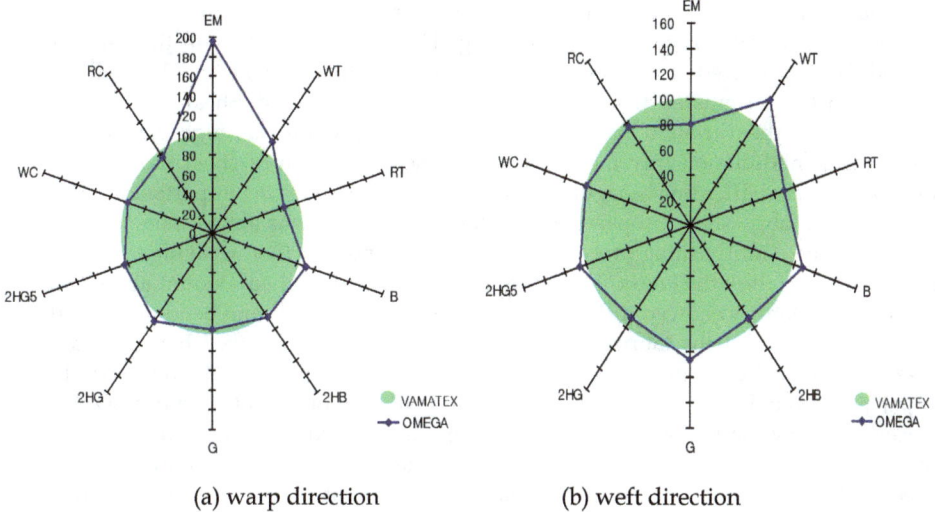

Fig. 41. The diagram of relative fabric mechanical properties between Vamatex and Omega looms.

As shown in Fig. 41 (a), tensile energy (WT), and extensibility (EM) of the fabric woven by Omega were much higher than those by Vamatex on the warp direction, which seems to be due to high warp tension of the Omega loom which was shown in Fig. 33. But, this phenomena was not shown on the weft direction as shown in Fig. 41 (b), which seems to be due to the same on the weft tension between Vamatex and Omega looms which was shown in Fig. 35. The tensile resilience of Vamatex was a little higher than that of Omega, which means that elastic recovery of fabric woven by Vamatex is better than that by Omega. It was shown that the bending rigidity of fabric woven by Omega was a little higher than that by Vamatex, which is also due to high weaving tension of Omega loom which was shown in Fig. 33, Any difference of fabric shear property between Omega and Vamatex was not shown and also was not shown in fabric compressional property. These results demonstrate that fabric tensile and bending properties are affected by warp yarn tension of loom, fabric shear and compressional properties are not affected by only warp tension, which properties are affected by both warp and weft tensions.

6. Conclusion

Linear relationship between warp yarn tension and shed amount of loom for the worsted fabric was shown. Warp yarn tension variation for the worsted fabric between edge sides of fabric and center of fabric was above about 20gf, the highest at center part and the lowest at the right side as viewed in front of loom. These shed amount and warp yarn tension affect extensibility and bending rigidity of finished fabrics, i.e. the higher warp yarn tension and the lower shed amount, the more extensible gray fabric. The warp extensibility of finished fabric for the continuous and discrete finishing showed big difference, the variation of warp extensibility among various looms by continuous finishing was smaller than that of discrete finishing. Warp bending rigidity of gray fabric woven under low warp yarn tension showed low values, after finished, the effects of high warp yarn tension during weaving were remained for the case of continuous finishing. The bending rigidity on the center of the finished fabrics showed the highest values comparing to the right and left sides on the fabrics, which is originated from high yarn tension for weaving. Shear modulus of gray fabrics showed the variation according to the weaving looms, i.e. high weaving tension makes shear modulus of gray fabric high. But, these variation of shear modulus of gray fabric disappeared after finishing, this phenomena demonstrates the importance of finishing process to the fabric shear property. Fabric surface property was almost same as the fabric shear property. And finishing process is much more important than weaving tension for the control of even fabric thickness. The warp and weft tensions, shed amount among various looms for PET fabrics showed different characteristics. The tensile and bending properties in the warp direction of the fabrics woven by low tension loom showed higher values than those of high tension loom owing to the high warp yarn tension, on the other hand, shear property showed lower value. On the weft direction, contrary phenomena was shown. Concerning the variation of the mechanical properties according to the fabric positions, the fabric woven by high tension loom showed more fluctuation than that of low tension loom. It seems that these results make fabric hand and garment tailorability deteriorating. The shed amount and warp tension for PET fabrics were also increased from front heald to back one like worsted fabrics. Warp tension variation according to the warp position showed same phenomena as the worsted fabrics.

7. References

Adanur, S., & Mohamed, M.H. (1988). Weft Insertion on Air-jet Looms: Velocity Measurement and Influence of Yarn Structure. Part I: Experimental System and Computer Interface, *J. Textile Inst.*, Vol. 79, No. 2, pp.(297-315), 0040-5000

Adanur, S., & Mohamed, M.H. (1991). Analysis of Yarn Tension in Air-jet Filling Insertion, *Textile Res. J.*, Vol. 61, No. 5, pp.(259-266), 0040-5175

Adanur, S., & Mohamed, M.H. (1992). Analysis of Yarn Motion in single Nozzle Air-jet Filling Insertion. Part I: Theoretical Models for Yarn Motion, *J. Textile Inst.*, Vol. 83, No. 1, pp.(45-55), 0040-5000

Basu, A.K. (1987). Effect of Different Let-off Mechanisms on Fabric Formation and Dimension of Fabric in the Loom. *Textile Res. J.*, Vol. 57, No. 7, (Jul, 1987), pp.(379-386), 0040-5175

Belforte, G., Mattiazzo, G., Viktorov, V., & Visconte, C. (2009). Numerical Model of an Air-jet Loom Main Nozzle for Drag Forces Evaluation. *Textile Res. J.*, Vol. 79, No. 18, (Nov 13, 2009), pp.(1664-1669), 0040-5175

Bilisik, K., & Mohamed, M.H. (2009). Multiaxis Three-dimensional Flat Woven Preform (Tube Rapier Weaving) and Circular Woven Preform (Radial Crossing Weaving)*. *Textile Res. J.*, Vol. 79, No. 12, (Jul 17, 2009), pp.(1067-1084), 0040-5175

Bilisik, K., & Demiryurek, O. (2011). Effect of Weaving Process on Tensile Characterization of Single and Multiple Ends of Air-Entangled Textured Polyester Yarns. *Fibers and Polymers*, Vol. 12, No. 3, pp.(376-383), 1229-9197

Gokarneshan, N., Jegadeesan, N., & Dhanapal, P. (2010). Recent innovations in loom shedding mechanisms. *Indian Journal of Fibre & Textile Research*, Vol. 35, (March, 2010), pp.(85-94), 0971-0426

He, X., Taguchi, Y., Sakaguchi, A., Matsumoto, Y.I., & Toriumi, K. (2004). Measuring Cloth Fell Fluctuation on a Weaving Machine. *Textile Res. J.*, Vol. 74, No. 7, (July 2004), pp.(576-580), 0040-5175

Islam, A.T.M.S., & Bandara, M.P.U. (1996). Yarn Spacing Measurement in Woven Fabrics with Specific Reference to Start-up Marks. *J. Textile Inst.*, Vol. 87, Part 1, No. 1, pp.(107-119), 0040-5000

Islam, A.T.M.S., & Bandara, M.P.U. (1999). Cloth Fell Control to Revent Start-up Marks in Weaving. *J. Textile Inst.*, Vol. 90, Part 1, No. 3, pp.(336-345), 0040-5000

Kim, S.J., Oh, A.G., Cho, D.H., Chang, D.H., & Song, J.S. (1995). Study on Correlation between Fabric Structural Parameter and Processing Shrinkage of Polyester Woven Fabric. *Journal of The Korean Fiber Society*, Vol. 32, No. 5, pp.(480-487), 1225-1089

Kim, S.J., Oh, A.G., Cho, D.H., Chang, D.H., & Song, J.S. (1995). Study on Correlation between Mechanical Properties and Warp Density of Polyester Woven Fabric. *Journal of The Korean Fiber Society*, Vol. 32, No. 5, pp.(488-493), 1225-1089

Kim, S.J., Sohn, J.H., Kang, J.M., & Park, M.H. (2004). Effects of Weaving Machine Characteristics on the Physical Properties of PET Fabrics (I). *Journal of The Korean Society of Dyes and Finishers*, Vol. 16, No. 4, pp.(206-215), 1229-0033

Kim, S.J., & Kang, J.M. (2004). Effects of Rapier Weaving Machine Characteristics on the Physical Properties of Worsted Fabrics for Garment (1). *Journal of The Korean Society for Clothing Industry*, Vol. 6, No. 6, pp.(765-771), 1226-2060

Kim, S.J., & Kang, J.M. (2004). Effects of Rapier Weaving Machine Characteristics on the Physical Properties of Worsted Fabrics for Garment (2). *Journal of The Korean Society for Clothing Industry*, Vol. 6, No. 6, pp.(772-777), 1226-2060

Kim, S.J., & Jung, K.J. (2005). Effects of the Projectile and the Air-jet Weaving Machine Characteristics on the Physical Properties of Worsted Fabrics for Garment (1). *Journal of The Korean Society for Clothing Industry*, Vol. 7, No. 1, pp.(101-105), 1226-2060

Kim, S.J., & Jung, K.J. (2005). Effects of the Projectile and the Air-jet Weaving Machine Characteristics on the Physical Properties of Worsted Fabrics for Garment (2). *Journal of The Korean Society for Clothing Industry*, Vol. 7, No. 1, pp.(106-110), 1226-2060

Kopias, K. (A2008). New Concept of Weaving Loom Construction. *FIBRES & TEXTILES in Esatern Europe January*, Vol. 16, No. 5(70), (December, 2008), pp.(74-76), 1230-3666

Lappage, J. (2005). End Breaks in the Spinning and Weaving of Weavable Singles Yarns: Part 2: End Breaks in Weaving. *Textile Res. J.*, Vol. 75, No. 6, (June 2005), pp.(512-517), 0040-5175

Lindberg, J., Westerberg, L., & Svennson, R. (1960). Wool Fabrics as Garment Construction Materials. *J. Textile Inst.*, Vol. 51, No. 12, pp.(1475-1493), 0040-5000

Ly, N.G., Tester, D.H., Buckenham, P., Roczniok, A.F., Adriaansen, A.L., Scaysbrook, F., & De Jong, S. (1991). Simple Instruments for Quality Control by Finishers and Tailors. *Textile Res. J.*, Vol. 61, No. 7, pp.(402), 0040-5175

Morooka, H., & Niwa, M. (1978). Physical Properties of Fabrics Relating to Marking Up and Good Appearance. *J. Text. Mach. Soc. Japan*, Vol. 24, No. 4, pp.(105-114), 0040-5043

Natarajan, V., Prasil, V., Egrt. F., & Hrus, M. (1993). New Air-jet Weft-insertion System without an Air-guide and with Double-holed Relay Nozzle: A Study of Double-holed Relay Nozzle. *J. Textile Inst.*, Vol. 84, No. 3, pp.(314-325), 0040-5000

Niwa, M., Nakanishi, M., Ayada. M., & Kawabata, S. (1998). Optimum Silhouette Design for Ladies' Garments Based on the Mechanical Properties of a Fabric. *Textile Res. J.,* Vol. 68, No. 8, pp.(578-588), 0040-5175

Oh, A.G., Kim, S.J., Cho, D.H., Chang, D.H., Kim, S.K., Kim, T.H., & Seo, M.H. (1993). Study on Correlation between Mechanical Properties and Processing Shrinkage of Polyester Woven Fabric. *Journal of The Korean Fiber Society,* Vol. 30, No. 11, pp.(803-816), 1225-1089

Oh, A.G., & Kim, S.J. (1993). Study on the Mechanical Properties of Polyester Woven Fabric (1) – Tensile Behavior under Low Sterss. *Journal of The Korean Fiber Society,* Vol. 30, No. 9, pp.(641-651), 1225-1089

Oh, A.G., & Kim, S.J. (1993). Study on the Mechanical Properties of Polyester Woven Fabric (2) – Nonlinearity of Shear Properties. *Journal of The Korean Fiber Society,* Vol. 30, No. 10, pp.(719-730), 1225-1089

Pavlinic, D.Z., Gersak, J., Demsar, J., & Bratko, I. (2006). Predicting Seam Appearance Quality. *Textile Res. J.,* Vol. 76, No. 3, (May 4, 2006), pp.(235-242), 0040-5175

Postle, R., & Dhingra, R.C. (1989). Measuring and Interpreting Low-Stress Fabric Mechanical and Surface Properties: Part III: Optimization of Fabric Properties for Men's. Suiting Materials. *Textile Res. J.,* Vol. 59, No. 8, pp.(448-459), 0040-5175

Seyam, A.M. (2003). Weavability Limit of Yarns with Thickness Variation in Shuttleless Weaving. *Fibers and Polymers,* Vol. 4, No. 4, pp.(176-181), 1229-9197

Shishoo, R.L. (1989). Interactions between Fabric Properties and Garment Making-up Processes. *Lenzinger Berichte,* Vol. 67, pp.(35-42), 0024-0907

Simon, D.M., Githaiga, J., Lieva, V.L., Hung, D.V., & Puissant, P. (2005). Simulation of the Dynamic Yarn Behavior on Airjet Looms. *Textile Res. J.,* Vol. 75, No. 10, (Oct 1, 2005), pp.(724-730), 0040-5175

Simon, D.M., Patrick, P. & Lieva, V.L. (2009). Three-dimensional Simulation of the Dynamic Yarn Behavior on Air-jet Looms. *Textile Res. J.,* Vol. 79, No. 18, (Nov 13, 2009), pp.(1706-1714), 0040-5175

Yokura, H., & Niwa, M. (1990). Durability of Fabric Handle and Shape Retention During Wear of Men's Summer Suits. *Textile Res. J.,* Vol. 60, No. 4, pp.(194), 0040-5175

4

An Integration of Design and Production for Woven Fabrics Using Genetic Algorithm

Jeng-Jong Lin

Department of Information Management, Vanung University

Taiwan, R.O.C.

1. Introduction

Nowadays, the enterprises all over the world are approaching toward globalizing in design and production in order to be more sustainable. Integration of interior divisions in a company or cooperation among different companies worldwide is of great importance to the competence enhancement for entrepreneurs. There have been a variety of developed applications to integrating different divisions (Cao et al., 2011) (Yamamoto et al., 2010). Moreover, the range of R&D cycle for textiles is much narrowed than ever. It is necessary for an enterprise to afford the demand of marketing change in small quantity and large variety for the commodity. Thus, it is crucial for textile manufacturer to integrate the design and production processes.

Generally speaking, at the very beginning a piece of fabric appeals to a consumer by its appearance, which is related to the weave structure and the colors of warp and weft yarns. Next, the characteristics, e.g., the permeability, the thickness, the tenacity, the elongation et al. of the fabric are required. Finally, the price of the fabric is used as an evaluation basis, by comparing which to the above-mentioned items (i.e., the outlook and the characteristics), the value of the fabric can thus be defined and determined. If the value is satisfactory, the fabric will be accepted by the consumer. Otherwise, it will become a slow-moving-item commodity.

It is essential for the fabric with good quality to be of appropriate weaving density except being equipped with satisfactory pattern. If the weaving density is too less, the fabric will seem obviously too sparse to have good enough strength. The more weight consumption of the material yarns is, the higher cost needed for the manufacturing of a piece of fabric is. Thus, it is a crucial issue for a designer to make a good balance between the cost and the essential consumption of the material yarns during woven fabric manufacturing.

Woven fabric is manufactured through the interlacing between the warp and weft yarn. The pattern of the woven fabric is illustrated through the layout of the different colors of the warp and weft yarn. Therefore, the application of computer-aided design (CAD) (Dan, 2011) (Wang et al., 2011) (Liu et al., 2011) (Gerdemeli et al., 2011) (Mazzetti et al., 2011) to simulated woven-fabric appearance and to the other aspects has been a major interesting research in recent years and various hardware and software systems are now available on the market for widely commercial applications. Until these systems became available, a considerable amount of time and money had been needed to show designers' ideas of fabric design in pattern (fabric sample) form. Probably only 15-20% of the patterns produced would have been approved for production by the sales department or the customers or both. Thus design can be a very expensive exercise for manufacturers engaged in the fancy woven fabric market. The introduction of CAD to textiles has provided a major breakthrough in multicolor weave design. With the help of CAD, designers can display, examine, and modify ideas very quickly on the color monitor before producing any real fabrics. Thus CAD allows a greater scope for free creative work on the part of designers without incurring a large cost increase. CAD allows a greater flexibility in the designer's work, and the designer's creativity is more effectively used.

A designer can do a weave structure design by using his/her inspiration. However, for a designer it can happen to run out of his/her creativity for pattern design from time to time. Though the CAD is becoming more and more applicable to the pattern design (Hu, 2009) (Zhang et al., 2010) (Penava et al.,2009) it has not yet become a complete tool to the textile designer because of limits to the function of color and material yarn selecting that can be created automatically. Up to the present, designers have got to be satisfied with a limited function of their own chosen color and material yarn recently available to display the simulation of the fabrics.

There are huge amount of researches on the weave structure of woven fabric. Griswold (Griswold, 2011) proposed algorithms on using Boolean operations in weave pattern design. Rasmussen (Rasmussen, 2008) discusses the theory of binary representation of fabric structures and the possibilities of weave category in order to design families of weave patterns. Rao et al. (Rao et al., 2009) developed 3-D geometric models for the morphological construction of fabrics with the unit-cells of four harness, five harness, and eight harness. Shinohara et al. (Shinohara et al., 2008) proposed a novel automatic weave diagram construction method from yarn positional data of woven fabric. Ozdemir et al. (Ozdemir et al., 2007) developed a method to obtain computer simulations of woven fabric structures based on photographs taken from actual yarns along their lengths. On the other hand, weave pattern design will benefit from some theoretical studies on binary matrices, e.g., pattern mining techniques from binary data. Ma et al. (Ma et al., 2011) proposed an encoding algorithm to reveal the hidden information in the binary matrix of a weave pattern so as to obtain a solution to determine features of the weave pattern. It enables the possibility to quickly produce required weave geometries and weave textures at different levels of detail.

In order to go beyond the simulation function of a conventional CAD system, a design system, which can generate a variety of patterns for a designer to evaluate each of them and scoring them by preference, is of great value to be created and developed. Such a system is developed by using genetic algorithm in this study. Genetic algorithm (GA) is powerful and broadly applicable stochastic search and optimization techniques based on principles from evolution theory. GA has widely been applied in varieties of fields, e.g., CAD/CAM integration (Ahmad et al., 2011), electromechanical product design (Yang et al., 2011), Process planning (Salehi et al., 2009), information management optimization (Wei et al., 2009), and manufacturing cycle cost finding (Deiab et al., 2007). Through the assistance of the GA-developed system in this study, a fabric designer can proceed with the design process of weave structure more flexibly and effectively. With the help of GA-based CAD, a satisfactory creative pattern, which is of a specific weave structure with certain colors of warp and weft yarn, can be obtained.

Once the pattern design is determined, the characteristics (e.g., the thickness, the permeability et al.) need to be set for the next. Another Search system based on GA is developed for acquire weaving parameters as well. With this system, a fabric designer can efficiently determine weaving parameters to help manufacture the fabrics of required characteristics.

With GA-based design and production system, a fabric designer can efficiently determine the warp (or weft) yarn color, weave structure and the required weaving parameters (e.g., yarn count N_1, N_2, and weaving density n_1, n_2 of warp and weft yarn) for manufacturing satisfactory fabric. Thus, the design and production divisions can be integrated together. The running out of creative inspiration for a designer can be eliminated. The system can provide several appropriate combination sets of layout parameters which can meet a designer's satisfaction on the appearing pattern of the fabric and the demand of weaving parameters which can produce the fabric on expected material cost without lab manufacturing in advance. The construction of the integrated system for design and production is described as follows.

2. Framework of integrated system for design and production

As shown in Figure 1, the system consists of two major components, i.e., the search mechanisms for weaving parameter and weave structure, and the user interface. Each of them is described briefly as follows.

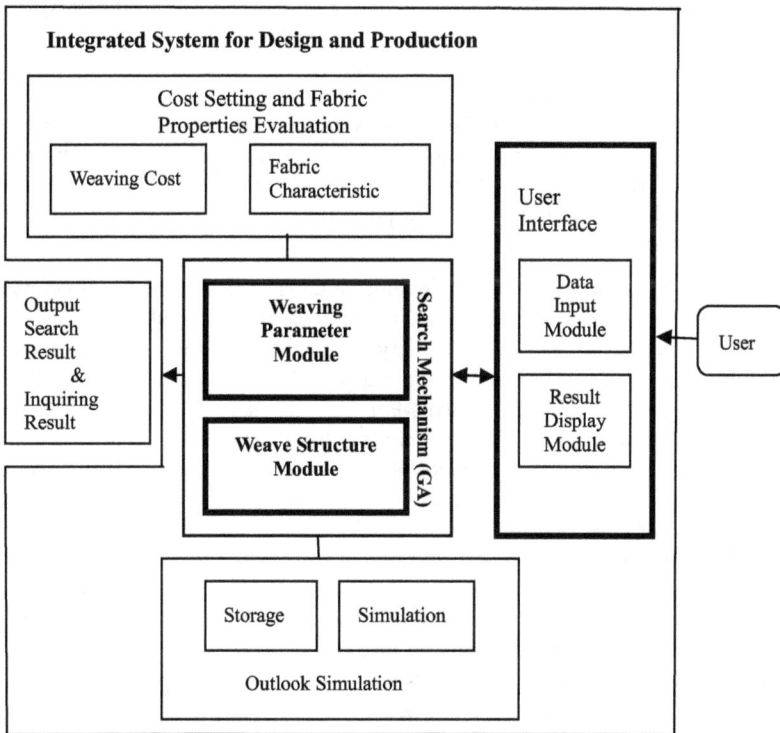

Fig. 1. Scheme of integrated system

2.1 Search mechanism

There are two search mechanisms included in the system; one is developed for the search of weave structure pattern, the other is for that of weaving parameters. The search mechanism is based on genetic algorithm (GA), which is an optimization technique inspired by biological evolution (Karr, 1999). Base on the natural evolution concept, GA is computationally simple and powerful in its search for improvement and is able to rapidly converge by continuously identifying solutions that are globally optimal with a large search space. By using the random selection mechanism, the GA has been proven to be theoretically robust and empirically applicable for search in complex space.

2.2 User interface

The user interface allows the user to set the basic parameters, e.g., population size, crossover rate, mutation rate, and the number of maximum generations, for GA to run. Besides, the user interface is of a function to display the searched results of the weave structure and the weaving parameters, on the monitor for the user to refer.

3. Search module for weaving parameters

3.1 Weaving parameters in production

Recent application of computer technology in the textile field (Inui et al., 1994) (Liu et al., 1995) (Ohra et al., 1994) (Hu, 2009) (Zhang et al., 2010) (Penava et al.,2009) , e.g., simulation systems for color matching, computer aided design (CAD) systems for static and dynamic states, and semantic color-generating systems for garment design. In this study, we propose an intelligent searching system theory based on a genetic algorithm to search for weaving parameters. There are five weaving parameters, i.e., warp yarn count, weft yarn count, warp yarn density, weft yarn density, and total yarn weight, which are all correlated to one another in weaving. If two or more than two parameters are unknown among them, there will be many available combinations.

Let's suppose there is a weaving mill that develops a fabric whose total weight consumption is preset as 5.6×10^{-7} (lb) per square inch. For simplification, the shrinkage of the fabric during weaving is neglected. There exist many combinations of weaving parameters (i.e., both yarn count and weaving density of the warp and weft), that can be used for preset weight consumptions of the material yarns. For instance, samples A, B, and C, shown in Table 1, all answer these demands. The areas of these three fabric pieces are similar- 1 square inch- but they have different yarn counts and weaving densities. Now the question is how a designer can easily and immediately obtain a lot of available combination sets of these four weaving parameters. In other words, it's difficult for a designer to acquire all the possible combination sets of weaving parameters simply through common sense. In addition, in order to speed up the production rate, the weft yarn count used in weaving is usually smaller than the warp yarn count. Thus the weaving density of the warp yarns is usually larger than that of weft yarns during implementation. Sample D's weaving density of warp yarn is smaller than that of its weft yarn. Sample E's warp yarn count is smaller than its weft yarn count. Sample F's weaving density of warp yarn is smaller than its weft yarn, and its warp yarn count is smaller than its weft yarn count. Therefore, Samples D-F shown in Table 1 are not available for practical use in weaving engineering.

Samples	N_1, 840yd/lb	N_2, 840yd/lb	n_1, ends/in	n_2, ends/in	W, lb	Size, Width × length
✓ A	1771.6 $N_1 \geq N_2$	885.8	10 $n_1 \geq n_2$	10	5.6 × 10⁻⁷	1 in × 1 in
✓ B	1328.8 $N_1 \geq N_2$	1063.0	10 $n_1 \geq n_2$	10	5.6 × 10⁻⁷	1 in × 1 in
✓ C	664.4 $N_1 \geq N_2$	531.5	5 $n_1 \geq n_2$	5	5.6 × 10⁻⁷	1 in × 1 in
× D	885.8 $N_1 \geq N_2$	885.8	5 $n_1 < n_2$	10	5.6 × 10⁻⁷	1 in × 1 in
× E	664.4 $N_1 < N_2$	5315.0	10 $n_1 \geq n_2$	10	5.6 × 10⁻⁷	1 in × 1 in
× F	664.4 $N_1 < N_2$	1063.0	5 $n_1 < n_2$	10	5.6 × 10⁻⁷	1 in × 1 in

a ✓ available for practical application, × unavailable for practical application, N_1, N_2 yarn count of warp and weft yarns, respectively, n_1, n_2 weaving density of warp and weft yarns, respectively, and W total weight of woven fabrics.

Table 1. Combination sets of weaving parameters for samples A-F.[a] (Lin, 2003)

3.2 Genetic operators

A genetic algorithm (GA) (Gen et al., 1997) (Goldberg, 1989) (Karr et al., 1999) is a search method based on the mechanism of genetic inheritance. A genetic algorithm maintains a set of trial solutions, called a population, and operates in cycles called generations. Each individual in the population is called a chromosome, representing a solution to the problem at hand. A chromosome is a string of symbols, usually, but not necessarily, a binary bit string.

We adopted a search method, genetic algorithm (GA), to the combination sets. A genetic algorithm maintains a set of trial solutions, called a population, and operates in cycles called generations. Each individual in the population is called a chromosome, representing a solution to the problem at hand. A chromosome is a string of symbols, usually, but not necessarily, a binary bit string.

During each generation, three steps are executed.

Step 1. Each member of the population is evaluated and assigned a fitness value which serves to provide a ranking of the member.
Step 2. Some members are selected for reproductions.
Step 3. New trial solutions are generated by applying the recombination operators to those members which construct the new population after reproduction.

The genetic algorithm is shown in Figure 2 and a brief discussion on the three basic operators of GA is made as below.

a. Crossover
   Crossover is the main genetic operator. It operates on two chromosomes at a time and generates offspring by combining both chromosomes' features. A simple way to achieve crossover would be to choose a random cut-point and generate the offspring by combing the segment of one parent to the left of the cut-point with the segment of the other parent to the right of the cut-point. This method of genetic algorithms depends to a great extend on the performance of the crossover operator used.
b. Mutation
   Mutation is a background operator, which produces spontaneous random changes in various chromosomes. A simple way to achieve mutation would be to alter one or more

genes. In genetic algorithms, mutation serves the crucial role of preventing system from being struck to the local optimum.

c.   Reproduction

Each set is evaluated by a certain evaluation function. According to the value of evaluation function, the number which survives into the next generation is decided for each set of strings. The system then generates sets of strings for the next generation.

3.3 Chromosome representation

A main difference between genetic algorithms and more traditional optimization search algorithms is that genetic algorithms work with a coding of the parameter set and not the parameters themselves. Thus, before any type of genetic search can be performed, a coding scheme must be determined to represent the parameters in the problem in hand. In finding solutions, consisting of proper combination of the four weaving parameters, i.e., warp yarn count (N_1), weft yarn count (N_2), weaving density of warp yarn (n_1), and weaving density of weft yarn (n_2), a coding scheme for three parameters must be determined and considered in advance. A multi-parameter coding, consisting of four sub-strings, is required to code each of the four variables into a single string. In a direct problem representation, the transportation variables themselves are used as a chromosome. A list of warp yarn count/weft yarn count/weaving density of warp yarn/weaving density of weft yarn was used as chromosome representation. The structure of a chromosome is illustrated in Table 2.

Parameter	Gene (bits)	Order	Layout of Chromosome
Warp count (N_1)	4	1~4	
Weft count (N_2)	4	5~8	
Warp Density(n_1)	4	9~12	
Warp Density(n_2)	4	13~16	

16 15 14 13 12 11 10 9 8 7 6 5 4 3 2 1
4 bits   4 bits   4 bits   4 bits
$\longleftarrow n_2 \longrightarrow\!\longleftarrow n_1 \longrightarrow\!\longleftarrow N_2 \longrightarrow\!\longleftarrow N_1 \longrightarrow$

Table 2. Structure of a chromosome (Lin, 2003)

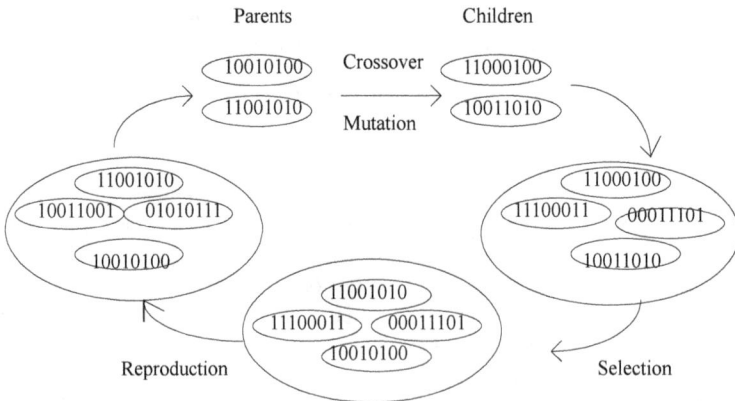

Fig. 2. Flow chart of genetic algorithm

3.4 Encoding and decoding of chromosome

The domain of variable x_i is $[p_i, q_i]$ and the required precision is dependent on the size of encoded-bit. The precision requirement implies that the range of domain of each variable should be divided into at least $(q_i-p_i)/(2^n-1)$ size ranges. The required bits (denoted with n) for a variable is calculated as follows and the mapping from a binary string to a real number for variable x_i is straight forward and completed as follows.

$$x_i = p_i + k_i(q_i-p_i)/(2^n-1) \tag{1}$$

where k_i is an integer between $0 \sim 2n$ and is called a searching index.

After finding an appropriate k_i to put into Equation 1 to have a x_i, which can make fitness function to come out with a fitness value approaching to '1', the desired parameters can thus be obtained.

Combine all of the parameters as a string to be an index vector, i.e. $X=(x_1,x_2,....,x_m)$, and unite all of the encoder of each searching index as a bit string to construct a chromosome shown as below.

$$P=b_{11}...b_{1j}b_{21}...b_{2j}........b_{i1}...b_{ij} \qquad b_{ij} \in \{0,1\} \; ; \; i=1,2,...,m; \; j=1,2,...,n; \tag{2}$$

Suppose that each x_i was encoded by n bits and there was m parameters then the length of Equation 2 should be a N-bit $(N=m \times n)$ string. During each generation, all the searching index k_is of the generated chromosome can be obtained by Equation 3.

$$k_i = b_{i1}*2^{n-1} + b_{i2}*2^{n-2}+...+ b_{in}*2^{n-n} \qquad i=1,2,...,m; \tag{3}$$

Finally the real number for variable x_i can thus be obtained from Equations 1-3. The flow chart for the encoding and decoding of the parameter is illustrated in Figure 3.

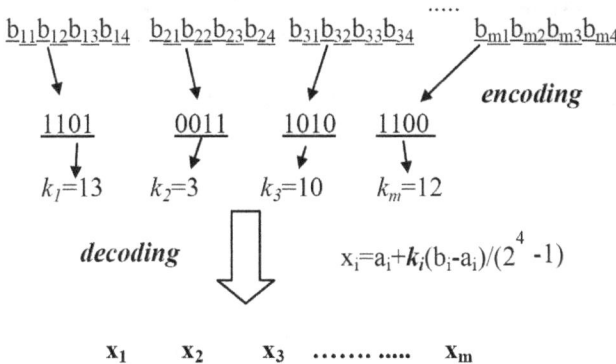

Fig. 3. Flow chart for encoding and decoding of the parameter with 4-bit precision

3.5 Object function

The fitness of the GA used in the system is shown in Equation 4. This approach will allow the GA to find the minimum difference between W and W_g when the fitness function value is maximized.

$$Fitness(W_g) = 1 - \frac{|W - W_g|}{W} \tag{4}$$

Where W_g (lb) is the decoded weight of the yarn for each generation, W(lb) is the target weight of the yarn and can be calculated by using Equation 5.

$$W(lb) = \frac{n_1 \times L \times Width}{(1 - S_1) \times N_1 \times 840} + \frac{n_2 \times L \times Width}{(1 - S_2) \times N_2 \times 840} \tag{5}$$

Where n_1 is the density of the warp (ends/inch), n_2 is the density of the weft (picks/inch); N_1 is the warp yarn count (840 yd/lb), N_2 is the weft yarn count (840 yd/lb), S_1 is the shrinkage of the warp yarn (%),S_2 is the shrinkage of the weft yarn (%), L (yd) is the length of the fabric, and Width (inches) is the width of the fabric.

3.6 Necessary to set constrained conditions

Generally speaking, while expecting to increase the production rate, a designer often leaves the weft yarn thicker than the warp yarn. Furthermore, the number of interlacing points will be different depending on the weave structure. There exists a maximum warp or weft weaving density for woven fabrics during weaving (Lin, 1993) (Pon, 1992). As a practical consideration, the weaving mill always adopts 90% of the maximum weaving density for warp and weft yarns to prevent jammed fabrics, which have a bad hand. In order to make the searching mechanism of the system more realistic and approvable, it is necessary to consider the constrained condition mentioned above. The conditions that essentially need to be considered during weaving can be illustrated as follows: (!)$n_1 \geq n_2$, (2) $N_1 \geq N_2$ (3) $n_1 \leq X \times 0.9$, (4) $n_2 \leq Y \times 0.9$, where X is the maximum weaving density of the warp yarn (end/inch), Y is the maximum weaving density of the weft yarn (picks/inch), n_1 is the density of the warp (ends/inch), n_2 is the density of the weft (picks/inch), N_1 is the warp yarn count (840 yd/lb), and N_2 is the weft yarn count (840 yd/lb). If the acquired chromosome's decoded part is not live up to the above-mentioned constrained conditions, it is essential for the system to set the fitness value at zero. Thus, the goal of preventing the system from deriving unfit solutions for the designer during the search is achieved.

3.7 Example

Table 3 is a set of conditions for a manufacture of fabrics. Table 4 is the results of the tenth generation searched by using GA. With the assistance of this system, many solution sets, consisting of weaving parameters (e.g., N_1,N_2,n_1,n_2), are obtained in a short time to help the designer make a decision more easily when exploring innovative fabrics. Moreover, the system will figure out the fractional cover (i.e., C) (Lin, 2003) of each solution set depending on the combination of weaving parameters generated after the interactive operation of the GA. For instance, the example shown in Table 3, has a GA whose operation conditions of crossover probability, mutation probability, and initial population are set to 0.6, 0.033, and 30, respectively. The decoded value of the ninth chromosome (i.e., 0100010010010100) from right to left per four bits is 30.7 (=N_1, 0100), 44.0 (N_2, 1001), 70.7 (=n_1, 0100), and 70.7 (=n_2, 0100), respectively. By putting these four decoded values into the calculation equation (Lin, 2003) of cover factor, a value of fractional cover can be obtained as 0.6394, yet it conflicts

Example	Known condition and set target			Constrained conditions
Searching for weaving parameters	Length(L)	constant	120 yds	(1)$n_1 \geq n_2$
	Width(width)	constant	64 in	(2)$N_1 \geq N_2$
	Shrinkage of warp yarn (S_1)	constant	6.3%	(3)$n_1 \leq X*0.9$
				(4)$n_2 \leq Y*0.9$
	Shrinkage of weft yarn(S_2)	constant	6.3%	where $X=a \times b \times a'/(b \times b'+a \times c)$
	Warp yarn per inch(n_1)	variable	60-100 ends/in	$Y=b \times a \times b'/(a \times a'+b \times c')$ X,Y:maximum weaving density
	Weft yarn per inch(n_2)	variable	60-100 picks/in	of warp and weft yarns a,b:maximum number of warp
	Count of warp yarn(N_1)	variable	20-60 Ne	and weft yarn capable of being laid out per inch
	Count of weft yarn(N_2)	Variable	20-60 Ne	a',b': number of warp and weft yarns per unit weave structure
	Weight(W)	constant	58 lb	c,c': number of interlacing point
	Weave structure	constant	plain 1/1	in warp and weft directions per unit weave structure
	Material yarn (Warp/Weft)	constant	Cotton/cotton	

Table 3. Set target, known conditions, and constrained conditions (Lin, 2003)

with the constrained conditions mentioned above that N_1 (=30.7) be smaller than N_2 (=44.0). Thus, the fitness of this solution is set at zero. Among the thirty chromosomes illustrated in Table 4, a fabric manufacturer can easily choose several solution sets, whose fitness values are closer to 1 and are of appropriate fractional cover. Thus, the manufacturer can avoid designing a woven fabric that cannot be manufactured by the production division. Furthermore, a designer can achieve the goal of considering many essential design factors such as cost, functionality (e.g., hand, air permeability, and heat retaining properties, et al.), and the possibility of weaving during design stage.

4. Search module for weave structure

4.1 Weave structure in pattern design

Nowadays, the range of R&D cycle for textiles is much narrowed than ever before. Moreover, it is necessary for the enterprise to afford the demand of marketing change in small quantity and large variety for the commodity (Chen, 2005). Application of computer technology in the textile field is widely spreading. For instance, computers are used for the control of processing machine in the apparel production process. Grading or marking of the cutting machine is one of the successful applications of computer technology. The computer has enhanced a lot not only the functions of the hardware but the applications of software. However, most of its applications in textile industry focus on manufacturing processes and quality improvement (Ujevic et al.,2002), . Some of them are applied to the computer aided design (CAD) systems (Rodel, 2001) (Sano, 2001) (Inui, 2001) (Luo, 2005) (Cho, 2005). In the past, the traditional drawing tools and skills were used to transform designers' ideas into

Population	Chromosome	N_1	N_2	n_1	n_2	C	Fitness
1	0100010000011100	52.0	22.7	70.7	70.7	0.6946	0.7531
2	0100010000111001	44.0	28.0	70.7	70.7	0.6760	0.6948
3	0100010001011110	57.3	33.3	70.7	70.7	0.6247	0.5640
4	0100010100111001	44.0	28.0	73.3	70.7	0.6835	0.7050
5	0001010101101111	60.0	36.0	73.3	70.7	0.6165	0.5359
6	0100010101001110	57.3	30.7	73.3	70.7	0.6440	0.6029
7	0100010101011111	60.0	33.3	73.3	70.7	0.6275	0.5623
8	0001010100111011	49.3	28.0	73.3	62.7	0.6381	0.6266
9	0100010010010100	30.7	44.0	70.7	62.7	0.6394	0.0000
10	0100010000011100	52.0	22.7	70.7	70.7	0.6946	0.7531
11	0100010001111111	60.0	38.7	70.7	70.7	0.5995	0.5056
12	0001010101001110	57.3	30.7	73.3	62.7	0.6103	0.5590
13	0100110001111111	60.0	38.7	92.0	62.7	0.6314	0.0000
14	0100110001011110	57.3	33.3	92.0	70.7	0.6814	0.0000
15	0100011001101110	57.3	36.0	76.0	70.7	0.6283	0.5532
16	1101011001111111	60.0	38.7	76.0	94.7	0.7036	0.0000
17	0100010001010101	33.3	33.3	70.7	94.7	0.7667	0.0000
18	0100010000011111	60.0	22.7	70.7	70.7	0.6832	0.7226
19	0100010001011010	46.7	33.3	70.7	70.7	0.6451	0.6114
20	0100010000011110	57.3	22.7	70.7	70.7	0.6867	0.7319
21	0100110000011100	52.0	22.7	92.0	70.7	0.7442	0.0000
22	1101110001101110	57.3	36.0	92.0	94.7	0.7529	0.0000
23	0100010001111111	60.0	38.7	70.7	94.7	0.6924	0.0000
24	0100010000011110	57.3	22.7	70.7	70.7	0.6867	0.7319
25	0001011000011011	49.3	22.7	76.0	70.7	0.7117	0.0000
26	0001010100111100	52.0	28.0	73.3	62.7	0.6325	0.6138
27	0100010001111001	44.0	38.7	70.7	62.7	0.6035	0.5428
28	0100010100111111	60.0	28.0	73.3	70.7	0.6538	0.6302
29	0001010101001110	57.3	30.7	73.3	70.7	0.6440	0.6029
30	0100010001111111	60.0	38.7	70.7	70.7	0.5995	0.5056

Population:30, chromosome 16 bits, generation 10, crossover rate 0.6, mutation rate 0.033, N_1 840 yds/lb, N_2 840 yds/lb, n_1 ends/in, n_2 picks/in

Table 4. Result of the tenth generation (Lin, 2003)

concrete works. Due to the limitations above, a fabric design could not proceed more easily and effectively. Plenty of time was wasted on repeated paper drawing of the same types of different materials' colors and patterns. Besides, the different color combinations' outlooks of warp and weft yarn for a piece of fabric could only be obtained through a designer's imagination. Now, a simulation system (Ujevic, 2002) for color matching has already been put to practical use. The user of this system can confirm color matching of yarns by changing colors or patterns of weave structure displayed on the computer monitor of a computer with the system.

Therefore, the application of CAD to simulated woven-fabric appearance has been a major interesting research in recent years and various hardware and software systems are now available on the market for widely commercial applications (Denton et al., 1989) (Inui, 1994) (Pon, 1992). Until these systems became available, a considerable amount of time and money had been needed to show designers' ideas of fabric design in pattern (fabric sample) form. Probably only 15-20% of the patterns produced would have been approved for production by the sales department or the customers or both. Thus design can be a very expensive exercise for manufacturers engaged in the fancy woven fabric market. The introduction of CAD to textiles has provided a major breakthrough in multicolor weave design. With the help of CAD, designers can display, examine, and modify ideas very quickly on the color monitor before producing any real fabrics. Thus CAD allows a greater scope for free creative work on the part of designers without incurring a large cost increase. CAD allows a greater flexibility in the designer's work, and the designer's creativity is more effectively used.

However, the inspiration of a designer can do a weave structure designing, there is time for the designer to run out of his creativity for pattern design. Though the CAD is becoming more and more applicable to the pattern design, it has not yet become a complete tool to the textile designer because of limits to the function of color and material yarn selecting that can be created automatically. Up to the present, designers have got to be satisfied with a limited function of their own chosen color and material yarn recently available to display the simulation of the fabrics. Besides, due to the limits to the creative inspiration each designer has to face everyday, an intelligent design system is developed in this study to help a fabric designer with the creative weave structure design.

A piece of fabric is woven through the interlacing between the warp and weft yarn. The pattern of the woven fabric is illustrated through the layout of the different colors of the warp and weft yarn (Pon, 1992). The developed system can provide several appropriate combination sets of layout parameters that can meet the designer's satisfaction on the appearing pattern of the fabric without the need of advance lab manufacturing. With this system, a fabric designer can efficiently determine the warp (or weft) yarn color and weave structure to manufacture his satisfying fabric. Thus, the design and production division can be integrated. Furthermore, the running out of creative inspiration for a designer can thus be eliminated through the assistance of this developed system.

4.2 Encoding

Woven fabrics consist of warp and weft (filling) yarns, which are interlaced with one another according to the class of structure and the form of design desired (Hearle, 1969) (Shie, 1984) (Tsai, 1986). A concrete way of encoding textile weave is illustrated as Figure 4(B), whose weave structure is shown as. Figure 4(A). The unit of weaves was restricted to 8 by 8 for the sake of simplicity. The encoding value of '1' on the weave structure indicates 'warp float', the warp is above the weft and the encoding value of '0' indicates 'weft float', the warp is below the weft. The encoded result of a weave structure can be saved as a two-dimensional matrix and can be transformed into a bit string as a chromosome to proceed with crossover, mutation, and reproduction. After finishing the evolution, we can directly apply the obtained chromosome, i.e., bit string result ('1' denotes warp float, '0' denotes weft float) to plot the weave structure. The color of warp and weft yarn is encoded with 4 bits. There are totally 16 (= 24) kinds of color for each warp or weft yarn.

Weft ⟶

Warp ↓

$$\begin{matrix}
0 & 0 & 1 & 1 & 0 & 0 & 1 & 1 \\
0 & 1 & 1 & 0 & 0 & 1 & 1 & 0 \\
1 & 1 & 0 & 0 & 1 & 1 & 0 & 0 \\
1 & 0 & 0 & 1 & 1 & 0 & 0 & 1 \\
0 & 0 & 1 & 1 & 0 & 0 & 1 & 1 \\
0 & 1 & 1 & 0 & 0 & 1 & 1 & 0 \\
1 & 1 & 0 & 0 & 1 & 1 & 0 & 0 \\
1 & 0 & 0 & 1 & 1 & 0 & 0 & 1
\end{matrix}$$

(A)   (B)

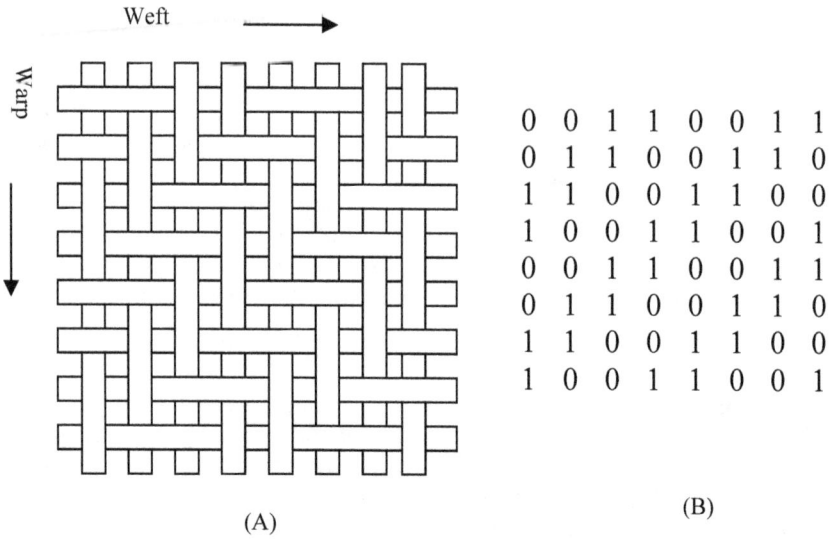

Fig. 4. (A) The schematic weave for the twill weave (B) The strings of the 8-harness twill weave

4.3 Chromosome

A main difference between genetic algorithms and more traditional optimization search algorithms is that genetic algorithms work with a coding of the parameter set and not the parameters themselves (Goldberg, 1989). Thus, before any type of genetic search can be performed, a coding scheme must be determined to represent the parameters in the problem in hand. In finding weave pattern solutions, consisting of proper combination of the three variables, including weave structure (i.e., the combination of warp float and weft float), warp yarn color, and weft yarn color. A multi-parameter coding, consisting of three sub-strings, is required to code each of the three variables into a single string. In this study, a binary coding is utilized and the bit-sizes of the encoding for the three variables are as follows. In a direct problem representation, the weave pattern variables themselves are used as a chromosome. A list of weave structure/warp color/weft color is used as chromosome representation, which represents the permutation of patterns associated with assigned weave structure, warp color and weft color. A gene is an ordered triple (weave structure, warp color, and weft color). This representation belongs to the direct way, which is sketched in Figure 5.

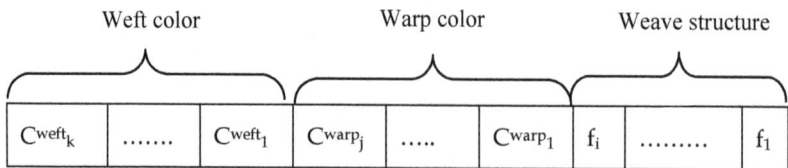

Weft color   Warp color   Weave structure

| C^{weft}_k | .... | C^{weft}_1 | C^{warp}_j | ..... | C^{warp}_1 | f_i | .... | f_1 |

Fig. 5. Representation Scheme of Chromosome

4.4 Fitness function

In this study, the fitness function (i.e., evaluation function) is the user's preference. They are completely free for an operator to give a mark to the generated pattern without any restrictions depending on his/her being satisfied with it. GA is of an evolution capability based on the fitness value of each chromosome. The bigger the fitness of a chromosome is, the more probability it has to survive (be chosen). In other words, the gene (the feature) of the chromosome, which has bigger fitness than the others, will survive (be chosen) more easily to proceed with the operators, such as crossover, mutation, and reproduction, to create a brand new chromosome. Thus, the feature of the new generated chromosome will be inherited from the old chromosome composed of the required gene (i.e., feature). Moreover, each result of the generated chromosome for every generation can be reevaluated according to user's preference after examining the displayed pattern of weave structure decoded from generated parameters of W_S, C_{warp}, and C_{weft1}~C_{weft8}. By judging from the status of the displayed pattern, the user gives a mark (i.e., 0~1) to each of the pattern according to the satisfaction degree of each of them to the user. Thus, the fitness function is the user's preference and can be formed as Equation 6.

$$\text{Fitness}(\text{pattern}_i)=\text{User's preference}(\text{pattern}_i) \tag{6}$$

4.5 Example

In this study, we use this system to search pattern parameters afforded to the predetermined specifications set as in Table 5, i.e., unit weave structure: 8 ends × 8 picks, the layout of warp yarn color is adopted as one color for simplification (if necessary, it can be set as various colors), and that of weft yarn is adopted as various colors. The desired pattern style is something like being of both the features of regular grid and interlacing twist.

4.5.1 GA coding scheme and chromosome

A main difference between genetic algorithms and more traditional optimization search algorithms is that genetic algorithms work with a coding of the parameter set and not the parameters themselves. Thus, before any type of genetic search can be performed, a coding scheme must be determined to represent the parameters in the problem in hand. In finding solutions, consisting of proper combination of the three pattern parameters mentioned above, a coding scheme for three variables (i.e., the weave structure, warp yarn color, and weft yarn color) must be determined and considered in advance. A multi-parameter coding, consisting of three sub-strings, is required to code each of the three variables into a single string.

In this study, a binary coding (Gen et al., 1997) (Goldberg, 1989) is utilized and the bit size of encoding for the three variables, i.e., the warp and the weft yarn color were all set as 4 bits, and the weave structure were set according to the size. For instance, the bit size of a weave structure consisting of 8 ends × 8 picks is set as 64 bits. In spite of the same weave structure, the pattern of the fabrics can be various a lot due to the different layout of the yarn color. Therefore, the layout of the warp (or weft) yarn color is a crucial factor for the woven fabric design. The searched result for the pattern of weave fabric is various with the layout of the yarn's color. For simplification, the 16-color (4 bits) layout resolution is applied in this study.

Example	Known condition and set target			Constrained conditions
Searching for pattern parameters	(1) Unit weave structure		8 ends × 8 picks	(1)There is none of "1" successively in a column for an unit weave structure
	(2) Layout of color yarn	Warp	Same color	
		Weft	Different colors	
	(3) Desired pattern		The desired pattern style is of both the features of "regular grid" and "interlacing twist".	(2)There is none of "0" successively in a column for an unit weave structure
				(3) There is none of "1" successively in a row for an unit weave structure
				(4) There is none of "0" successively in a row for an unit weave structure

Table 5. Set target, known conditions, and constrained condition (Lin, 2008)

If necessary, it is available for the system to adopt more than 16 colors (4 bits) layout resolution. The coding and decoding methods for the color resolution are briefly discussed as follows. For instance, in case of the searching range of yarn's color ranges between 0 and 15 (i.e., 16 colors), 4 bits are needed for encoding. Thus Equation 1 can be reformed as follows.

$$
\left.
\begin{aligned}
k = 0 \quad & x = 0 + 0 * (15 - 0) / 15 = 0 \\
k = 1 \quad & x = 0 + 1 * (15 - 0) / 15 = 1 \\
k = 2 \quad & x = 0 + 2 * (15 - 0) / 15 = 2 \\
\cdots\cdots\cdots & \cdots\cdots\cdots\cdots\cdots\cdots\cdots\cdots \\
k = 15 \quad & x = 0 + 15 * (15 - 0) / 15 = 15
\end{aligned}
\right\} \tag{7}
$$

The chromosome of weave pattern consists of 3 parts, i.e., (1) the gene of weave pattern (1~64 bits) (2) the gene of the layout of the color of warp yarn (65~68 bits) (3) the genes of layout of the color of weft yarn (69~100 bits). For simplification, colors for all the warp yarns are set as the same in this study, the bit size is set to 4 bits. On the other hand, the colors for all the weft yarn (i.e., 8 picks) were set various to one another. Therefore, the size of the first sub-bit-string, representing the weave structure, is 64 bits, that of the second sub-bit-string, representing the color of warp yarn, is 4 bits, and that of the third sub-bit-string, representing all the different weft yarn colors, is 32 bits (=8 picks × 4 bits). Therefore, a chromosome string consisting of 100 bits can be formed and its layout can be shown as Table 6. Besides, if necessary, it is available for the system to set the colors of the warps (8 ends) to be various to one another. In other words, the second sub-bit-string (i.e., the second gene), representing the colors of the warps can be increased as 32 bits (= 8 ends × 4 bits), and the number of bits for the chromosome is reformed as 128 bits as well.

4.5.2 Fitness function and solution search

GA is of an evolution capability based on the fitness value of each chromosome. The bigger the fitness is, the bigger probability to survive (be chosen) is. In other words, the gene (the

Parameters		Gene size (bits)	Sequence of layout	Chromosome
Weave structure (8warpx8weft) (WS)		1 bit× 64 float	1~64	69~100 | 65~68 | 1~64
Layout of color yarns	Warp color (C$_{warp}$) 8 ends (same)	4 bits× 1 end	65~68	←— 32 bits —→|←4 bits→|←— 64 bits —→|
	Weft color (C$_{weft1}$) 8 picks (different)	4 bits× 8 picks	69~100	C$_{weft1}$ ~C$_{weft8}$ C$_{warp}$ WS

Table 6. Layout of chromosome (Lin, 2008)

feature) of the chromosome, which has bigger fitness than the others, will survive (be chosen) more easily to proceed with the operators, such as crossover, mutation, and reproduction, to create a brand new chromosome. Thus, the feature of the new generated chromosome will be inherited from the old one composed of the required gene (i.e., feature). Moreover, each result for every generation can be reevaluated by examining the pattern display of weave structure decoded from generated parameters of W$_S$, C$_{warp}$, and C$_{weft1}$~C$_{weft8}$. By judging from the status of the displayed pattern, the user gives the mark to each of the pattern according to the satisfaction degree for each of them to him/her. In this study, the object function (i.e., fitness function) is the user's preference shown as Equation 6.

The simulation mechanism of the system can illustrate patterns on the monitor to help a designer give each a weighting value (i.e., fitness value). The user interface of the system is developed for helping a designer give each generated chromosome a specific weighting value (i.e., fitness value) ranging from '0' (denotes completely unsatisfied) to '1' (denotes completely satisfied) depending on the degree of his preference through comparing displayed pattern on the screen. Once each generated pattern is given a mark according to the user's preference, search mechanism of system can thus proceed with the operations such as crossover, mutation, and reproduction to produce the next generation.

Next the user can continue with the evolution of GA to search for the genuine design fabric pattern, with which he/she is satisfied. The user expects a satisfying pattern can finally be obtained. For an instance, eight weave structure patterns are illustrated on the monitor as shown in Figure 6A, the designer can give each of them a fitness value (i.e., weighting value of satisfaction) such as 0.8, 0.2, 0.8, 0.4, 1.0, 0.2, 0.4, and 0.6. After proceeding with several generations of evolution (rate of crossover: 0.6; probability of mutation: 0.033; Initial population:8), there comes up with a satisfying solution for this design case. Figure 6B shows the result of weave pattern after one generation. The designer can obtain a satisfying pattern among the obtained patterns listed on Figure 6B. In case none of them is satisfying, the designer can continue proceeding with the procedure illustrated as Figure 2 for another several generations till there is one can be obtained.

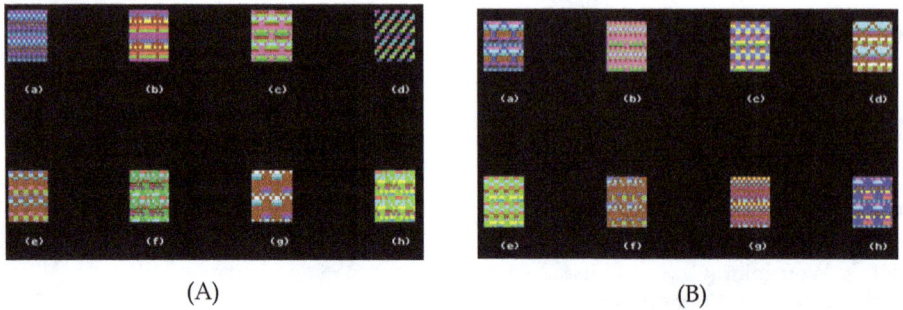

(A)   (B)

Fig. 6. (A) Weave pattern decoded from strings in the initial state,(a)~(h): denotes the
1st~8th population (B) Weave pattern after 1 generation of GA, (a)~(h): denotes the 1st~8th
population

4.5.3 Necessity to set constrained conditions

Regarding the gene of the weave structure pattern (1~64 bits), one should check if the weave
structure of 8 ends × 8 picks happens that there is complete warp or weft float in a unit
weave structure (i.e., There are eight bits of "1" (or "0") successively in a column (or row)
for a unit weave structure as shown in Figure 7.). If the above mentioned case does happen,
the weft or warp yarn will not be fixed to the fabric's surface because none of crossing point.
Thus, the genes (1~64 bits) of the chromosome should be set to "0" to avoid the bad
evolution for the next other generations. Thus, the outcome gained after evolutions can be fit
for the practical application.

4.5.4 Experiment results

With the assistance of this system, many solution sets, consisting of weave structure pattern
parameters (e.g., W_S, C_{warp}, C_{weft1}~ C_{weft8}) are obtained in a short time to help the designer
create a satisfying weave structure pattern more easily during exploring innovative fabrics.
The example shown in Table 5, has a GA whose operation conditions of crossover
probability, mutation probability, and initial population are set to 0.6, 0.033, and 8,
respectively. The desired pattern style for the fabric is something like both the features of
regular grid and interlacing twist.

Firstly, the system displays eight default patterns of weave structures on the monitor
waiting for the user to give each of them a mark depending on his preference. Then the user
can find out the monitor-displayed patterns closely similar to the feature of desired pattern
such as Figure 6A(a), (c), (e), (h), which are of both the feature grid and interlacing twist
pattern look. Judging from the eight patterns of weave structures illustrated on the monitor,
the user can give each of them (i.e., a~h) a fitness value (i.e., weighting value of preference)
such as 0.8, 0.2, 0.8, 0.4, 1.0, 0.2, 0.4, and 0.6 respectively.

Finally, the search mechanism proceeds with crossover, mutation, and reproduction to create
new chromosomes (i.e., bits strings) fit for his demand. The first chromosome (i.e., Figure 6B

(a)) of survey results after one generation is shown as in Table 7. In this instance, if the user is not yet satisfied with the searched result illustrated on the monitor, he/she can just proceed the next generation by giving each pattern another weighting value of preference to search for another combination of bit strings. He/She expects that there exists a chromosome with a much better pattern for his/her referring to during design stage by following the processing procedure: Start → Show eight patterns on the screen → Give each pattern a mark by user → System proceed with crossover, mutation, and reproduction → End.

The results of the first generation are shown in Table 7. The decoded value of the first chromosome (i.e., 0011100001010...10101), from right to left, the first 64 bits string (i.e., 10000...1010101), which can be reformed as a two dimensional matrix to represent the weave structure of fabrics, shown as the solution row in Table 7. The next 4 bits string (i.e., 0001), which denotes the color of the warp, is decoded as 1 (i.e., blue). The left 32 bits string (i.e., 0011100...11101), which can be decoded from right to left per four bits as 13 (=C_{weft1}=light Purple, 1101), 11 (=C_{weft2}=light Cyan, 1011), 3 (=C_{weft3}=Cyan, 0011), 6 (=C_{weft4}=Brown, 0110), 6 (=C_{weft5}=Brown, 0110), 5 (=C_{weft6}= Purple, 0101), 8 (=C_{weft7}=Gray, 1000), and 3 (=C_{weft8}=Cyan, 0011), respectively for each color of the eight picks of weft yarn. According to these obtained weave structure parameters (W_S, C_{warp}, and C_{weft1}~C_{weft8}), the simulation mechanism of the system can display the simulation image of the weave structure pattern, which is illustrated as in Figure 6B(a) with an expected look of the mixed feature of both grid and interlacing twist pattern.

Items	Pattern of weave structure																
Chromosome	0011 1000 0101 0110 0110 0011 1011 1101 0001 10000001 01000010 00100100																
	00011000 00011000 00100100 01000010 01010101																
Pattern parameters	Unit weave structure	Layout of yarn colors															
		Warp (the same)								Weft (the various)							
		1	2	3	4	5	6	7	8	1	2	3	4	5	6	7	8
Solution	10000001←8th weft 01000010←7th weft 00100100←6th weft 00011000←5th weft 00011000←4th weft 00100100←3rd weft 01000010←2nd weft 01010101←1st weft ↑↑↑↑↑↑↑↑ 12345678 warp	0001	0001	0001	0001	0001	0001	0001	0001	1101	1011	0011	0110	0110	0101	1000	0011
		1	1	1	1	1	1	1	1	13	11	3	6	6	5	8	3

Table 7. Searched results of desired pattern

It is available for the simulation mechanism of the system to display the amplified simulation appearance of each the generated weave structure on the screen. With the assistance of the simulation mechanism, the designer can more precisely give the mark to each generated chromosome as a fitness value (i.e., weighting value of preference) depending on his own preference. Figure 7 is the pattern for the amplified weave structure of Figure 6B(a). There is none of complete warp (or weft) float in the horizontal or vertical direction of the weave structures as shown in Figure 6B. Thus, the practical use of these generated weave structures can be ensured. Furthermore, in case none of them is satisfied, the designer can still continue proceeding with the GA for another generation till there is a desired one can be obtained.

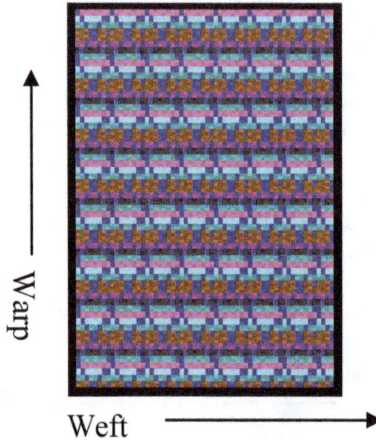

Fig. 7. Simulation pattern of the 1st chromosome for the first generation

5. Implementation of integration

At the very beginning, a consumer is usually attracted by the appearance of a fabric, which is related to the weave structure and the colors of the warp and weft yarns. Next, the characteristics, e.g., the permeability, the thickness, the tenacity, the elongation et al. of the fabric are required to be taken into consideration. Finally, the price of the fabric is used as an evaluation basis, by comparing which to the above-mentioned items (i.e., the outlook and the characteristics), the value of the fabric can thus be defined and determined.

Figure 8 illustrates the integration schemes of design and production for woven fabric. There are two search mechanisms included in the developed system, one is used to search for the weave structure, and the other is for the weaving parameters. The desired outlooks presented by the weave structure and the layout of color yarns can be generated and determined by using the weave structure module. The search for unknown combination of weaving parameters (i.e., N_1, N_2, n_1, and n_2) based on a preset cost consideration (i.e., total weight of material yarn) can proceed under both known width and length of a loom.

For instance, the outlook demand for weave structure pattern (i.e., a pattern style of regular grid and interlacing twist) is listed as shown in Table5 and that for production cost (i.e., fabric weight = 58 lb) is illustrated in Table 3. Firstly, in terms of designing a required

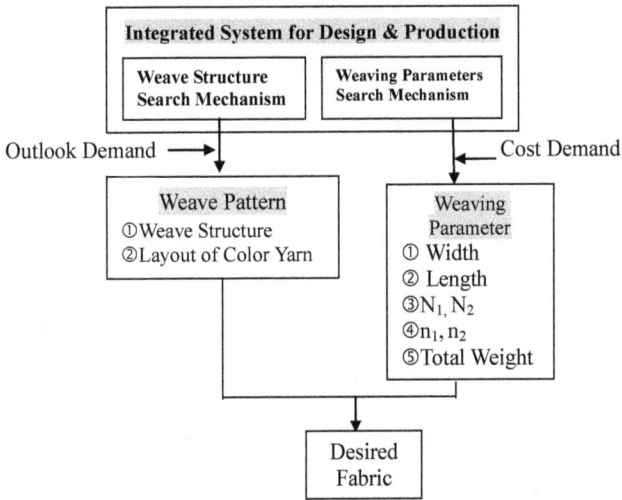

Fig. 8. Integration schemes of design and production for woven fabric

innovative weave structure, in order to obtain an integrated pattern of the appearance of above-mentioned characteristics, three basic patterns (as shown in Figure1a, 1c, and 1e) of the eight ones provided by the system, which are of some sorts of required characteristics and are more similar to the desired pattern, are given higher scores (i.e., fitness values) as 0.8, 0.8, and 1.0 respectively than the other five ones. After proceeding with several generations of evolution, there comes up with a satisfying solution as shown in Figure 6(B)a, the amplified of which is illustrated in Figure 7, for the desired pattern style. Secondly, as for finding a manufacturing solution to meet the cost demand (i.e., desired weight of 58 lb) of the fabric, the search mechanism for weaving parameters can help determine the combinations of weaving parameters (i.e., N_1, N_2, n_1, and n_2). The searched results of tenth generation are listed in Table 4, from which a designer can easily pick out the solution (i.e., N_1=52.0, N_2=22.7, n_1=70.7, n_2=70.7) of maximum fitness 0.7531, which can closely meet the weight demand of 58 lb while manufacturing. Through the assistance of this system, the design and production divisions can thus be integrated together.

6. Conclusions

In this study, an integrated system of design and production for woven fabric is proposed. There are two search mechanisms included in the integrated system. One is for the search of weaving parameters and is of an excellent search capacity to allow the fabric designer to obtain the best combinations of weaving parameters during manufacturing, considering costs. The other is for that of weave structure and can efficiently find appropriate combination sets of the pattern parameters, such as the weave structure (i.e., W_S), the layout of colors for warp yarns (i.e., C_{warp}), and that for weft yarns (i.e., C_{weft1}~C_{weft8}), during pattern design. A fabric designer can efficiently determine what the colors of the warp (weft) yarn and the weave structure should be adopted to manufacture satisfying fabric without the advance sample manufacturing. Both the time and cost consuming for sample manufacturing during design stage can be eliminated now. Moreover, the problem of running out of inspiration for a designer can be solved through the system's assistance as well. With the assistance of the developed integration system proposed in this study, the integration between design division and production one can be achieved. Thus, the competence of an enterprise can increase in the mean time.

7. References

Ahmad, N. ; Tanaka, T. & Saito, Y. (2011). Optimization of multipass turning parameters by genetic algorithm, *Advanced Materials Research*, Volume 264-265, 2011, pp. 1545-1550

Cao, W.; Zhang, F.; Chen, S.-L.; Zhang, C. & Chang, Y.A. (2011). An integrated computational tool for precipitation simulation, *JOM*, Volume 63, Issue 7, July , pp. 29-34

Chen, Z. (2005), Fashion clothing supply chain management, Britain to China, *Textile Asia*, Vol. 36, , pp. 53-56

Cho, Y.; Okada, N.; Park, H.; Takatera, M.; Inui, S. & Shimizu, Y. (2005). An interactive body model for individual pattern making, *International Journal of Clothing Science and Technology*, Vol. 17, pp. 100-108

Dan, L. (2011). On computer aided design college art design thinking course, *Proceedings of 2011 International Conference on Electronic and Mechanical Engineering and Information Technology*, EMEIT, Volume 6, 2011, Article number 6023661, pp. 2928-2930

Deiab, I.M.& Al-Ansary, M.D.R. (2007). GA based optimization of manufacturing cycle cost, *Proceedings of the IASTED International Conference on Modelling and Simulation*, pp. 532-537

Denton, M.J. & Seth, A.K. (1989). Computer Simulation of the Appearance of Fabric Woven from Blended-fibre Yarns, *Journal of Textile Institute*, Vol.80, pp. 415-440

Geerts, F.; Goethals, B. & Mielikainen, T. (2004). Tiling data-bases. *Discovery Sci Berlin: Springer*, pp. 278-289

Gen, M. & Cheng, R. (1997). Genetic Algorithms & Engineering Design, *JOHN WILEY & SONS, INC.* NY

Gerdemeli, I.; Cotur, A.E.; Kayaoglu, E. & Candas, A. (2011). Computer aided valve design of shock absorbers used in vehicles, *Key Engineering Materials*, Volume 486, 27 July, pp. 270-273

Goldberg, D.E. (1989). Genetic Algorithms in Search, Optimization & Machine Learning, *Addision-Wesley Publish Co.*, NY

Griswold R.E. (June 2010). Designing weave structures using Boolean operations, Part 1,2,3, Available from
        http://www.cs.arizona.edu/patterns/weaving/webdocs/gre_bol{1,2,3}.pdf.

Hearle, J. W. S.; Grosberg, P. & Baker, S. (1969). Structure Mechanics of Fibers, Yarns, and Fabrics, Vol. 1, *John Wiley & Sons*, NY

Hu, Z.-H. (2009). A hybrid system based on neural network and immune co-evolutionary algorithm for garment pattern design optimization, *Journal of Computers*, Volume 4, Issue 11, November, pp. 1151-1158.

Inui, S. (1994). A Computer Aided System Using a Genetic Algorithm Applied to Apparel Design, *Sen'I Gakkaishi* , 50(12), pp. 593-598

Inui, S. (2001). A preliminary study of a deformable body model for clothing simulation, *International Journal of Clothing Science and Technology*, Vol. 13, pp. 339-350

Karr, C.L. & Freeman, L. M. (1999). Industrial Applications of Genetic Algorithms, *CRC Press LLC*, NY

Lai, T.-P. (1985). Practical Fiber Physical Chemistry, *Tai-Long Publishing Co.*, Taipei, R.O.C.

Lin, C.A. (1993). Report on Formulation of Weight Consumption of Material Yarn for Woven Fabrics, *Foundation of Textile Union*, R.O.C.

Lin, J.J. (2003). A Genetic Algorithm for Searching Weaving Parameters for Woven Fabrics, *Textile Research Journal*, Vol.73, pp. 105-112

Lin, J.-J. (2008). A GA-based search approach to creative weave structure design, Journal of Information Science and Engineering, Vol.24, No.3, pp. 949-963

Lin, Z. & Harlock, S. C. (1995). A Computer-Aided Grading System for Both Basic Block and Adapted Clothing Patterns, Part III: The Auto Grader System, *Textile Research Journal*, Vol.65, No.3, pp. 157-162

Liu, N.B.; Wu, Z.Y.; Li, G.Y. & Weng, N. (2011). Research on the CAD technology of resin-bonded diamond abrasive tool based on quality control, *Key Engineering Materials*, Vol. 487, pp. 204-208

Luo, Z.G. & Yuen, M.M.F. (2005). Reactive 2D/3D garment pattern design modification, *CAD Computer Aided Design*, Vol.37, pp. 623-630

Ma, L.; Baciu, G.; Hu, J.L. & Zhang, J. (2011). A novel weave pattern encoding method using neighbor information and its applications, *Textile Research Journal*, Vol. 81, No. 6, April, PP 632-648

Mazzetti, S.; De Luca, M.; Bracco, C.; Vignati, A.; Giannini, V.; Stasi, M.; Russo, F. & Regge, D. (2011). A CAD system based on multi-parametric analysis for cancer prostate detection on DCE-MRI, *Progress in Biomedical Optics and Imaging - Proceedings of SPIE 7963*, art. no. 79633Q

Ohta, K.; Shibuya, A.; Imaoka, H. & Shimizu, Y. (1994). Rewiew for CAD/CAM Technology of Textile Products, *Sen'i Gakkaishi*, Vol.50, No.9, pp. 510-523

Özdemir, H. & Baser, G. (2007). Computer simulation of woven structures based on actual yarn photographs. In *Computational Textile*, Zeng, X., Li, Y., Ruan, D. & Koehl, L. (Eds), 75-91, Springer, Berlin, Germany

Penava, Z.; Sukser, T. & Basch, D. (2009). Computer aided construction of reinforced weaves using matrix calculus, *Fibres and Textiles in Eastern Europe*, Vol. 76, No. 5, pp. 43-48

Pensa, R. & Boulicaut, J.-F. (2005). Towards fault-tolerant formal concept analysis. *Proc. 9th Congress of the Italian Association for Artificial Intelligence*, September, pp. 212-223

Pon, E.-S. (1992). Computerized Fabrics Design, *Gou-Tzan Publishing Co.*, Taichung, R.O.C.,.

Rao, M.P.; Pantiuk, M. & Charalambides, P.G. (2009). Modeling the geometry of satin weave fabric composites, *Journal of Composite Material*, Vol. 43, pp. 19-56

Rasmussen, S.L. (2008). On 3-harness weaving: cataloguing designs generated by fundamental blocks having distinct rows and columns, Electronic Journal of Combinatorics, Vol. 15, pp. 1-39

Rödel, H.; Krzywinski, S.; Schenk, A. & Herzberg, C. (2001). Links between design, pattern development and fabric behaviours for cloths and technical textiles, *Tekstilec*, Vol. 44, pp. 197-202

Salehi, M. & Tavakkoli-Moghaddam, R. (2009). Application of genetic algorithm to computer-aided process planning in preliminary and detailed planning, *Engineering Applications of Artificial Intelligence*, Vol. 22, No. 8, December, pp. 1179-1187

Sano, T. & Yamamoto, H. (2001). Computer aided design system for Japanese kimono, *IEEE Instrumentation and Measurement Technology Conference*, Vol. 1, pp. 326-331

Shie, P. (1984). Weaving Engineering, *Su-Hern Publishing Co.*, Taichung, R.O.C.

Shinohara, T.; Takayama, J.-Y.; ohyama, S. & Kobayashi, A. (2008). Automatic weave diagram construction from yarn positional data of woven fabric, *Textile Research Journal*, Vol.78, pp. 745-751

Grundler, D. & Rolich, T. (2009). Evolutionary algorithm for matching the weave and color of woven fabrics, *Tekstilec*, Vol. 58, No. 10, October, pp. 473-484

Tsai, C. M. (1986). Weaving Engineering for Shuttleless Loom, *Zuah-Wei Publishing Co.*, Taichung, R.O.C.

Ujević, D.; Rogale, D. & Trzic, D. (2002). Development and application of computer support in garment and technical textile manufacturing processes, *Tekstilec*, Vol. 51, pp. 224-229

Wang, X.; Yu, L. & Xie, S. (2011). Research on micro-electro-mechanical system computer aided design, *Proceedings of 2011 International Conference on Electronic and Mechanical Engineering and Information Technology*, EMEIT 2011, Vol. 3, Article number 6023393, pp.1556-1559

Wei, S. & Baodong, L. (2009). Study on information management optimization for CAPP by GA, *2009 2nd International Conference on Information and Computing Science*, ICIC, Vol. 4, 2009, Article number 5169142, pp. 128-130

Yamamoto, H. & Qudeiri, J.A. (2010). A concurrent engineering system to integrate a production simulation and CAD system for FTL layout design, *International Journal of Product Development*, Vol. 10, No. 1-3, pp. 101-122

Yang, X.A.; Sun, G.L.; Chen, H.; Liu, X.H. & Zhang, W.S. (2011). Research on design method of electromechanical product form based on MA-GA, School of Science, Chang'an University, South Er Huan Road, Xi'an, 710064, China; email: yang5336@126.com.cn, Copyright 2011 Elsevier B.V., All rights reserved.

Zhang, J.; Baciu, G. & Liang, S. (2010). A creative try: Composing weaving patterns by playing on a multi-input device, *Proceedings of the ACM Symposium on Virtual Reality Software and Technology*, VRST.

5

Polyester Microfilament Woven Fabrics

Hatice Kübra Kaynak[1] and Osman Babaarslan[2]
[1]*Gaziantep University, Textile Engineering Department*
[2]*Çukurova University, Textile Engineering Department*
Turkey

1. Introduction

Synthetic fiber industry has been enforced to make developments due to the increasing performance demand for textile products. One of the most important developments in synthetic fiber industry, is absolutely producing extremely fine fibers which are named as microfibers and nanofibers (Kaynak & Babaarslan, 2010). Until today, there is no exact definition for microfibers. But common opinion is defining a fiber finer than 1 dtex or 1 denier as microfiber (Leadbetter & Dervan, 1992; Bianchi & Maglione, 1993; Purane & Panigrahi, 2007; Basu, 2001; Mukhopadhyay, 2002; Falkai, 1991; Rupp & Yonenaga, 2000). 1 dtex polyester fiber has a fiber diameter of approximately 10 μm (Falkai, 1991). On the other hand, nanotechnology refers to the science and engineering concerning materials, structures and devices which at least one of the dimensions is 100 nanometers (0.1 μm) or less (Ramakrishna, et al., 2005).

Fabrics produced from microfilaments are superior to conventional fiber fabrics, due to their properties such as light weight, durability, waterproofness, windproofness, breathability and drapeability. Tightly woven fabrics produced from microfilament yarns have a very compact structure due to small pore dimensions between the fibers inside the yarns and between yarns themselves. These fabrics provide very good resistance against wind for different end uses such as parachutes, sails, wind-proof clothes, tents while serving light weight and high durability properties (Babaarslan & Kaynak, 2011). Wind resistance is usually assessed by measuring air permeability. This is the rate of air flow per unit area of fabric at a standard pressure difference across the faces of the fabric (Horrocks & Anand, 2004). Airflow through textiles is mainly affected by the pore characteristics of fabrics. The pore dimension and distribution in a fabric is a function of fabric geometry (Bivainyte & Mikucioniene, 2011). So, for woven fabrics, number of yarns per unit area, yarn linear density, weave type, fabric weight and fabric thickness are the main fabric parameters that affect air permeability (Fatahi & Yazdi, 2010; Çay & Tarakçıoğlu, 2007; Çay & Tarakçıoğlu, 2008; Turan & Okur, 2010). On the other hand, considering the yarn structure; yarn production technology, yarn diameter, yarn twist, hairiness, being staple or filament yarn, fiber fineness, fiber cross-section and yarn packing density are also important parameters (Turan & Okur, 2010). The pores of a fabric can be classified as pores between the fibers inside the yarns and between yarns themselves. The dimensions of the pores between the yarns are directly affected by the yarn density and yarn thickness. By increasing of the yarn density, the dimensions of the pores become smaller, thus the air permeability decreases.

The dimensions of pores between the fibers inside the yarns (micro voids) are generally affected by fiber fineness, yarn count, yarn twist and crimp and also the deformation and flattening of the yarns (Çay & Tarakçıoğlu, 2008).

In an earlier study (Varshney et al., 2010), the effect of filament fineness on air permeability of polyester and polyester/viscose woven fabrics was observed. It was seen that decreasing filament fineness has a decreasing effect on air permeability. An another study, (Laourine & Cherif, 2011) was performed on the effect of filament fineness and weave type on air permeability of polyester woven fabrics for surgical protective textiles. The study showed that air permeability can be reduced by decreasing the filament fineness. Also, for woven fabrics with nearly same degree of cover factor, lower levels of air permeability can be reached with plain weaves than those of twill weaves. Kaynak & Babaarslan (2011), investigated the filament fineness on polyester woven fabrics for plain weave. As a result of this study, lower air permeability values were obtained by decreasing the filament fineness. The present study investigates the effects of filament fineness, weft sett and weave type on air permeability of polyester microfilament woven fabrics in a more detailed manner, aiming to determine the proper construction parameters of fabric. In addition regression analyses were conducted to estimate the air permeability before production.

2. General properties of microfibers

Potentially, any man-made fiber could be made into a microfiber (Smith, n.d.). Microfibers are most commonly found in polyester and nylon (Smith, n.d.; Purane & Panigrahi, 2007; Anonymous, 2000). Trevira Finesse, Fortrel Microspun, DuPont Micromattique and Shingosen are all trade names for various polyester microfibers, whereas Supplex Microfibre, Tactel Micro and Silky Touch are some of the trade names for nylon microfibers (Mukhopadhyay & Ramakrishnan, 2008). Nylon is claimed to have advantages over polyester in having a better cover, plus lower density, higher strength and abrasion resistance. Polyester is easier to spin and is available in finer filaments than nylon. Polyester raised fabrics are easier to produce. This has given polyester an economic advantage in apparel and sportswear markets (Anonymous, 2000). However, micro-denier versions of rayon, acrylic and polypropylene products are available to consumers (Purane & Panigrahi, 2007; Smith, 2011; Anonymous, 2000). Microfibers can be used alone or blended with conventional denier man-made fibers as well as with natural fibers such as cotton, wool, viscose and silk (Smith, n.d.; Anonymous, 2000). This enhances the appearance, hand, drape and performance properties of the fabrics (Basu, 2001).

Ultra-fine fibers were produced in the late 1950s which were not of continuous filament type but were fine staple fibers of random length and found no application except for being processed into nonwoven sheets immediately after spinning. In 1961 a petal-shaped ultra fine filament type fiber described in a patent was probably the first example of a potential ultra-fine filament. Another patent issued simultaneously described splitting two component conjugate fibers of non-circular cross section into the two separate components after weaving. No attention was paid at that time for combining these technologies to produce ultra-fine fibers (Okamoto, 2000). The first microfibers were invented in the mid sixties by Dr. Miyoshi Okamoto, a chemist in the Toray industries textile research laboratory. In the beginning, the single fine man-made fibers found scarcely any appropriate application. The breakthrough came with the success of imitation game leather (Rupp &

Yonenaga, 2000). This was the first attempt to produce an ultra fine fiber intentionally. Matsui et al. of Kanebo also tried multi layer conjugate spinning in 1968 for the production of ultra fine filaments. Since no application of ultra-fine fibers was foreseen in the 1960s, there had been no technical or commercial interest in them until Toray put the new suede-like material on the market in 1970 (Okamoto, 2000).

Microfibers are being increasingly used throughout the world for various end uses due to their fineness, high performance characteristics and their unique ability to be engineered for a specific requirement (Anonymous, 2000). Fundamental characteristics of microfibers are as follows;

- Since bending and torsional stiffness are inversely proportional to fiber diameter, ultrafine fibers are extremely flexible (Okamoto, 1993) and for the same reason microfiber yarns impart excellent drapeability to the fabric (Basu, 2001; Purane & Panigrahi, 2007; Rupp & Yonenaga, 2000)
- Yarn strength is high due to the high number of fiber per cross sectional area (Basu, 2001). Microfiber fabrics are also relatively strong and durable in relation to other fabrics of similar weight (Purane & Panigrahi, 2007).
- The yarns made from micro denier fiber contain many more filaments than regular yarns producing fabrics with water tightness and windproofness but improved breathability (Basu, 2001).
- More filaments in yarn result in more surface area. This can make printed fabrics more clear and sharp as compared to normal fabrics (Basu, 2001). Greater fiber surface area results making deeper, richer and brighter colors possible (Smith, n.d.).
- Microfiber fabrics are very soft and have a luxurious hand with a silken or suede touch (Rupp & Yonenaga, 2000; Purane & Panigrahi, 2007).
- Microfibers have a quick stress relief so microfiber fabrics resist wrinkling and retain shape (Purane & Panigrahi, 2007; Okamoto, 2000).
- Microfiber fabrics are washable and dry-cleanable
- Microfiber fabrics insulate well against wind, rain and cold and also they are more breathable and more comfortable to wear
- Microfibers are super-absorbent, absorbing over 7 times their weight in water
- Microfiber dries in one-third of the time of ordinary fibers (Purane & Panigrahi, 2007).
- A large ratio of length to diameter resulting in easy entanglement
- Good interpenetrating capacity to other materials
- Bio-singularity to living tissues and fluids (Okamoto, 2000).

One caution related to synthetic microfibers is heat sensitivity. Because the fiber strands are so fine, heat penetrates more quickly than thicker conventional fibers. So, microfiber fabrics are vulnerable to damage from careless ironing. They will scorch or glaze if too much heat is applied for a too long period (Smith, n.d.; East, 2005).

They also have a tendency to snag easily and, as with all fine fabrics, they need to be handled with care. Jewelry cause pulls, snags or general abrasion to garments (Smith, n.d.; East, 2005).

Since microfibers have an increased surface area, resulting in a dyeing rate four times higher than that of normal which can cause unlevelness in dyeing. They require more dyestuff than normal fibers to attain a given shade depth (Jerg & Baumann, 1990; Anonymous, 2000; Falkai, 1991).

3. Production of microfibers

There are various methods of producing microfibers. All three conventional spinning methods, namely melt spinning, wet spinning, and dry spinning can be employed to manufacture microfibers (Purane & Panigrahi, 2007). Although, it is possible to produce microfibers through conventional melt spinning, to create such fine filaments requires very strict process controls and a uniformly high quality of polymer (Tortora & Collier, 1997).

Ultra-fine fibers are classified into two types: filament type, and staple type. Recent developments in the field of ultra-fine fibers have focused on the filament type (Okamoto, 2000).

Ultra fine fiber of the filament type is produced by the methods including:

1. Direct spinning (conventional extrusion)
2. Conjugate spinning
   a. Islands in a sea type
   b. Separation or splitting type
   c. Multi-layer type (Okamoto, 2000).

3.1 Production of filament type microfibers

3.1.1 Direct spinning

In this method microfiber is directly manufactured by melt spinning (Purane & Panigrahi, 2007). With this method, the fineness of the microfibers produced is limited to 0.1 dtex because of the tendency of the individual filaments to stick together. Improvements in processing conditions and finishing, such as more accurate spinnerets and strictly controlled cooling conditions after extrusion, together with lower polymer viscosity, can however make the production of microfiber yarns possible (Rupp & Yonenaga, 2000).

3.1.2 Conjugate spinning

The technical problems in direct spinning can be solved by conjugate spinning, which yields homogenous ultra-fine fibers. Okamoto et al. and Matsui et al. investigated the extrusion of conjugate fibers with a cross section consisting of highly dispersed conjugate components by modifying the spinneret structure. Conjugate spinning is classified into two types from technical viewpoint: the islands in a sea type and separation or splitting type. In either case, the microfiberization is performed in the form of fabrics. No special technical problems arise in later processing, compared with conventional spinning (Okamoto, 2000).

3.1.2.1 Islands in a sea type

In islands in a sea method, a number of continuous very fine filaments are extruded in a matrix of another polymer. In the spinneret a number of bi-component sheath-core polymer flows are combined into a single flow and extruded. The islands in the sea fiber is then quenched and drawn in the usual way (Richards, 2005). Polyester, nylon, polypropylene, polyethylene and polyphenylene sulfide are the polymers employed as island components (Okamoto, 2000). The various combinations of polymers to form fibers by this method successfully are polystyrene/polyamide and polystyrene/polyester (Purane & Panigrahi, 2007). The sea component is removed by dissolving it in a solvent after conventional processing into woven, knitted or nonwoven fabrics. This technology provided a means of industrial production of suede type artificial leather, silk like fabrics, wiping cloths and fine filters. Since the ultra fine filaments (the island component) are sheathed by the sea component, they are protected from damage during later processing (Okamoto, 2000).

Three component spinning can be carried out with two island components by designing a three component spinneret assembly. The sea component can be reduced to 2-10% of the total components, but the space between the ultra fine filaments is also reduced and this may lead to poorer handle of the products. When the sea component is small in amount and not miscible with the island component, the splitting can be carried our mechanically (Okamoto, 2000).

3.1.2.2 Separation or splitting type

This type of spinning aims to utilize the second component in the final product by splitting the two components instead of removing the second component by dissolving. The ultra fine fiberization is performed by a mechanical or chemical process in the splitting and separation types of spinning (Okamoto, 2000). This method of microfiber production involves extruding a bicomponent fiber which two polymers with poor adhesion to each other are used (Richards, 2005). Applications of these fibers are suede for clothing and upholstery, silk like fabrics, wiping cloths, wall coverings, automobile trims, golf gloves and moisture-permeable and water-repellent fabrics (Okamoto, 2000; Richards, 2005). In this method the overall shape of the fiber determines the ease of splitting. If the components are in a radial configuration then splitting is more difficult than if one polymer is located at the ends of the lobes in a multilobal shape (Richards, 2005). Suitable polymer combinations for splittable bi-component filament spinning are polyamide/polyester and polyester/polyolefin (Purane & Panigrahi, 2007).

3.1.2.3 Multi-layer type

Two components are spun into a conjugate fiber of multilayered structure with an oval-shaped cross section, which is microfiberized into filaments of 0.2-0.3 denier (Okamoto, 2000).

3.2 Production of staple type microfibers

Ultra fine fiber of the staple type is produced by the methods including:

1. Melt blowing or jet spinning
2. Flash spinning
3. Polymer-blend spinning
4. Centrifugal spinning
5. Fibrillation
6. Turbulent flow-moulding
7. Bursting (Okamoto, 2000).

3.2.1 Melt blowing or jet spinning

This method is employed for the production of nonwoven fabrics of polypropylene ultra fine fibers. The polymer melt is blown apart immediately after extrusion by an air jet stream in this method, so it is sometimes termed "jet spinning". Thus, this method is an application of spraying technology rather than true spinning (Okamoto, 2000). It finds applications in an increasing number of fields, such as filtration, absorbency, hygiene and apparel (Mukhopadhyay & Ramakrishnan, 2008).

3.2.2 Flash spinning

The polymer is dissolved in transparent solutions at high temperature under high pressure. The spinning solution is jetted out of a nozzle into the air to form a fibrous network. A fiber network is obtained by spreading a single stream of fiber spun from one spinneret hole. The filament thickness varies from 0.01 denier to 10 denier (Okamoto, 2000).

3.2.3 Polymer blend spinning

In this method the conjugate fiber is produced by extruding and drawing a blended polymer melt of two components. The fiber fineness can not be controlled and the fiber often breaks during spinning, although the spinning stability is strongly dependent on the combination of polymers. Since the dispersed polymer phase is drawn to yield ultra-fine fibers, no filament type of ultra fine fiber is produced at present by polymer blend spinning (Okamoto, 2000).

4. Applications of microfibers

At the start of the development, the researchers searched intensively for suitable fields of application for their microfibers, since they had not yet existed in previous clothing and technical textile concepts (Rupp & Yonenaga, 2000). Microfibers offer a great variety of applications in fashion clothing owing to their extra softness, full handle, drape, comfort and easy care properties (Anonymous, 2000). One general point that should be mentioned is that the desired properties (i.e. sophisticated handle, pleasant silky appearance, leather look-alike, good filtration properties, etc.) are only obtained when a suitable fabric construction is produced. As well as fineness, the material combination, cross-section of the elementary filaments and their effect when used in combinations are extremely important and can offset negative parameters (i.e. proneness to creasing, somewhat lower absolute tenacity) (Falkai, 1991). Items of polyester microfibers, both 100 percent polyester and blends with other fibers, include coats, suits, blouses, dresses, wall coverings, upholstery, sleeping bags, tents, filters and toweling (Kadolph et al., 1993).

4.1 Protection against the weather

Woven fabrics for protection against the weather were previously coated with polyvinyl chloride in most cases (Rupp & Yonenaga, 2000). But today, closely woven microfilament fabrics offer a new era for protection against weather. As the number of filaments in a yarn of given linear density increases, then the surface area of all the fibers increases and, in a fabric of close construction, the gaps between the fibers become smaller (Richards, 2005).

They are only very tiny gaps for air to blow through. So, closely woven fabrics are constructed from microfilament yarn which results in small size of gaps to give maximum protection against wind and rain. The use of microfibers ensures that gaps in the fabric are very small even when dry (Mukhopadhyay & Midha, 2008; East, 2005). Moreover, the fabric and fiber surface is enlarged; therefore moisture is transported via more channels, since a better capillary effect is achieved (Rupp & Yonenaga, 2000). This type of weaving results in a windproof fabric with an excellent water vapor permeability compared with laminates and coatings (Mukhopadhyay & Midha, 2008).

Tightly woven microfilament fabrics exhibit an exceptional property of obstructing water droplets from penetrating. Liquid water is prevented by surface tension from penetrating the fabric, which will have a degree of water repellency (Richards, 2005; East, 2005; Anonymous, 2000). These fabrics exhibit an exceptional property of obstructing water droplets from penetrating while allowing water vapor to escape resulting in increased comfort (Anonymous, 2000; Falkai, 1991; Rupp & Yonenaga, 2000). Microfilaments make it possible to structure fabrics so that they repel wind and rain without loosing their textile character. In this respect the low water absorption of polyester plays a part (Basu, 2001). Functionality of densely woven microfilament fabrics with respect to waterproofness and windproofness can be reinforced by hot calendering (Rupp & Yonenaga, 2000). Their improved water impermeability and lower air permeability make microfiber fabrics highly suitable for waterproof and windproof application area such as sportswear, rain clothes, and tents.

4.2 Clothing

Microfibers are used in a variety of fabrics, but most commonly in dress and blouse weight garments. Suit jackets and bottom weights are becoming available. (Smith, n.d.). As the color, appearance resembles silk, these fabrics can be used for blouses, dresses, tailored suits, hosiery and evening wear (Basu, 2001). Fabrics produced with microfiber yarns are consequently softer and drape better than those made with normal fiber yarns. Even if tightly woven, microfiber yarns have a low weight per unit area and are not stiff. Polyester microfibers can be used to create fashionable women's outerwear fabrics with a new hand and smooth drape (Jerg & Baumann, 1990).

4.3 Synthetic game leather

Synthetic game leather and leather products are today produced industrially in Japan by impregnating nonwovens produced from PET, PA, PAN microfibers with polyurethane. These products offer outstanding advantages as against natural leather and game leather in terms of uniformity, dimensional stability, ease of care, color fastness and low mass (Rupp & Yonenaga, 2000). Microfiber suede fabrics are used in upholstery due to their elegant look, comfortable feel and easy care and clean properties.

4.4 Filtration

Owing to their fine, compact structure, microfiber textiles offer excellent filtration effects for both air and fluid filtration. Ultra-fine microfiber products such as 0.05 dtex PP microfiber nonwovens, in combination with a high electrical voltage, which will provide permanent polarization to the nonwoven, attract and absorb charged dust particles. Microfiber textiles can produce excellent filtration effects in the process of filtering solid or liquid materials. Microfiber liquid filters offer; high water passage speed, high extraction performance and ease of cleaning micro-particles from the filter (Rupp & Yonenaga, 2000). The extremely fine diameter of splittable synthetic microfibers makes them suitable for filter applications by increasing the filtration performance. The splittable microfibers are more suited as flex-resistant materials. In pulsing applications where the filter medium is continuously flexed but also requires stiffness, splittable synthetic fibers add a high degree of reinforcement to the filter medium. Because there are at least 16 times the numbers of fibers available for reinforcement when they are split for segmented fibers (Mukhopadhyay & Ramakrishnan, 2008).

4.5 Microfibers for cleaning

The microfiber fabric can be used for producing cleaning cloths. Most of the stick dirt is caused by dust accumulating on thin layers of fat which merely spreads and barely touched by conventional wiping cloths. This is because the fiber of these wiping cloths is normally 10 mm thick and is unable to capture 1 μm thick oil layers (Basu, 2001). Unlike ordinary cleaning fabrics that move or push dirt and dust from one place to another, the microfilaments can penetrate into the thin fatty layer of dirt and trap it within the micropockets among the filaments and then store the dirt particles in the fabric until it is washed. They are perfect for asthma and allergy sufferers, as they remove dust and dust mites without chemicals. They are also excellent at removing fingerprints from any surface. Grease, tar, splattered bugs come off with the cleaning cloth (Purane & Panigrahi, 2007). The cleaning properties of the microfibers are further enhanced because they have a cationic (positive) charge due to the presence of the polyamide in the microfibers. Most dirt and dust particles, bacteria, pollen, oxidation on metals, etc., have an anionic (negative) charge. Thus, the microfibers naturally attract negatively charged particles, bacteria, etc. (Mukhopadhyay & Ramakrishnan, 2008). The dirt trapped in the micropockets can be removed by washing. These wiping cloths are used for cleaning car mirrors, computers, jewels and noble metals, fingerprints from photos and films (Basu, 2001). Microfiber cleaning towels are used for cleaning floors, windows, furnishings and interior and exterior of cars.

4.6 Medical applications

When compared to commonly textiles to microfiber nonwovens, they are lower in cost, easier to use, more versatile, safer, and features of better disposability. With this in mind, it is no wonder that microfiber nonwovens are found in hospital surgical drapes and gowns, protective face masks, gloves, surgical packs, and bedding (Purane & Panigrahi, 2007). Fabrics from microfibers have excellent breathability and have been used for wound care. Polypropylene microfiber spunbonds have application in wound-care, where they are used as hydrophobic backings. At the same time, the air permeability and breathability of these nonwovens promote healing and their softness and flexibility allow excellent adaptation to the skin. In addition, polypropylene microfiber spunbonds have potential application in disposable surgical gowns and masks where spunlaced fabrics are widely used. The barrier properties of these spunbonds are more than 25% better than the spunlaced fabrics at about half their weight. Their softness, high permeability and breathability guarantee a high level of comfort in wearing when used as surgical gowns and for application as surgical face masks (Mukhopadhyay & Ramakrishnan, 2008).

The biocompatibility of microfibers offers great potential and manifold possibilities of use in the medical field. Microfiber filter produced from melt blown polyester are sold to hospitals for use in blood transfusion and blood donation. Polyester microfibers are very compatible with human organs and have proved themselves in use of vascular prothesis (Rupp & Yonenaga, 2000).

4.7 Construction applications

Polypropylene and bi-component microfibers can be very important components of fiber-reinforced composites, as they function not only as a reinforcing element, but also as a binder fiber between the individual layers. Polypropylene and bi-component microfibers are used in many different composite products: Microfiber reinforced concrete (to reinforce and prevent cracks), insulation material (to avoid the use of chemical binders), multifunctional liquid transport media (acquisition and distribution layers), woven fabrics (as a dimensional stability network), and laminated products (lamination between textiles and boards) (Purane & Panigrahi, 2007).

5. Weaving of microfibers

Microfilaments demand extraordinarily high quality weaving warps (Rupp & Yonenaga, 2000). In the case of warping microfiber yarns versus warping conventional fiber yarns, it must be considered that the smaller amount of force is needed to break a microfiber filament during warping. In addition, the eyelets used in the tension device in the creel of warping machine must be made of a low friction material. The surfaces must be free of cuts and snags. Also, the reed blades must be absolutely free of any snags or bars (Basu, 2001).

Due to their fineness the total surface area of microfiber yarn is far greater than ordinary fibers (Anonymous, 2000). The surface of microfiber yarns is 10-15% larger than conventional fiber yarns. Due to the higher surface area of microfiber smoothed yarn absorbs approximately 10% more sizing agent and textured yarns approximately 15% more sizing agent than conventional yarn. By reducing the viscosity of the sizing agent distribution of the sizing agent may be improved. The squeezing pressure must be adapted to the speed to achieve regular sizing. As a result of the larger surface and the small mass of the individual filaments the microfiber can be heated faster as drying also occurs faster. So cooling must be effected in a few seconds. The drying temperature can be lowered by 10% in the case of microfilaments (Basu, 2001). Since microfibers have very small interstices, desizing become quite difficult and costly. Desizing must always be clean to prevent problems in dyeing. Knowledge of the type of size used is very important to optimize the desizing process. Pretreatment must be done either on tensionless open width washers or in the overflow or jet dyeing machine. Control of PH is important for optimum size removal (Anonymous, 2000; Rupp & Yonenaga, 2000).

Many machine manufacturers recommend the use of air jet and rapier looms for microfilament woven fabric production. They also recommend the use of prewinding units and yarn brakes, which are suitable for processing filament yarns for weft insertion.

Basically weft yarns must be inserted with the greatest care. In this respect, processing should be effected with at least two yarn storage units in order to keep taking off speed as low as possible. Warp ends should be fed by a finely controlled warp let off system with an absolute tension sensor and a positively controlled rotating back rest roller. Abrasion resistant materials must be employed to prevent filament breakage in the case of the winding unit. For this the use of vulcanized or rubberized materials are suggested. For double width weaving machines, an additional pressure roller should be used for fabric slippage. Textured filament yarns should be more intensively intermingled in order to ensure good running characteristics. This may however have a rather detrimental effect on air consumption in the case of air jet looms (Rupp & Yonenaga, 2000). The harness with drop wires of warp stop motion, the reed and the healds come into particularly intimate contact with microfiber yarns. The surfaces of these items, therefore, need to have particularly low roughness. Fabrics woven with microfiber yarns are often densely constructed; beat up must also be relatively severe. If the reed wires have sharp edges, they can easily cut the individual filaments and thus damage the yarn. Weft accumulators must provide sensitive gentle tensioning of fine yarns. Two or more weft accumulators must be used with weaving machines to reduce the withdrawal of weft insertion. Temples recommended for silk and silk like fabrics must be used for microfiber yarn fabrics in order to prevent the fabric bowing out in the selvedge zone (Basu, 2001).

6. Experimental

This chapter is focused on the effect of filament fineness, weft sett and weave type on air permeability of 100% polyester microfilament woven fabrics. Fabrics were made in three weave types; 1/1 Plain, 2/3 Twill (Z) and 1/4 Satin. The highest and the lowest weft sett values for the weave types were determined by production trials. For each weave type four different weft sett values were applied considering the weaveability limitations. For plain 30, 32, 34, 36 wefts/cm, for twill 41, 43, 45, 47 wefts/cm and for satin 43, 45, 47, 49 wefts/cm were determined as proper values. Polyester microfilament textured yarns of 110 dtex with 0.33, 0.57 and 0.76 dtex filament finenesses and conventional polyester textured yarns of 110 dtex with 1.14, 3.05 dtex filament finenesses were used as weft. For warp yarn 83 dtex polyester yarn with 1.14 dtex filament fineness was used. Warp set was 77 warps/cm for plain weave types and 85 warps/cm for twill and satin weave types. By this way 60 woven fabric samples were produced. Sample fabrics were woven by a loom with an electronic Dobby shedding mechanism and rapier weft insertion at a loom speed of 420 rev/min. Before desizing, to obtain dimensional stability, samples were applied to thermal fixation at 195°C with 25m/min process speed.

Yarn linear density was measured according to ISO 2060, yarn tenacity and elongation were measured according to ISO 2062, shrinkage was determined according to DIN EN 14621, crimp contraction, crimp module and crimp stability were determined according to DIN 53840-1 standards. The properties of weft yarns are given in Table 1.

Figure 1, exhibits the microscopic views of cross sections of weft yarns used in the study. As seen from Figure 1, as the filament fineness decreases, the number of filament in yarn cross section increases. Thus, total void area between the filaments is smaller for finer filaments than that of coarser filaments. Furthermore, number of pores between the filaments and total surface area of the filaments is increased. The cross sections of the filaments are essentially round. But after the texturizing process the view of the filament cross section was changed to cornered shape.

	Weft yarn filament fineness, dtex				
	3.05	1.14	0.76	0.57	0.33
Yarn linear density, dtex	110				
Yarn tenacity, cN/Tex	3.0	3.6	3.6	3.0	3.0
Yarn breaking elongation, %	21	27	25	24	26
Shrinkage , %	3.0	2.6	3.6	4.0	4.5
Crimp contraction, %	15	11	8	8	5
Crimp module, %	10	6	4	4	3
Crimp stability, %	81	82	79	77	72
Oil content, %	3	3	3	3	3
Intermingling frequency, points/meter	90	100	90	70	60
Intermingling retention (Stability at 3% elongation)	45	50	50	45	60

Table 1. Weft yarn properties used in the study

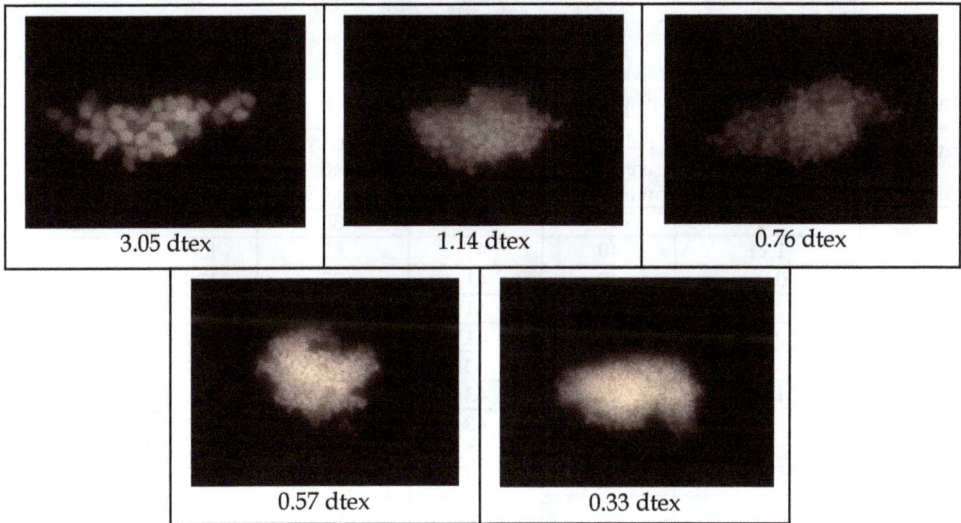

Fig. 1. Microscopic views of weft yarn cross sections by magnification X200

Structural properties of sample fabrics after thermal fixation and desizing processes were determined according to following standards and results were given in Tables 2, 3 and 4.

- ISO 7211-2 Textiles -- Woven fabrics -- Construction -- Methods of analysis -- Part 2: Determination of number of threads per unit length
- ISO 3801 Textiles -- Woven fabrics -- Determination of mass per unit length and mass per unit area
- ISO 5084 Textiles -- Determination of thickness of textiles and textile products

Weave type	Weft yarn filament fineness, dtex	Adjusted weft sett, wefts/cm	Fabric weft sett, wefts/cm	Fabric warp sett, warps/cm	Fabric weight, g/m²	Fabric thickness, mm
Plain	3.05	28	30	77	117	0.25
	1.14	28	30	77	112	0.23
	0.76	28	30	77	111	0.23
	0.57	28	30	77	113	0.24
	0.33	28	30	77	114	0.24
	3.05	30	32	77	118	0.24
	1.14	30	32	77	116	0.23
	0.76	30	32	77	115	0.22
	0.57	30	32	77	116	0.22
	0.33	30	32	77	117	0.22
	3.05	32	34	77	119	0.24
	1.14	32	34	77	120	0.22
	0.76	32	34	77	118	0.21
	0.57	32	34	77	120	0.22
	0.33	32	34	77	121	0.22
	3.05	33	36	77	121	0.23
	1.14	33	36	77	122	0.22
	0.76	33	36	77	120	0.21
	0.57	33	36	77	121	0.22
	0.33	33	36	77	123	0.21

Table 2. Structural properties of plain weave sample fabrics

Weave type	Weft yarn filament fineness, dtex	Adjusted weft sett, wefts/cm	Fabric weft sett, wefts/cm	Fabric warp sett, warps/cm	Fabric weight, g/m²	Fabric thickness, mm
2/3 Twill (Z)	3.05	40	41	85	125	0.22
	1.14	40	41	85	128	0.22
	0.76	40	41	85	126	0.22
	0.57	40	41	85	129	0.22
	0.33	40	41	85	130	0.22
	3.05	42	43	85	128	0.22
	1.14	42	43	85	130	0.22
	0.76	42	43	85	130	0.22
	0.57	42	43	85	132	0.22
	0.33	42	43	85	133	0.22
	3.05	44	45	85	131	0.22
	1.14	44	45	85	133	0.22
	0.76	44	45	85	132	0.22
	0.57	44	45	85	135	0.22
	0.33	44	45	85	137	0.23
	3.05	46	47	85	134	0.23
	1.14	46	47	85	136	0.22
	0.76	46	47	85	136	0.22
	0.57	46	47	85	138	0.23
	0.33	46	47	85	141	0.23

Table 3. Structural properties of twill weave sample fabrics

Weave type	Weft yarn filament fineness, dtex	Adjusted weft sett, wefts/cm	Fabric weft sett, wefts/cm	Fabric warp sett, warps/cm	Fabric weight, g/m²	Fabric thickness, mm
	3.05	42	43	85	129	0.23
	1.14	42	43	85	129	0.22
	0.76	42	43	85	128	0.22
	0.57	42	43	85	130	0.22
	0.33	42	43	85	130	0.22
	3.05	44	45	85	132	0.23
	1.14	44	45	85	131	0.22
	0.76	44	45	85	132	0.22
	0.57	44	45	85	133	0.22
1/4 Satin	0.33	44	45	85	133	0.23
	3.05	46	47	85	133	0.23
	1.14	46	47	85	132	0.23
	0.76	46	47	85	134	0.23
	0.57	46	47	85	136	0.22
	0.33	46	47	85	136	0.23
	3.05	48	49	85	136	0.23
	1.14	48	49	85	139	0.22
	0.76	48	49	85	138	0.23
	0.57	48	49	85	140	0.23
	0.33	48	49	85	138	0.23

Table 4. Structural properties of satin weave sample fabrics

Figure 2 shows the SEM views of three of fabric samples which have the highest weft sett and lowest weft yarn filament fineness.

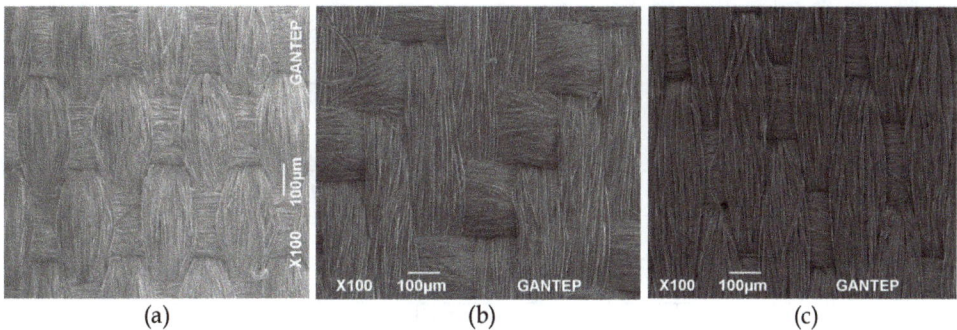

(a)   (b)   (c)

Fig. 2. SEM views of fabric samples X 100 with 0.33 dtex weft yarn filament fineness and 1.14 dtex warp yarn filament fineness (a) Plain, 36 wefts/cm, 77 warps/cm, (b) Twill, 47 wefts/cm, 85 warps/cm, (c) Satin, 49 wefts/cm, 85 warps/cm

All yarn and fabric samples were conditioned according to ISO 139 before tests and tests were performed in the standard atmosphere of 20±2°C and 65±4% humidity. The air permeability of sample fabrics was measured by SDL Atlas Digital air permeability test device according to ISO 9237 with 20 cm² test head and 200 Pa air pressure drop. Each sample was replicated twice and ten repeated measurements were done for each replication. The mean values of the test results were used in graphical representation.

Design-Expert statistical software package was used to interpret the experimental data and to compose the regression models. Regression models were formed to define the relationship between independent variables (filament fineness and weft sett) and response variable (air permeability) for plain, twill and satin weave types. General Factorial Design was selected to compose regression models. The air permeability test results of samples were used to analyze the general factorial design. The analysis of variance, lack of fit tests and residual analysis were performed to select the proper model for the air permeability.

7. Results and discussion

Tightly woven fabrics produced from microfilament yarns have a very compact structure due to small pore dimensions between the fibers inside the yarns and between yarns themselves. These fabrics provide very good resistance against wind for different end uses such as parachutes, sails, wind-proof clothes, tents while serving light weight and high durability properties (Kaynak & Babaarslan, 2011). Wind resistance is usually assessed by measuring air permeability. Air permeability is the rate of air flow per unit area of fabric at a standard pressure difference across the faces of the fabric (Horrocks & Anand, 2000). The passage of air is of importance for a number of fabric end uses such as industrial filters, tents, sail cloths, parachutes, raincoat materials, shirting, down proof fabrics and airbags (Saville, 2002). Airflow through textiles is mainly affected by the pore characteristics of fabrics (Bivainyte & Mikucioniene, 2011). As fabric interstices increase in number and size, air permeability increases. In other words as fabric porosity increases, air permeability increases (Collier & Epps, 1999).

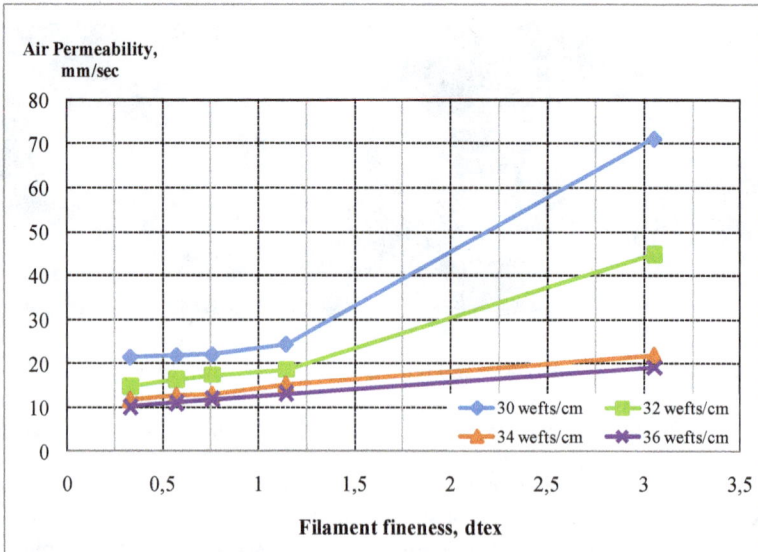

Fig. 3. Air permeability of plain weave type samples

Figure 3 exhibits the air permeability of plain weave samples for different filament fineness and weft sett. The highest air permeability value is 71.2 mm/sec and the lowest value is 10.2 mm/sec. As seen from Figure 3, increasing weft sett values cause a decrease of air permeability values for plain weave samples. Higher weft sett values provide the gaps between the yarns which the air pass through to become smaller thus lead to lower air permeability. The literature survey shows that some former studies on this topic (Fatahi & Yazdi, 2010; Çay &Tarakçıoğlu, 2008; Çay &Tarakçıoğlu, 2007) are agree with our work. Nevertheless, it must be considered that the effect of weft sett on air permeability is more obvious for coarser filaments. In other words the influence of weft sett on air permeability decreases as the filament fineness decreases. On the other hand lower filament finenesses cause lower air permeability. Because, lower filament fineness results in higher number of filament in yarn cross section. Thus, dimensions of gaps between the filaments within the yarns decreases. This is an expected result, since micro voids between the fibers become smaller as the fiber diameter decreases, thus the air permeability decreases as pointed out in an earlier study (Varshney, 2010).

The statistical analyses show that the best fitting model is the cubic model for plain weave type. ANOVA results for air permeability of plain weave type samples is given in Table 5.

Source	Sum of squares	Degree of freedom	Mean Square	F value	P value	Significance
Model	7541.65	7	1077.38	637.04	< 0.0001	Significant
A	635.42	1	635.42	375.72	< 0.0001	Significant
B	261.96	1	261.96	154.89	< 0.0001	Significant
A²	90.57	1	90.57	53.55	< 0.0001	Significant
B²	280.35	1	280.35	165.77	< 0.0001	Significant
AB	1174.29	1	1174.29	694.34	< 0.0001	Significant
A²B	118.57	1	188.57	70.11	< 0.0001	Significant
AB²	138.09	1	138.09	81.65		
Residual	54.12	32	1.69			
Lack of Fit	47.25	12	3.94	11.46	< 0.0001	Significant
Pure Error	6.87	20	0.34			
Cor. Total	7595.77	39				

Table 5. ANOVA results for air permeability of plain weave type samples

ANOVA results in Table 5 show that the effect of filament fineness (A) and weft sett (B) on air permeability is significant for plain weave samples from a statistical approach.

The regression equation of the cubic model for plain weave type is as follows:

$$\text{Air Permeability (mm/sec)} = 36.02279 + 450.27578\ A + 1.132469\ B + 47.20693\ A^2 + 0.053972\ B^2 - 29.40924\ AB - 1.35053\ A^2B + 0.47562\ AB^2 \tag{1}$$

In this equation (1); A and B are the filament fineness (dtex) and weft sett (wefts/cm) independent variables respectively. The air permeability of plain weave polyester microfilament woven fabrics can be predicted by this equation. Mean Square Error (MSE), Mean Absolute Error (MAE), Mean Absolute Percent Error (MAPE%), R-square predicted (R^2predicted) and R-squared (R^2) values which contribute the performance of the statistical model for plain weave type samples are seen in Table 6.

According to model performance values, the correlation coefficient between predicted and observed air permeability values is 0.9854 indicating a strong predictive capability of the regression model for plain weave types. Also, this regression model can predict the air permeability with 95.47% accuracy. So it can be said that the regression equation gave satisfactory results.

Performance parameter	Value
R^2	0.9936
R^2predicted	0.9854
MSE	1.35
MAE	0.89
MAPE, %	4.53

Table 6. Performance of the model for plain weave type

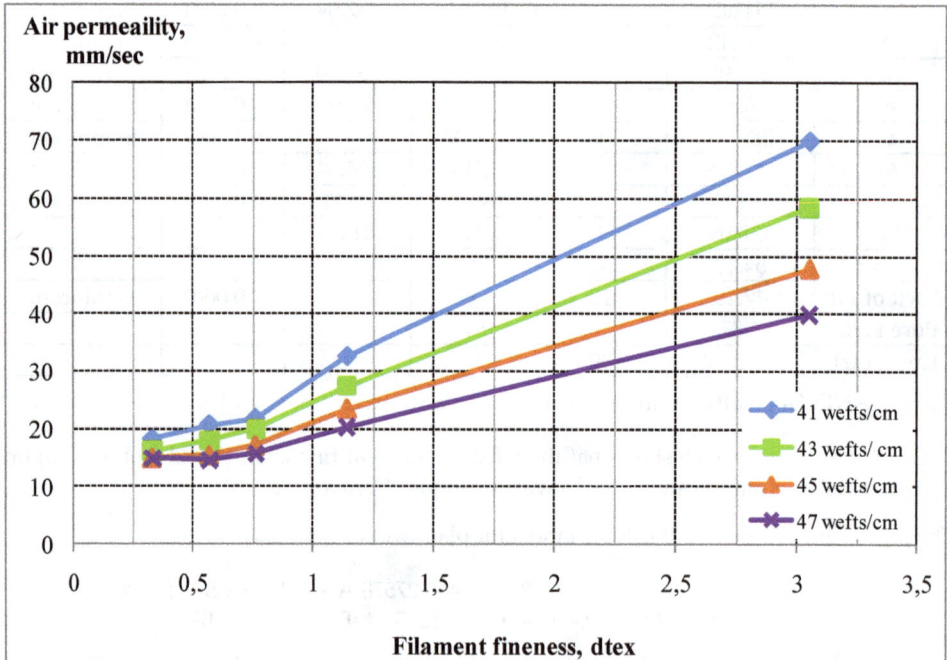

Fig. 4. Air permeability of twill weave type samples

Figure 4 illustrates the air permeability versus filament fineness of twill weave samples for different weft sett values. The highest air permeability value is 69.9 mm/sec while the lowest value is 14.5 mm/sec. It must be emphasized that reducing the air permeability from 69.9 mm/sec to 14.5 mm/sec is an important result showing the effect of filament fineness and weft sett on the air permeability. It is clear from Figure 4 that decreasing the filament fineness has a decreasing effect on the air permeability. A direct linear relationship between filament fineness and air permeability is also obvious. This relationship is similar for all weft sett values.

Furthermore, increasing weft sett causes a decrease of air permeability for twill woven fabrics. But the magnitude of this effect is decreased as the filament fineness decreased. The difference of air permeability value between the highest and the lowest weft sett are; 30.1 mm/sec, 12.3 mm/sec, 6.1 mm/sec, 6.1 mm/sec and 3.3 mm/sec for 3.05, 1.14, 0.76, 0.57 and 0.33 dtex filament finenesses respectively. These findings suggest that, total void volume has the vital importance with respect to air permeability and total void volume which the air flow pass through is affected by filament fineness much more than weft sett.

The statistical analyses show that the best fitting model is the cubic model for twill weave type. ANOVA results for air permeability of twill weave type samples are given in Table 7.

Source	Sum of squares	Degree of freedom	Mean Square	F value	P value	Significance
Model	9399.30	6	1566.55	3897.76	< 0.0001	Significant
A	190.33	1	190.33	473.56	< 0.0001	Significant
B	1232.42	1	1232.42	3066.40	< 0.0001	Significant
A²	11.58	1	11.58	28.80	< 0.0001	Significant
B²	10.10	1	10.10	25.13	< 0.0001	Significant
AB	526.39	1	526.39	1309.71	< 0.0001	Significant
A³	34.22	1	34.22	85.15	< 0.0001	Significant
Residual	13.26	33	0.40			
Lack of Fit	8.28	13	0.64	2.55	0.0289	Significant
Pure Error	4.98	20	0.25			
Cor. Total	9412.56	39				

Table 7. ANOVA results for air permeability of twill weave type samples

ANOVA results in Table 7 show that the effect of filament fineness (A) and weft sett (B) on air permeability is significant for twill weave samples from a statistical approach.

The regression equation of the cubic model for twill weave type is as follows:

$$\text{Air Permeability (mm/sec)} =$$
$$261.98458 + 55.06852\,A - 11.05995\,B + 27.54636\,A^2 + 0.12562\,B^2 - 1.66115\,AB - 5.87156\,A^3 \quad (2)$$

In this equation (2); A and B are the filament fineness (dtex) and weft set (wefts/cm) independent variables respectively. The air permeability of polyester microfilament woven fabrics can be predicted by this equation. Mean Square Error (MSE), Mean Absolute Error (MAE), Mean Absolute Percent Error (MAPE%), R-square predicted (R^2predicted) and R-squared (R^2) values which contribute the performance of the statistical model are seen in Table 8.

According to model performance values, the correlation coefficient between predicted and observed air permeability values is 0.9979 indicating a strong predictive capability of the regression model for twill weave types. Also, this regression model can predict the air permeability with 98.4% accuracy. So it can be said that the regression equation gave satisfactory results.

Performance parameter	Value
R^2	0.9986
R^2 predicted	0.9979
MSE	0.33
MAE	0.41
MAPE %	1.60

Table 8. Performance of the model for twill weave type

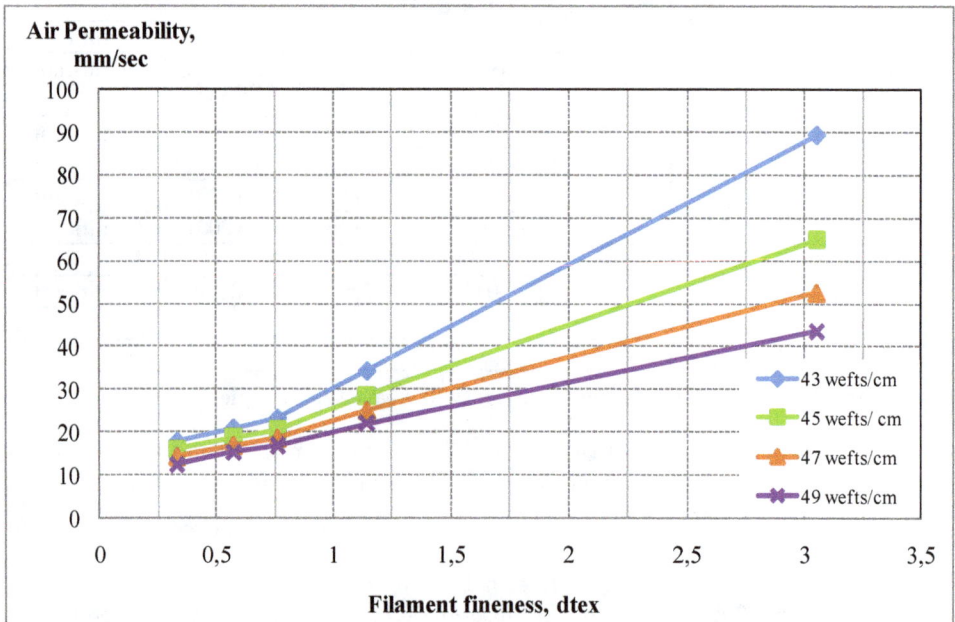

Fig. 5. Air permeability of satin weave type samples

The air permeability of satin weave type samples are shown in Figure 5. The highest and the lowest air permeability values are 89.3 mm/sec and 12.4 mm/sec respectively. The value of

76.9 mm/sec as the difference between these values is considerable. This value (76.9 mm/sec) is higher than the value (61 mm/sec) obtained for plain weave samples and the value (55.4 mm/sec) obtained for twill weave samples. Based on this result, it is clear that the effects of filament fineness and weft sett on satin weave type samples are higher than those of plain and twill weave type samples. Decreasing filament fineness causes a decrease on air permeability for satin weave samples. This is the result of reducing the void volume of the fabric due to decreasing filament fineness as seen before in plain and twill weave samples. A direct linear relationship between filament fineness and air permeability is also available for satin weave samples.

Besides, higher weft sett values resulted in lower air permeability for satin weave type samples. Similar with plain and twill weave samples, the influence of decreasing weft sett on air permeability decreases as the filaments become finer. The difference of the air permeability value between the highest and the lowest weft sett are; 45.7 mm/sec, 12.3 mm/sec, 6.4 mm/sec, 5.6 mm/sec and 5.5 mm/sec for 3.05, 1.14, 0.76, 0.57 and 0.33 dtex filament fineness respectively. This common tendency for both all weave types explains the effect of filament fineness on total void volume of fabrics. It should be noted that, for satin weave type there is a direct proportion between filament fineness and air permeability and a reverse proportion between weft sett and air permeability as seen in earlier studies (Fatahi & Yazdi, 2010; Çay &Tarakçıoğlu, 2008; Çay &Tarakçıoğlu, 2007; Varshney, 2010).

ANOVA results for air permeability of satin weave type samples are given in Table 9.

Source	Sum of squares	Degree of freedom	Mean Square	F value	P value	Significance
Model	14798.49	5	2959.70	778.77	< 0.0001	Significant
A	3831.95	1	3831.95	1008.28	< 0.0001	Significant
B	2267.35	1	2267.35	596.60	< 0.0001	Significant
B²	95.39	1	95.39	25.10	< 0.0001	Significant
AB	1259.27	1	1259.27	331.35	< 0.0001	Significant
AB²	81.93	1	81.93	21.56	< 0.0001	Significant
Residual	129.22	34	3.80			
Lack of Fit	76.20	14	5.44	2.05	0.0691	Not Significant
Pure Error	53.01	20	2.65			
Cor Total	14927.70	39				

Table 9. ANOVA results for air permeability of satin weave type samples

ANOVA results in Table 6 show that the effect of filament fineness (A) and weft sett (B) on air permeability is statistically significant for satin weave samples.

The regression equation of the cubic model for satin weave type is as follows:

$$\text{Air Permeability (mm/sec)} =$$
$$-400.27938 + 909.42821\,A + 17.24904\,B - 0.18175\,B^2 - 36.27354\,AB + 0.36635\,AB^2 \qquad (3)$$

In this equation (3); A and B are the filament fineness (dtex) and weft sett (wefts/cm) independent variables respectively. The air permeability of polyester microfilament woven fabrics can be predicted by this equation. Mean Square Error (MSE), Mean Absolute Error (MAE), Mean Absolute Percent Error (MAPE%), R-square predicted (R^2predicted) and R-squared (R^2) values which contribute the performance of the statistical model are seen in Table 10.

According to model performance values, the correlation coefficient between predicted and observed air permeability values is 0.9815 indicating a strong predictive capability of the regression model for twill weave types. Also, this regression model can predict the air permeability with 95.37% accuracy. So it can be said that the regression equation gave satisfactory results.

Performance parameter	Value
R^2	0.9913
R^2predicted	0.9815
MSE	3.23
MAE	1.22
MAPE %	4.63

Table 10. Performance of the model for satin weave type

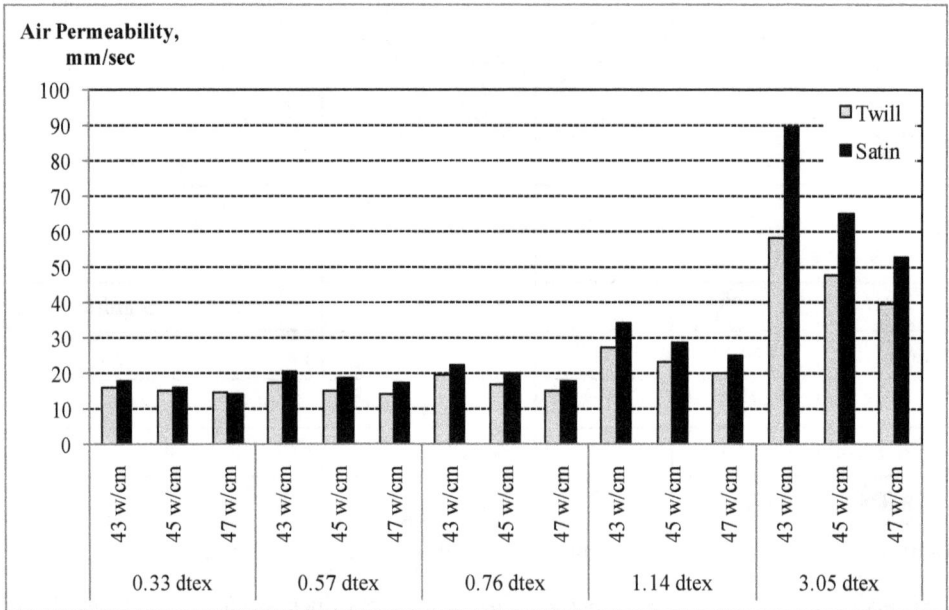

Fig. 6. The effect of weave type on air permeability

It is seen from Figure 6 that for all filament fineness and weft sett values, satin woven fabric samples show higher air permeability than twill fabric samples. The number of interlacing is lower in satin weave type than twill weave type. So the yarn mobility is higher for satin weave types. Yarn mobility is higher, thus the pore dimensions become larger because of the deformation during air flow (Çay & Tarakçıoğlu, 2008). So enlarging the pore dimensions cause the air flow to pass through the fabric more easily and air permeability of the fabrics increase. On the other hand, the effect of weave type on air permeability is considerable for 3.05 dtex filament fineness.

8. Conclusion

Wind resistance was achieved by coating fabrics formerly. But, it is already known that obtaining the wind proof fabrics with a better breathability is possible by weaving microfilament yarns with high densities. These types of fabrics provide a good thermal insulation in windy conditions in addition to submitting a comfortable wear by transporting the sweat vapor more easily than other counterparts. So, it is widely important to know how the woven fabric parameters affect wind resistance and to determine the proper values of these parameters for particular end uses. Consequently, air permeability of polyester microfilament and conventional filament fabrics is presented here. It is already known that a good wind resistance can be achieved by ensuring lower gaps in fabric structure with finer filaments and higher densities. A considerable effect of filament fineness on air permeability is seen, for all weave types used in this study. The experimental results showed that decreasing the filament fineness have a decreasing effect on fabric air permeability. This is not surprising, since the air gaps between the filaments within the yarns become smaller as the filament fineness decreases. Thus, air flow through the filaments is prevented. Furthermore, higher weft sett values provided lower air permeability values because of obtaining smaller air gaps between the yarns in fabric structure. This situation causes the air flowing through the fabric more difficultly and fabric air permeability decreases. It should be noted that our study was investigated the effect of filament fineness and fabric density by differentiating the parameters of weft direction solely. It may be concluded that by changing the parameters in the warp direction, lower air permeability can be achieved. From the point of weave type, it is observed that weave types with higher number of float or lower number of interlacing have higher air permeability values. Because this type of weaves provides better mobility for yarns in their structure and gaps between these yarns become larger with air flow more easily than others.

As mentioned earlier, very good resistance against wind can be achieved by tightly woven microfilament fabrics for different end uses. The most convenient fabric construction can be determined for a particular end use such as wind proof cloth, tent, e.t.c. with the aid of the results and regression analysis obtained from this experimental study. This study also lends assistance to decide the structural parameters for barrier fabrics in specialized applications such as surgical gowns which will be on horizon in near future. For further studies, fabric properties can be developed by using different fiber blends, applying fabrics mechanical surface treatments and special finishes. In the near future, it is expected to see nanofiber yarns for producing woven and knitted fabrics as well as microfiber yarns. Thus, several vast developments might be seen in fabric performance properties and fabric functionality.

9. Acknowledgement

This research is founded by "Scientific Research Projects Governing Unit of Çukurova University" with the project number of MMF2010BAP1.

10. References

Anonymous. (2000). Recent advancements in man-made textiles. *New Cloth Market*, Vol.14, No.4, pp.13-14, ISSN 0972-1711

Babaarslan, O. & Kaynak, H.K. (2011). A study on air permeability of windproof polyester microfilament fabrics. *ICONTEX 2011-International Congress of Innovative Textiles*, pp.418-427, ISBN 978 605 4265 14 5, İstanbul Turkey, October 20-22, 2011

Basu, A. (2001). Microfibers: Properties, processing and use. *Asian Textile Journal*,Vol.10, No.4, ISSN 1819-3358

Bianchi, R. & Maglione, R. (1993). Manufacturing of fine denier filaments, In: *Polyester: 50 Years of Achievement, Tomorrow's Ideas and Profits*, D. Brunnschweiler & J. Hearle (Eds), pp.196-198, Stanley Press, ISBN 1 870812 49 2, Dewsbury, England

Bivainyte, A. & Mikucioniene, D. (2011). Investigation on the air and water vapour permeability of double layered weft knitted fabrics. *Fibres and Textiles in Eastern Europe*, Vol.19, No.3 ,pp.69-73, ISSN 1230-3666

Collier, B.J. & Epps, H.H. (1999). *Textile Testing and Analysis*, Prentice Hall, ISBN 0-13-488214-8, Ohio, USA

Çay, A. & Tarakçıoğlu I. (2008). Relation between fabric porosity and vacuum extraction efficiency: Energy issues. *The Journal of The Textile Institute*, Vol. 99, No.6, pp.499-504, ISSN 0040-5000

Çay, A. & Tarakçıoğlu, I. (2007). Prediction of the air permeability of woven fabrics using neural networks. *International Journal of Clothing Science and Technology*, Vol.19, No.1, pp.18-35, ISSN 0955-6222

East, A.J. (2005). Polyester fibers. In: *Synthetic Fibers: Nylon, Polyester, Acrylic, polyolefin*, J.E. McIntyre (Ed), pp.95-166, Woodhead Publishing, ISBN 1 85573 588 1, Cambridge, England

Falkai, B.V. (1991). Production and properties of microfibers and microfilaments. *The Indian Textile Journal*, No.2, pp.62-70, ISSN 0019-6436

Fatahi, I. & Yazdi, A.A. (2010). Assessment of the relationship between air permeability of woven fabrics and its mechanical properties. *Fibres and Textiles in Eastern Europe*, Vol.18, No.6, pp.68-71, ISSN 1230-3666.

Horrocks, A.R. & Anand, S.C. (Eds). (2004). *Handbook of Technical Textiles*, CRC Press, ISBN 0-8493-1047-4, NewYork, USA

Jerg, G. & Baumann, J. (1990). Polyester microfibers: A new generation of fabrics. *Textile Chemists and Colorists*, Vol.22, No.12, pp.12-14, ISSN 0040-490X

Kadolph, S., Langford, A.L., Hollen, N., Saddler, J. (1993). *Textiles*, Macmillan Publishing, ISBN 0 02 362601 6, New York, USA

Kaynak, H.K. & Babaarslan O. (2011). Effects of microfilament fineness on woven fabric properties. *Electronic Journal of Textile Technologies*, Vol. 3, No.5, pp.30-39, e-ISSN 1309-3991

Kaynak, H.K. & Babaarslan O. (2010). Investigation of the effects of filament fineness on the performance properties of microfiber knitted sportswear fabrics, *4th International Technical Textiles Congress*, pp.51-52, ISBN 978 975 441 285 7, İstanbul-TURKEY, May 16-18, 2010

Laourine, E. & Cherif, C. (2011). Characterization of barrier properties of woven fabrics for surgical protective textiles. *AUTEX Research Journal*, Vol.11, No.2, pp.31-36, ISSN 1470 9589

Leadbetter, P. & Dervan, S. (1992). The microfiber step change. *Journal of the Society of Dyers and Colourists*, Vol. 108, No. 9, pp.369-371, ISSN 0037-9859

Mukhopadhyay, S. (2002). Microfibers – An Overview. *Indian Journal of Fiber & Textile Research*. Vol.27, No.3, pp.307-314, ISSN 0971-0426

Mukhopadhyay, A. & Midha, V.K. (2008). A Review on designing the waterproof breathable fabrics Part I: Fundamental principles and designing aspects of breathable fabrics. *Journal of Industrial Textile*, Vol.37, No.3, pp.225-262, e-ISSN 1530-8057

Mukhopadhyay, S. & Ramakrishnan, G. (2008). Microfibers. *Textile Progress*, Vol.40, No.1, pp.1-86, e-ISSN 1754 2278

Okamoto, M. (2000). Spinning of Ultra-fine Fibers, In: *Advanced fiber spinning technology*, T. Nakajima (Ed), pp.187-207, Woodhead Publishing, ISBN 1 85573 182 7, Cambridge, England

Okamoto, M. (1993). Ultra-fine Fibers: A new dimension for polyester, In: *Polyester: 50 Years of Achievement, Tomorrow's Ideas and Profits*, D. Brunnschweiler & J. Hearle (Eds), pp.108-111, Stanley Press, ISBN 1 870812 49 2, Dewsbury, England

Purane, S.V. & Panigrahi, N.R. (2007). Microfibers, microfilaments and their applications. *AUTEX Research Journal*, Vol.7, No.3, pp.148-158, ISSN 1470 9589

Ramakrishna, S., Fujihara, K., Teo, W.E., Lim, T.C., & Ma, Z. (2005). *An introduction to electrospinning and nanofibers*, World Scientific Publishing, ISBN 978-981-256-415-3, Singapore

Richards, A.F. (2005). Nylon fibers, In: *Synthetic Fibers: Nylon, Polyester, Acrylic, polyolefin*, J.E. McIntyre (Ed), pp.20-94, Woodhead Publishing, ISBN 1 85573 588 1, Cambridge, England

Rupp, J. & Yonenaga, A. (2000). Microfibers-The new man made fiber image. *International Textile Bulletin*, No.4, pp.12-24, ISSN 1029-8525

Saville, B.P. (2002). *Physical Testing of Textiles*, CRC Press. ISBN 0-8493-0568-3, NewYork, USA

Smith, J.A. (n.d.). Microfibers: Functional Beauty, In: *Ohio State University Extension Fact Sheet*, 27.10.2011, Available from:
<http://ohioline.osu.edu/hyg-fact/5000/5546.html>

Tortora, P.G. & Collier B.J. (1997). *Understanding Textiles*, Prentice Hall Press. ISBN 0 13 439225 6, New Jersey, USA

Turan, B.R. & Okur, A. (2010). Air permeability of fabrics. *The Journal of Textiles and Engineers*, No.72, pp.16-25, ISSN 1300-7599

Varshney, R.K., Kothari, V.K., Dhamija, S. (2010). A study on thermophysiological comfort properties of fabrics in relation to constituent fibre fineness and cross-sectional shapes. *The Journal of The Textile Institute*; Vol. 101, No.6, pp.495-505, ISSN 0040-5000

6

Woven Fabrics Surface Quantification

Jiří Militký

Faculty of Textile Engineering, Technical University of Liberec, LIBEREC,
Czech Republic

1. Introduction

It was revealed (Kawabata, 1980) that surface roughness is one of the main characteristics of fabric responsible for hand feeling. On the other hand it was found (Militký & Bajzík, 2000) that paired correlation between subjective hand ratings and surface roughness is statistically not significant. Anisotropy of mechanical and geometrical properties of textile fabrics is caused by the pattern and non-isotropic arrangement of fibrous mass. Fabric structural pattern characteristics are important from point of view of fabric appearance uniformity and have huge influence on the surface roughness, which is an important part of mechanical comfort (Militký & Bajzík, 2001). The complex structural pattern depends on the appearance of warp and weft on their surface, very often, one group of threads dominates. Typical examples of patterned fabric are cords where the so called "rows" parallel with machine direction are created. In the so called non-patterned fabrics the surface appearance or roughness is usually dependent on the weave and uniformity of fabric creation (Militký & Bleša, 2008). From a general point of view, the fabrics rough surface displays two basic geometrical features:

1. Random aspect: the rough surface can vary considerably in space in a random manner, and subsequently there is no spatial function being able to describe the geometrical form,

2. Structural aspect: the variances of roughness are not completely independent with respect to their spatial positions, but their correlation depends on the distance. Especially surface of woven fabrics is characterized by nearly repeating patterns and therefore some periodicities are often identified.

Periodic fluctuations of surface roughness can be spatially dependent due to arrangements of weft and warp yarns. Non-periodic type of spatial dependence is subtler.

The main aims of this chapter are:

- Characterization of fabrics surface profile i.e. "surface height variation (SHV) trace" by using of techniques based on the standard roughness evaluation, spatial analysis, Fourier regression, power spectral density (PSD) and utilization of fractal dimensions.
- Indication of micro and macro roughness by using of aggregation principle or by using of selected frequencies in PSD.
- The characterization of roughness anisotropy by using of profile spectral moments.

- Description of approach for contact-less evaluation of surface relief (macro roughness). This approach is based on the image analysis of especially prepared fabric images.

The simulated "teeth" profiles with variable height and thickness is used for identification of all kind of roughness parameters capability. These parameters are applied for characterization of some real patterned and non-patterned fabrics surface roughness.

2. Measurement of surface profiles

Surface irregularity of planar textiles can be identified by contact and contact-less techniques. For **contact measurements** the height variation (as thickness meter) or measurement of force needed for tracking the blade on the textile surface is applied (Ajayi, 1992, 1994; Militký & Bajzík, 2001, 2004). **Contacts less measurements** are usually based on the image analysis of fabric surfaces (Militký & Bleša, 2008). The subjective assessment of the fabrics roughness can be investigated as well (Stockbridge H.C. et. al, 1957).

KES for hand evaluation (Kawabata 1980) contains measuring device for registration the surface height variation (SHV) trace. This device (shown in the fig. 1) is a part of system KES produced by company KATO Tech.

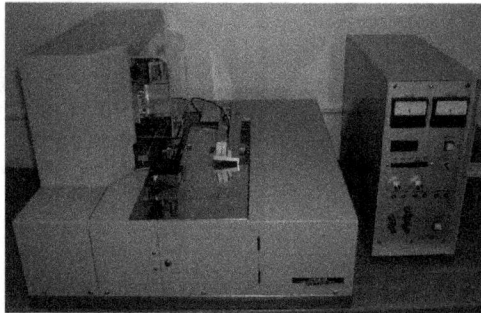

Fig. 1. Device for measurement of fabric surface characteristics

The main part of this device is contactor (see. fig. 2) in the form of wire having diameter 0.5 mm. The contact force 10 g is used.

Fig. 2. KES contactor for measuring of surface roughness

This contactor touches on the sample under the standardized conditions. The up and down displacement of this contactor caused by surface roughness is transduced to the electric signal by a linear transformer put at up ends of the contactor. The signal from the transducer is passed to the high pass digital filter having prescribed frequency response (wavelength being smaller than 1 mm). The sample is moved between 2 cm interval by a constant rate 0.1 cm/sec on a horizontal smooth steel plate with tension 20g/cm and SHV is registered on paper sheet. The SHV corresponds to the surface profile in selected direction (usually in the weft and warp directions are used for SHV creation).

The preprocessing of SHV traces, from images of paper sheet resulting form KES can be divided into the two phases.

- Digitalization of trace picture by image analysis system
- Removing parasite objects (grid, axes, base line etc.)

First of all the low ω_L and high ω_H surface frequency bands have to be specified. These cut-off frequencies are related to the wavelength limits l_L and l_H i.e. $l_L = 2\pi / \omega_L$ and $l_H = 2\pi / \omega_H$. The low pass cut-off is related to Nyquist criterion i.e. $l_L = dp / 2$ and the high pass cut-off is dependent on the maximum intersecting wavelength. For non-regular SHV, $l_H = L$ has to be selected. The results of digitalization and parasite object removing is set of "clean" heights $R(d_i)$ of fabric in places $0 < d < L$ (L is maximum investigated sample length and $i = 1...M$ is number of places). The distance between places $d_p = d_{i+1} - d_i$ is constant. For the case of Kawabata device $L = 2$ cm and $d_p = 2/(N-1)$ cm.

For deeper evaluation of SHV from KES device the rough signal from transducer has been registered and digitalized by using of LABVIEW system (Militký, 2007). Result is output voltage $U(d)$ in various distances d from origin of measurements. For calibration of this signal the mean value $E(U)$ and variances $D(U)$ were estimated. From KES apparatus the mean thickness $\bar{R}$ and corresponding standard deviation S_R were obtained. The transformation form voltage $U(d)$ to thickness $R(d)$ was realized by means of relation

$$R(d) = \bar{R} + S_R \left(\frac{U(d) - E(d)}{\sqrt{D(d)}} \right) \tag{1}$$

The result of this treatment is raw thickness $R(d)$ in various distances d from origin.

The technique of roughness evaluation will be demonstrated on the analysis of the surface trace SHV of twill fabric (see fig 3) in the machine direction.

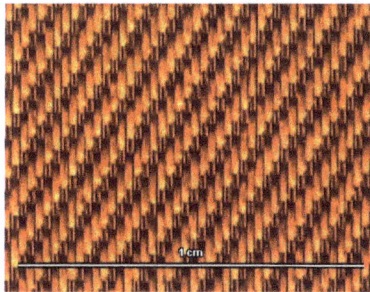

Fig. 3. Twill fabric used for roughness evaluation

The raw SHV trace of twill fabric is shown in the fig. 4.

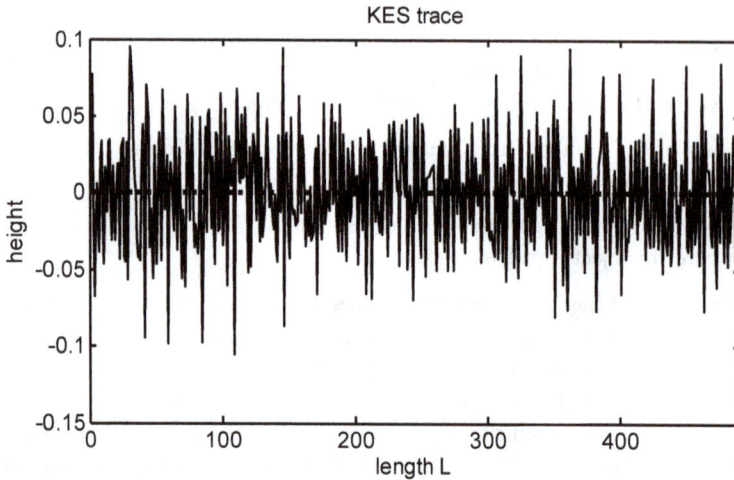

Fig. 4. SHV trace from LABVIEW

Similar approach is based on the measurement of R(d) by Shirley step thickness meter with replacement of measuring head by blade (Militký & Bajzík, 2001). The Shirley step thickness meter is shown in the fig. 5.

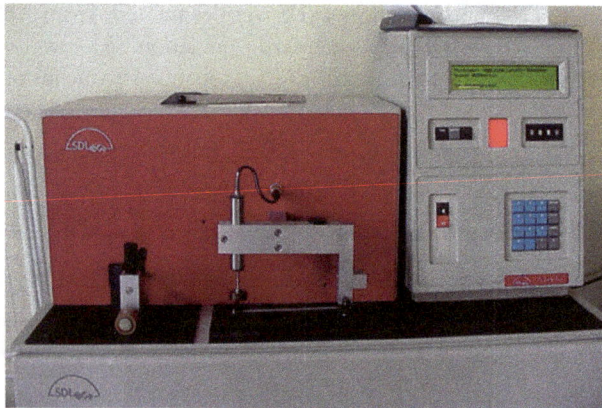

Fig. 5. The step thickness meter SDL M 034/1

The principle of profile roughness evaluation by the simple accessory to the tensile testing machine is registration of the force $F(d)$ needed for tracking the blade on the textile surface (Militký & Bajzík, 2004). Roughly speaking, the $F(d)$ should be proportional to the $R(d)$. In reality, the $F(d)$ profile is different due to small surface deformation caused by the tracked blade. Based on the complex testing the following working conditions have been selected:

Blade contact pressure 0.2 mN

Blade movement rate 0.6 mm/s
Sampling frequency 50 s^{-1} (length between samples Δd = 0.013 mm)
Investigated length T = 30 mm

The arrangement of this accessory to the tensile testing machine is shown in fig. 6.

Fig. 6. Accessory for roughness evaluation by the tensile testing machine

Output from measurements is sequence of loads $F(d_i)$.

Variation of thickness $R(d_i)$ or loads $F(d_i)$ can be generally assumed as combination of random fluctuations (uneven threads, spacing between yarns, non uniformity of production etc.) and periodic fluctuations caused by the repeated patterns (twill, cord, rib etc.) created by weft and warp yarns. For description of roughness the characteristics computed from $R(d)$ or $F(d)$ in places $0 < d < T$ (T is maximum investigated sample length and M is number of places) are used.

Profile of textile surfaces at given position along machine direction can be obtained by the analysis of especially prepared fabric images. The system RCM (Militký & Mazal, 2007) is composed from CCD camera, lighting system and special sample holder controlled by a personal computer (Fig. 7).

Fig. 7. Details of RCM apparatus

For good image creation the suitable lighting (laser from the top) and fabric arrangement (bend around sharp edge) were selected (see fig. 8).

Fig. 8. Details of lighting system

Result after image treatment is so called "slice" which is the roughness profile in the cross direction at selected position in machine direction (the line transect of the fabric surface). The system RCM offers reconstruction of surface roughness plane in two dimensions. For this purpose, the sample holder is step by step moved in controlled manner. From set of these profiles, it is possible to reconstruct the surface roughness plane (Militký & Mazal, 2007).

A finished cord fabric with relatively good structural relief was selected for demonstration of relief creation system capability. The original fabric surface is shown in the fig. 9.

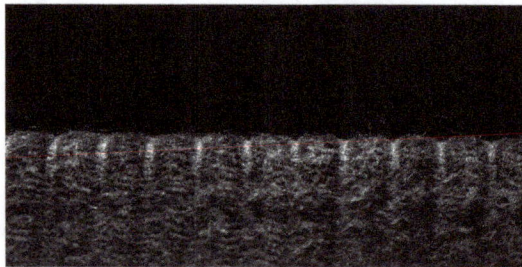

Fig. 9. Roughness profile in the cross direction of tested fabric

Individual relief slices were created by combination of threshold, set of morphological operations (erosion, dilatation) and Fourier smoothing to the 30 terms (Quinn & Hannan, 2001). Result of these operations is vector of surface heights in cross direction at specified machine direction (see fig. 10).

Fig. 10. Slice after morphological operations and cleaning

Output of data pre-treatment phase is array of slices i.e. array of vectors $R_{j(i)}$ where index i corresponds to the position in j[th] slice. From this array it is simple to reconstruct whole surface relief (see. fig. 11).

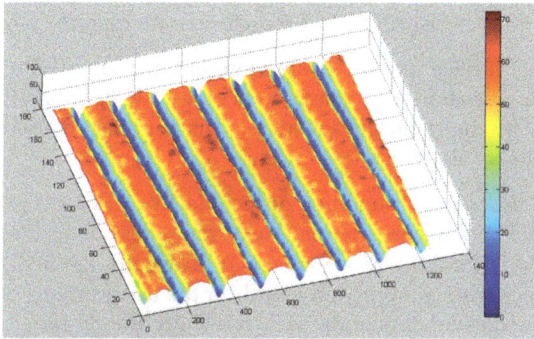

Fig. 11. Reconstructed roughness surface

Above described devices are based on the measurements of surface height variation (SHV) or force needed for tracking blade across of surface in given direction. The analysis of surface images in two dimensions is based on the different principles (Militký & Klička, 2007). These methods are not discussed in this chapter.

3. Surface roughness

There exists a vast number of empirical profile or surface roughness characteristics suitable often in very special situations (Quinn & Hannan, 2001; Zhang & Gopalakrishnan, 1996). Some of them are closely connected with characteristics computed from fractal models as fractal dimension and topothesy (Davies, 1999). A set of parameters for profile and surface characterization are collected in (Militký & Bajzík, 2003, Militký & Mazal, 2007). Parameters for profile and surface characterization can be generally divided into the following groups:

- Statistical characteristics of surface profile distribution (variance, skewness, kurtosis)
- Spatial characteristics as autocorrelation or variogram (denoted in engineering as structural function). Analysis is here in fact based on the analysis of random field moment characteristics of second order.
- Spectral characteristics as power spectral density or Fourier analysis.
- Characteristics of overall complexity based on random linear stationary processes, self-affined processes, long-range dependencies and on the theory of chaotic dynamics or nonlinear time series.

General surface topography is usually broken down to the three components according to wavelength (or frequency). The long wavelength (low frequency) range variation is denoted as form. This form component is removed by using of polynomial models or models based on the form shape. The low wavelength (high frequency) range variation is denoted as roughness and medium wavelength range variation separates waviness. The most common way to separate roughness and waviness is spectral analysis. This analysis is based on the Fourier transformation from space domain d to the frequency domain $\omega = 2\pi / d$.

Data from contact based measurements of roughness often represents height variation on line transects of the surface. Usually, it is possible to obtain structural data for one direction of the fabric, whereas the results on the other direction do not give clear information about the respective structural patterns. Some contacts-less methods based on the image analysis are found to be capable for measuring fabric structural pattern in the whole plane.

Standard methods of surface profile evaluation are based on the relative variability characterized by the variation coefficient - analogy with evaluation of yarns mass unevenness or simply by the standard deviation. This approach is used in Shirley software for evaluation of results for step thickness meter.

Common parameters describing roughness of technical surfaces are given in the ISO 4287 standard (Anonym, 1997). For characterization of roughness of textiles surfaces the mean absolute deviation MAD (SMD as per Kawabata) is usually applied (Meloun & Militký, 2011). The descriptive statistical approach based on the assumptions of independence and normality leads to biased estimators, if the SHV has short or long-range correlation. There is therefore necessity to distinguish between standard white Gauss noise and more complex models. For description of short range correlation the models based on the autoregressive moving average are useful (Maisel, 1971). The long-range correlation is characterized by the fractal models (Beran, 1984; Whitehouse, 2001). The deterministic chaos type models are useful for revealing chaotic dynamics in deterministic processes where variation appears to be random but in fact predictable. For the selection among above mentioned models the power spectral density (PSD) curve evaluated from experimental SHV can be applied (Eke, 2000; Quinn & Hannan, 2001).

Especially the fractal models (Mandelbrot & Van Ness, 1968) are widely used for rough surface description. For these models the dependence of log (PSD) on the log (frequency) should be linear. Slope of this plot is proportional to fractal dimension and intercept to the so-called topothesy. White noise has dependence of log (PSD) on the log (frequency), nearly horizontal plateau for all frequencies (the ordinates of PSD are independent and exponentially distributed with common variance). More complicated rough surfaces can be modeled by the Markov type processes. For these models the dependence of log (PSD) on the log (frequency) has plateau at small frequencies, then bent down and are nearly linear at high frequencies (Sacerdotti et. al, 2000). A lot of recent works is based on the assumption that the stochastic process (Brownian motion) can describe fabrics surface variation (Sacerdotti et. al, 2000). It is clear that for the deeper analysis of rough surface, the more complex approach should be used.

4. Simulated "teeth" profiles

The surface of cord fabrics has typical "teeth" in the machine direction. For indication of the influence of geometry of teeth on the values of the roughness characteristics the simulated roughness profile in the cross direction was created. The standard pattern is composed from two parts (see. fig. 12).

The top height of one tooth is selected as 1. Bottom height of one tooth is equal to the value of ym. The tooth size is therefore 1 – ym. The length of distance between teeth is equal to am and tooth thickness is equal 1- am. Total length of standard pattern equal to the 1 and 100 individual values are generated, i.e. the standard pattern is characterized by 100 points with constant increment. Teeth profile is then composed from 11 repetitions of standard pattern. The simulated teeth profiles were generated for the set of value $0.1 \leq am \leq 0.9$, $0.1 \leq ym \leq 0.9$ with increment 0.1. The tooth profile for the case $am = ym = 0.1$ is shown in the fig. 13a and for the case of $am = ym = 0.9$ is shown in the fig. 13b.

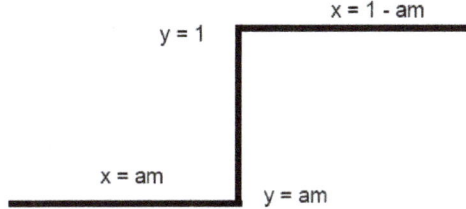

Fig. 12. Standard pattern

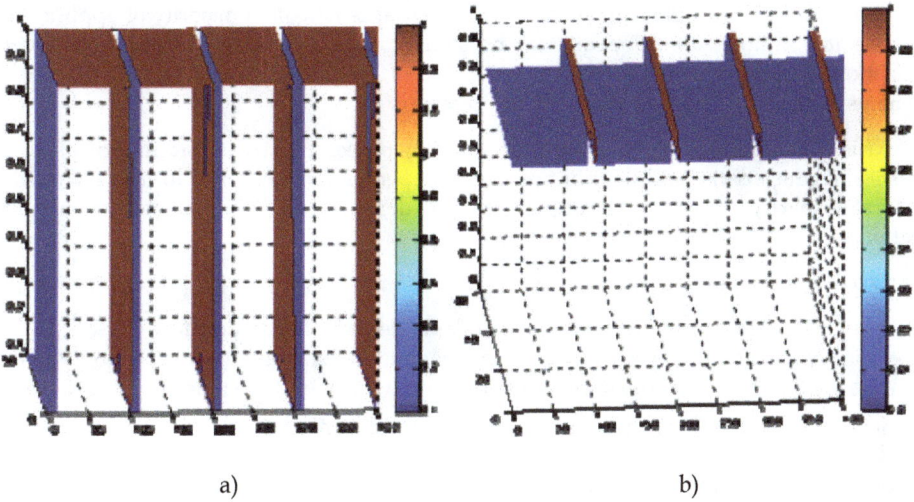

a)   b)

Fig. 13. Detail of teeth profile for a) $am = ym = 0.1$ and for b)

It can be easily derived that the mean height R_a of teeth profile is equal to the

$$R_a = 1 - am(1 - ym)$$ (2)

Corresponding standard deviation SD is equal to

$$SD = \sqrt{\frac{1}{100}\left([100ym\ am]^2 + [100(1 - ym)am]^2 - 100R_a^2\right)}$$ (3)

Some of the other characteristics can be analytically expressed as well but expressions are complicated. Generated teeth profiles were used for computation of profiles characteristics. The dependence of these characteristics on the am and ym are shown in subsequent paragraphs.

5. Basic assumptions testing

Consider a series $R_{(i)}$. Let this series is one "spatial" realization of random process $y = R(d)$. For analysis of this series it is necessary to know if some basic assumptions about behavior of underlying random process can be accepted or not. These basic assumptions are (Maisel, 1971):

- stationarity
- ergodicity
- independence

In fact the realizations of random process are $R_{j(i)}$, where index j correspond to individual realizations and index i corresponds to the distance d_i. In the case of ensemble samples there are values $R_{j(i)}$ for $i = const.$ and $j = 1$. M at disposal.

For these data there is no problem to use standard statistical analysis of univariate samples for creation of data distribution e.g. probability density function $p(R_{(i)})$ or computation of statistical characteristics as mean value $E(R_{(i)})$ or variance $D(R_{(i)})$. In majority of applications, the ensemble samples are not available and statistical analysis is based on the one spatial realization $R_{j(i)}$ for $j = 1$ and $i = 1.N$. For creation of data distribution and computation of moments, some additional assumptions are necessary.

The basic assumption is stationarity. The random process is **strictly stationary** if all the statistical characteristics and distributions are independent on ensemble location. The **wide sense stationarity** of g-th order implies independence of first g moments on ensemble location.

The second order stationarity implies that:

- mean value $E(R_{(i)}) = E(R)$ is constant (not dependent on the location d_i).
- variance $D(R_{(i)}) = D(R)$ is constant (not dependent on the location d_i).
- autocovariance, autocorrelation and variogram, which are functions of d_i and d_j are not dependent on the locations but on the lag $h = |d_i - d_j|$ only.

For example the covariance is $c(R(d_i) R(d_{i+h})) = c(h)$. For ergodic process the "ensemble" mean can be replaced by the average across distance (from one spatial realization) and autocorrelation $R(h) = 0$ for all sufficiently high h.

Ergodicity is very important, as the statistical characteristics can be calculated from one single series $R_{(i)}$ instead of ensembles which frequently are difficult to be obtained. Given a $R_{(i)}$ series, the selection of the appropriate approach for its analysis is not a trivial task because the mathematical background of the underlying process is unknown. Moreover, the $R_{(i)}$ are corrupted by noise and consist of finite number of sample values. The task to analyze real data is often to resolve the so-called inverse problem, i.e., given a series $R_{(i)}$, how to discover the characteristics of the underlying process. Three approaches are mainly applied:

- first based on random stationary processes,
- second based on the self affine processes with multiscale nature,
- third based on the theory of chaotic dynamics.

In reality the multiperiodic components are often mixed with random noise.

Before choosing the approach, some preliminary analysis is needed mainly to test the stationarity and linearity. This is important as some kind of stochastic (self affine) processes with power-law shape of their spectrum may erroneously be classified as chaotic processes on the basis of some properties of their non-linear characteristics, e.g., correlation dimension and Kolmogorov entropy. In this sense, the tests for stationarity and linearity may be regarded as a necessary preprocessing in order to choose an appropriate approach for further analysis. Prior to selecting any method for data analysis, some simple tests are useful to apply on the series $R_{(i)}$. The first one may be to observe the $R_{(i)}$ distribution e.g. via histogram as simple estimator of probability density function (pdf) or by using kernel density estimator (Meloun & Militký, 2011). The histogram of series $R_{(i)}$ corresponding to the raw SHV trace of twill weave fabric (shown in the fig. 4) is shown in fig. 14.

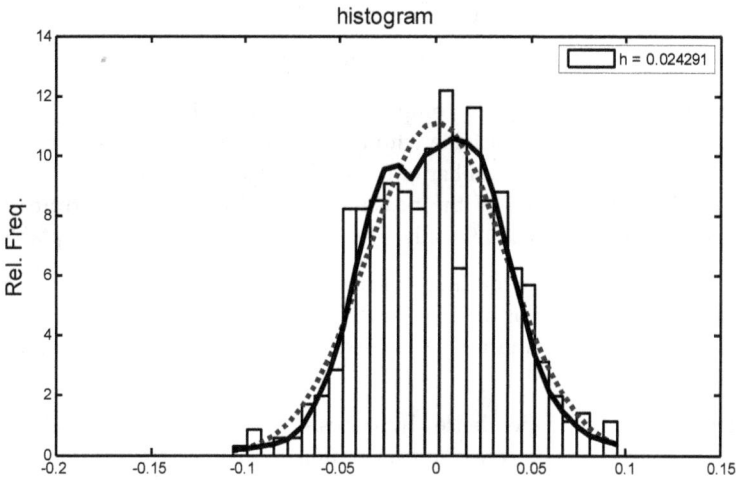

Fig. 14. Histogram and pdfs of raw SHV for twill fabric

In this figure, the solid line corresponds to the Gaussian pdf with parameters: mean = 0.000524 and standard deviation = 0.0358. The dotted line is nonparametric kernel density estimator wit optimal bandwidth h= 0.0243. The bimodality pattern is clearly visible.

In most of the methods for data processing based on stochastic models, normal distribution is assumed. If the distribution is proved to be non-normal (according to some test or inspection), there are three possibilities:

1. the process is linear but non-gaussian;
2. the process has linear dynamics, but the observations are as a result of non-linear "static" transformation (e.g. square root of the current values)
3. the process has non-linear dynamics.

It is suitable to construct the histograms for the four quarters of data separately and inspect non-normality or asymmetry of distribution. The statistical characteristics (mean and variances) of these sub series can support wide sense stationarity assumption (when their values are statistically indistinguishable).

The simple nonparametric test of stationarity uses the reverse arrangement evaluation. Test is based on the computation of times that $R_{(i)} > R_{(j)}$ with $i < j$ for all i. If the sequence of $R_{(i)}$ are independent identically distributed (i.i.d) random variables, the number of reverse arrangements NR is random variable with mean $E(NR) = N(N-1)/4$ and variance $D(NR) = N(2N+5)$ $(N+1)/72$. If observed number NR is significantly different from $E(NR)$, the non-stationarity (trend) is indicated. For rough SHV from fig. 4 reversation test statistic NT = 2.328 and upper limit for P=95%, is 1.96 only. The stationarity is therefore not acceptable.

The alternative "run test" can detect a monotonic trend in a series $R_{(i)}$ i = 1..N. A "run" is defined as a sequence of identical observations that is followed or preceded by a different observation or no observation at all. First the median med (R) of the observations $R_{(i)}$ is evaluated and the new series z(i) is derived from $R_{(i)}$ as

$$z\,(i) = 0 \text{ if } R_{(i)} < med\,(R)$$

$$z\,(i) = 1 \text{ if } R_{(i)} \geq med\,(R)$$

Then the member of runs in z(i) is computed. If $R_{(i)}$ is stationary random process, the number of runs NT is a random variable with mean $E(NT) = N/2 + 1$ and variance $D(NT) = (N(N - 2))\,/(4(N-1))$. As observed number of runs NT is significantly different from $E(NT)$. It indicates nonstationarity because of the possible trend. For rough SHV from fig. 4, NR= 18.14 and upper limit for P=95% is 1.96 only. The stationarity is here not acceptable.

Very simple check of presence of first order autocorrelation is creation of zero order variability diagram which is plot of $R_{(i+1)}$ on $R_{(i)}$. In the case of independence the random cloud of points appears on this graph. Autocorrelation of first order is indicated by linear trend.

For characterization of independence hypothesis against periodicity alternative the cumulative periodogram can be constructed. Cumulative periodogram is unbiased estimate of the integrated spectrum

$$CU(f_i) = \frac{\sum_{j=1}^{i} I(f_j)}{N\,s^2} \tag{4}$$

The function $C(f_i)$ is called the normalized cumulative periodogram (construction of $I(f_i)$ is described in par. 7). For white noise series (i.i.d. normally distributed data), the plot of $C(f_i)$ against f_i would be scattered about a straight line joining the points (0, 0) and (0.5, 1). Periodicities would tend to produce a series of neighboring values of $I(f_i)$ which were large. The result of periodicities therefore bumps on the expected line. The limit lines for 95% confidence interval of $C(f_i)$ are drawn at distances $\pm 1.36 / \sqrt{(N-2)/2}$. For rough SHV from fig. 4 cumulative periodogram is shown in fig. 15.

6. Aggregation principle

In the unevenness analysis, it is common to aggregate raw data. This is equivalent to cutting the material to pieces and measurement of variability between pieces only. In the case of

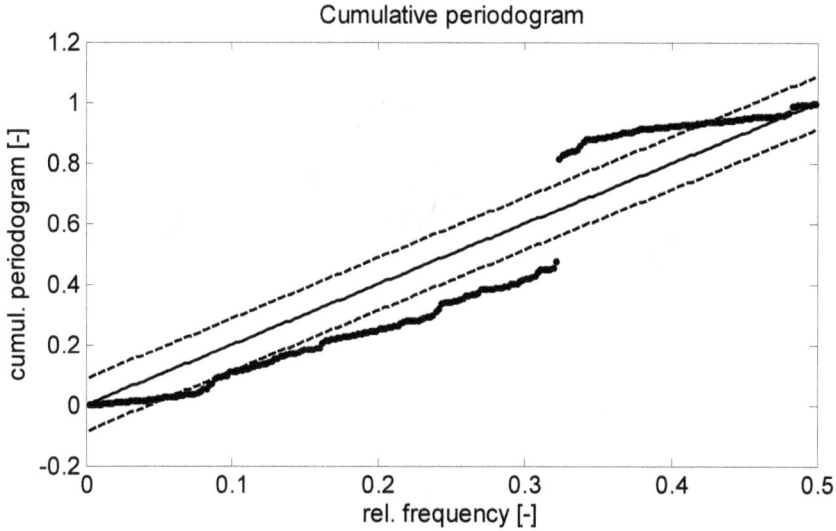

Fig. 15. Cumulative periodogram of raw SHV for twill fabric
It is visible that the raw SHV is approximately periodic.

roughness aggregation is tool for smoothing of roughness profiles and avoiding local (small scale) roughness. The principle of aggregation is joining of original data $R_{(i)}$ into non overlapping blocks or application of window of length L. By using of aggregation the resolution is decreased and roughness profile is created without local roughness variation. By averaging of original data $R_i = R_{(i)}$ in non overlapping blocks having L values the aggregated series are constructed. Aggregated series $R^{(L)}(i)$ are created according to relation

$$R^{(L)}(i) = \frac{1}{L}(R(i\ L - L + 1) + .. + R(i\ L))\quad L = 1,\ 2,\ 3..\tag{5}$$

For rough SHV from fig. 4, aggregate series for aggregation length $L = 2$ and 10 are shown in fig. 16.

It is known (Beran, 1984; Cox, 1984) that variance of aggregated series $v^{(L)}$ is connected with auto correlation structure of original series

$$v^{(L)} = \frac{v}{L} + \frac{2}{L^2}\sum_{s=1}^{L-1}\sum_{h=1}^{s} c(h)\tag{6}$$

Here $c(h)$ is autocorrelation function defined as $c(h) = \mathrm{cov}(R(i) * R(i - h))$ and $\log h = L * d_i$. Very important is lag one autocorrelation function for aggregated series

$$r^{(L)}(1) = 2\ \frac{v^{(2*L)}}{v^{(L)}} - 1\tag{7}$$

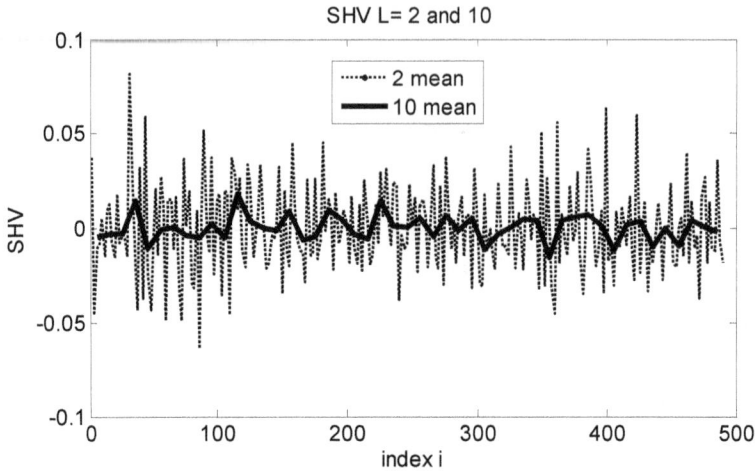

Fig. 16. Aggregate series (L =2 and 10) for twill fabric

The nature of original random series can be explained by using of characteristics of aggregated series. There are three main groups of series:

1. Series of random independent identically distributed (i.i.d.) variables. For this case are all $c(h)$ =0, for lags h = 1,2,.. and data are uncorrelated. This is ideal case for roughness analysis and it is implicitly assumed as valid in computation of basic geometric characteristics.
2. The short-range dependent stationary processes. In this case the sum of all $c(h)$ h= 1, 2, ... is convergent
3. The long-range dependent stationary processes. In this case the sum of all $c(h)$ h= 1, 2, ... is divergent

For short-range dependent stationary processes, the first order autocorrelation $r^{(L)}(1) = 0$ for $L \to \infty$. The same is valid for autocorrelation of all lags h. The aggregated series $R^{(L)}(i)$ therefore tends to the second order pure noise as $L \to \infty$. For large L variance $v^{(L)} = v / L$. The autocorrelation structure of aggregated series is decreased until limit of no correlation. Typical model of short-range processes are autoregressive moving average processes of finite order. For the higher L, data are approaching to the i.i.d. case.

For long-range dependent processes, variance $L\, v^{(L)} \to \infty$ as $L \to \infty$.

Then the autocorrelation structure is not vanishing. For these processes, it is valid that for sufficiently large L

$$c(h) \approx h^{-\beta} \text{ and } v^{(L)} \approx L^{-\beta} \tag{8}$$

where $0 < \beta < 1$ is valid for stationary series. For non-stationary case β can be outside of this interval. For the long-range processes correlation structure is identical for original and aggregate series. For strictly second order self-similar processes,

$$c(h) \approx \frac{v}{2}(1-\beta)\,(2-\beta)\,h^{-\beta} \qquad (9)$$

For the higher L the correlation structure remains the same and assumption of i.i.d. cannot be used. Instead of β the so-called Hurst exponent $H = 1 - 0.5 * \beta$ is frequently used. Where H = 0, this denotes a series of extreme irregularity and $H = 1$ denotes a smooth series.

For rough SHV from fig. 4 dependence of $\log\,v^{(L)}$ on aggregation length L shown in fig. 17.

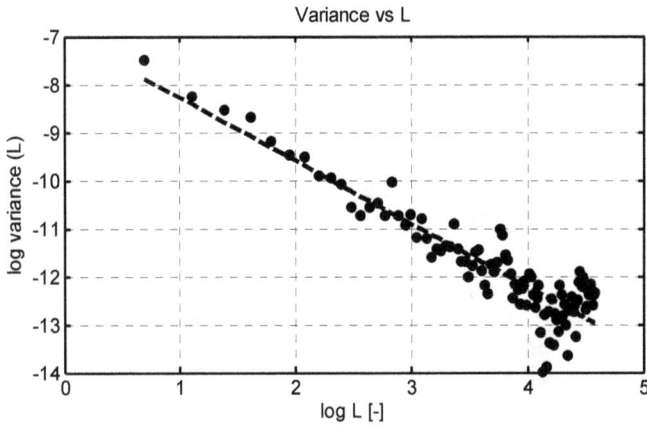

Fig. 17. Dependence of $\log\,v^{(L)}$ on L for twill fabric

It is clear that for higher L this dependence is scattered and corresponding slope is over 1.

The long range dependency is characteristic for self affine processes as well. Self similar processes are characterized by the fractal dimension FD. For self-affine processes, the local properties are reflected in the global ones, resulting in the well known relationship H + FD = 2. Long-memory dependence, or persistence, is associated with the case $H \in (0.5, 1)$ and linked to smooth curves with low fractal dimensions. Rougher curves with higher fractal dimensions occur for antipersistent processes with $H \in (0, 0.5)$.

If $1 - c(h) \approx h^{-\alpha}$ for $\alpha \to 0$, the process has called fractal dimension $FD = 2 - \alpha / 2$. Generally, the l-th central moment of aggregated long range dependent series is defined as

$$M_l^{(L)} = \frac{1}{N/L} \sum_{k=1}^{N/L} abs(y^{(L)}(k) - \overline{y})^l \qquad (10)$$

The $M_l^{(L)}$ asymptotically behaves like power function $M_l^{(L)} \approx L^{l(H-1)}$. If the series has finite variance and no long-range dependence, then H = 0.5 and the slope of the fitted line in log-log plot of $M_l^{(L)}$ on L should be – 1/2. It is assumed that both N and N/L are large. This ensures that both the length of each block and number of blocks is large. In practice the points at very low and high ends of the plot are not used for fitting least squares line. Indeed, short-range effects can distort the estimates of H if the low end of plot is used.

One of the best methods for evaluation of β or H is based on the power spectral density

$$g(\omega) = \frac{1}{2\pi} \int_{h=-\infty}^{\infty} c(h) \, \exp(-i \, h \, \omega) \, d\omega \quad -\pi < \omega < \pi \tag{11}$$

For small frequency range, it is valid that

$$g(\omega) \approx \omega^{-(1-\beta)} \quad \omega \to 0 \tag{12}$$

and for very high frequency range

$$g(\omega) \approx \omega^{-1-\alpha} \quad \omega \to \infty \tag{13}$$

The parameters β and α or are evaluated from empirical linear representation of dependence of the log of power spectral density (PSD) on log frequency in suitable range. The parameter β is often evaluated from empirical representation of the log of power spectral density

$$\log(g(\omega)) = -(1-\beta) \, \log(\omega) + a_0 + a_1 \, \omega + .. + a_p \, \omega^p \tag{14}$$

For long range processes, it is ideal to have all aj = 0, except a_0.

For rough SHV from fig. 4 dependence of $\log(g(\omega))$ on log frequency is shown in fig. 18.

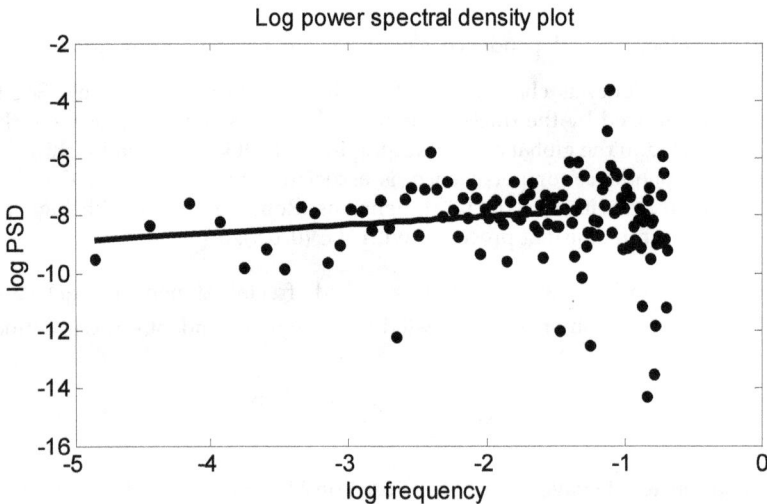

Fig. 18. Dependence of $\log(g(\omega))$ on log frequency for twill fabric

It is visible that the scatter of data is very big. The solid line in fig. 18 is regression line created for low frequency range data set. The slope is equal to - 0. 2831. Corresponding β = 0.7169 and Hurst exponent is 0.6416.

7. Classical roughness characteristics

Because the basic output form RCM is set of "slices" (roughness profiles in the cross direction at selected position in machine direction) it is possible to compute all profile roughness characteristics separately for each slice and show the differences between slices. Another possibility is to use the reconstructed surface roughness plane for evaluation of planar roughness.

There are two reasons for measuring surface roughness. First, is to control manufacture and is to help to ensure that the products perform well. In the textile branch the former is the case of special finishing (e.g. pressing or ironing) but the later is connected with comfort, appearance and hand.

From a general point of view, the rough surface display process which have two basic geometrical features:

- Random aspect: the rough surface can vary considerably in space in a random manner, and subsequently there is no spatial function being able to describe the geometrical form,
- Structural aspect: the variances of roughness are dependent with respect to their spatial positions and their correlation depends on the distance. Especially surface of textile weaves is characterized by nearly repeating patterns and therefore some periodicities are often identified.

The random part of roughness can be suppressed by proper smoothing. In this case the only structural part will be evaluated.

From the individual roughness profiles, it is possible to evaluate a lot of roughness parameters. Classical roughness parameters are based on the set of points $R(d_j)$ j =1.. N (SHV) defined in the sample length interval L_s. The distances d_j are obviously selected as equidistant and then $R(d_j)$ can be replaced by the variable R_j . For identification of positions in length scale, it is sufficient to know that sampling distance $d_s = d_j - d_{j-1} = L_s/N$ for $j>1$.The standard roughness parameters used frequently in practice are (Anonym, 1997):

i.   Mean Absolute Deviation MAD. This parameter is equal to the mean absolute difference of surface heights from average value (R_a). For a surface profile this is given by,

$$MAD = \frac{1}{N}\sum_j \left| R_j - \bar{R} \right| \tag{15}$$

This parameter is often useful for quality control and textiles roughness characterization (called SMD (Kawabata, 1980)). However, it does not distinguish between profiles of different shapes. Its properties are known for the case when R_j's are independent identically distributed (i. i. d.) random variables. For rough SHV from fig. 4, dependence of SMD on aggregation length L is shown in fig. 19.

ii.   Standard Deviation (Root Mean Square) Value SD. This characteristics is given by

$$SD = \sqrt{\frac{1}{N}\sum_j (R_j - \bar{R})^2} \tag{16}$$

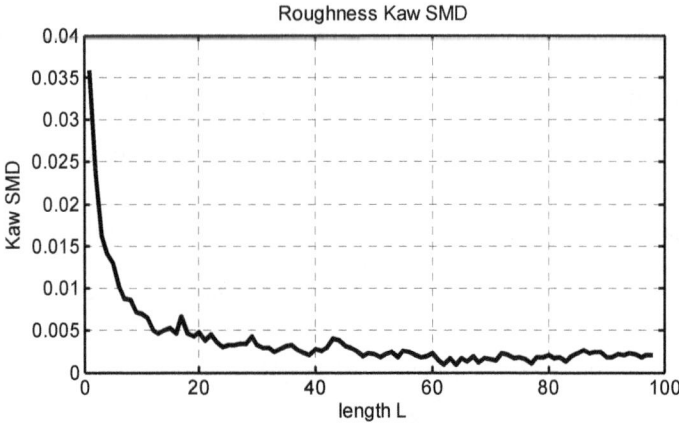

Fig. 19. Dependence of SMD on the aggregation length L for twill fabric

The influence of teeth profile parameters am and ym on MAD is shown in the fig. 20.

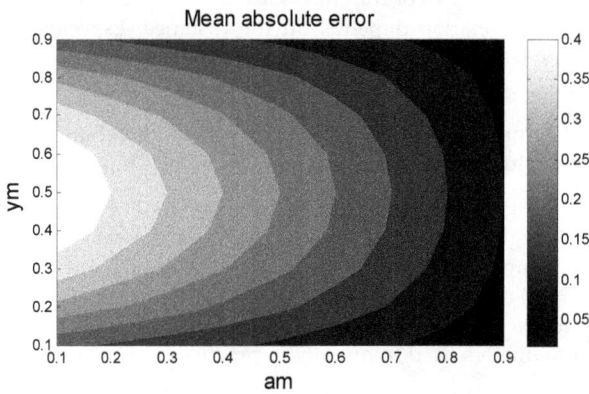

Fig. 20. Influence of am and ym on the MAD

Its properties are known for the case when R_j's are independent identically distributed (i.i.d.) random variables. One advantage of SD over MAD is that for normally distributed data, it can be simple to derive confidence interval and to realize statistical tests. SD is always higher than MAD and for normal data $SD = 1.25 \, MAD$. It does not distinguish between profiles of different shapes as well. The parameter SD is less suitable than MAD for monitoring certain surfaces having large deviations (corresponding distribution has heavy tail).

The influence of teeth profile parameters am and ym on the square of SD (i.e. variance) is shown in the fig. 21.

It is visible that the MAD and SD have similar dependence on the tooth parameters am and ym. SD is always higher than MAD and for normal data SD = 1.25 MAD. It does not distinguish between profiles of different shapes as well. The parameter SD is less suitable than MAD for monitoring certain surfaces having large deviations (corresponding distribution has heavy tail).

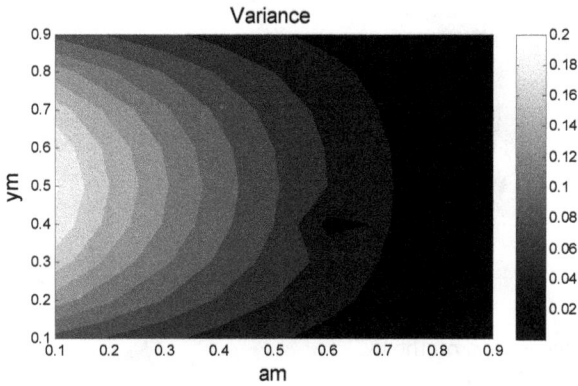

Fig. 21. Influence of *am* and *ym* on the variance (square of SD)

iii. The Standard Deviation of Profile Curvature *PC*. This quantity called often as waviness is defined by the relation

$$PC = \sqrt{\frac{1}{N}\sum_j \left(\frac{d^2 R(x)}{dx^2}\right)^2_j} \qquad (17)$$

The curvature is characteristics of a profile shape. The *PS* parameter is useful in tribological applications. The lower the slope, the smaller will be the friction and wear. Also, the reflectance property of a surface increases in the case of small *PC*.

For rough SHV from fig. 4 dependence of PC on aggregation length L is shown in fig. 22.

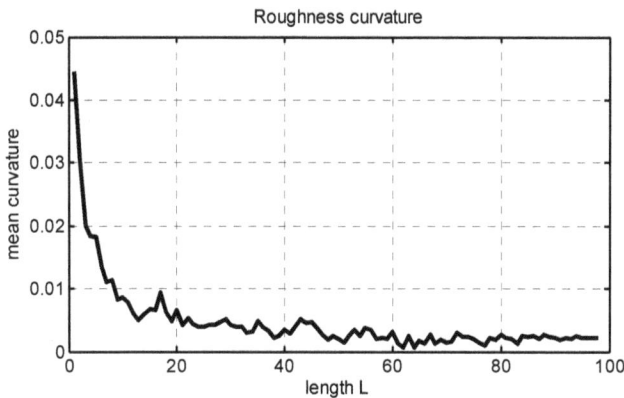

Fig. 22. Dependence of SMD on aggregation length *L* for twill fabric

The influence of teeth profile parameters *am* and *ym* on the PS are shown in the fig. 23.

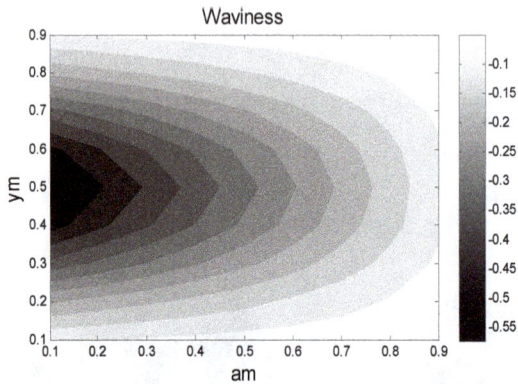

Fig. 23. Influence of *am* and *ym* on the PC

It is visible that the growing of *am* (i.e. decreasing of tooth thickness) leads to the increase of PC. Lowest values of PC are around *ym* equal to the 0.5. This behavior is "inverse" to the behavior of MAD and SD. The PC parameter is useful in tribological applications. The lower the slope the smaller will be the friction and wear. Also, the reflectance property of a surface increases in the case of small PC.

The *MAD* and *PC* characteristics for all slices for cord fabric (see fig. 11) are shown in the fig. 24.

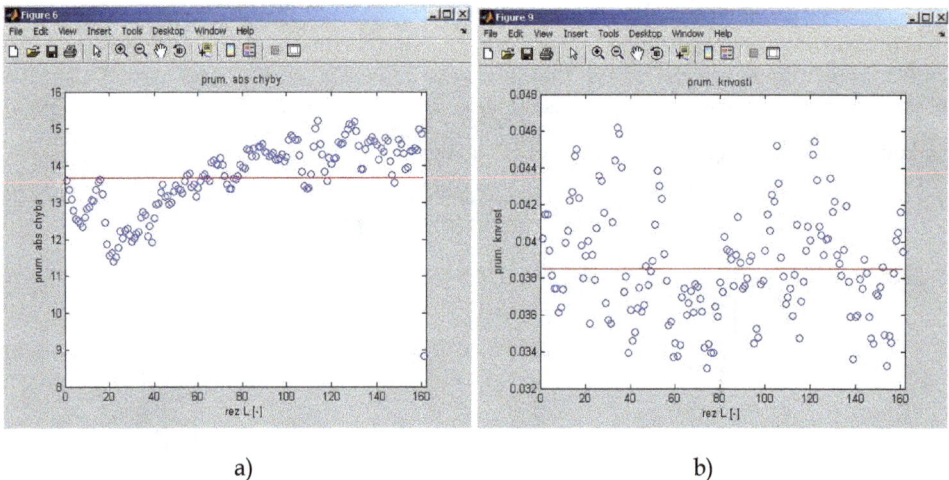

a)   b)

Fig. 24. The a) MAD and b) PC values for all slices of cord fabric (fig. 11)

In the case of MAD, a systematic trend is visible. The variation of PC is nearly random.

For the characterization of hand, it will be probably the best to use waviness *PC*. The characteristics of slope and curvature can be computed for the case of fractal surfaces from power spectral density, autocorrelation function or variogram.

8. Spectral analysis

The primary tool for evaluation of periodicities is expressing of signal $R(d)$ by the Fourier series of sine and cosine wave. It is known that periodic function given by equally spaced values R_i, $i = 0, ..., N - 1$ can be generally expressed in the form of Fourier series at Fourier frequencies $f_j = j/N$, $1 \leq j \leq [N/2]$. If N is odd with $N = 2m + 1$, the Fourier series has form (Quinn & Hannan, 2001)

$$R_i = a_0 + \sum_{k=1}^{m} \left(a_k \, \cos(\omega_k \, i) + b_k \, \sin(\omega_k \, i) \right) \quad i = 0, ..\, N - 1 \tag{18}$$

where $\omega_k = 2 \, \pi \, f_k = 2 \, \pi \, k / N$ $k = 1, ..\, m$ $k = 1, ..., m$ are angular frequencies. The eqn. (17) is for known frequencies harmonic linear regression model with $2m + 1$ parameters (intercept and $2m$ sinusoids amplitudes at the m Fourier frequencies). The sinusoid with the j-th Fourier frequency completes exactly j cycles in the span of the data. Due to selection of Fourier frequencies all regressors ($sin(.)$ and $cos(.)$ terms) are mutually orthogonal, so that standard least-squares method leads to estimates $a_0 = \bar{R}$ and

$$a_k = \frac{2 \sum_{i=0}^{N-1} R_i \, \cos(\omega_k \, i)}{N} \qquad b_k = \frac{2 \sum_{i=0}^{N-1} R_i \, \sin(\omega_k \, i)}{N} \qquad k = 1, ..m \tag{19}$$

Basic statistical characteristic in the frequency domain is power spectral density PSD defined as Fourier transform of covariance function.

The simple estimator of power spectral density is called periodogram. The periodogram of an equally spaced series R_i, $i = 0, ..., N - 1$ is defined by equation

$$I(\omega) = \frac{1}{N} \left(\sum_{i=0}^{N-1} R_i \, \cos(\omega \, i) \right)^2 + \frac{1}{N} \left(\sum_{i=0}^{N-1} R_i \, \sin(\omega \, i) \right)^2 \tag{20}$$

and can be expressed in the alternate form

$$I(\omega_k) = \frac{N}{4} (a_k^2 + b_k^2) \quad k = 1, ..m \tag{21}$$

For rough SHV from fig. 4 periodogram is shown in fig. 25.

The periodogram ordinates correspond to analysis of variance decomposition into m orthogonal terms with 2 degrees of freedom each because,

$$\sum_{k=1}^{m} I(\omega_k) = 0.5 \sum_{i=0}^{N-1} \left(R_i - \bar{R} \right)^2 \tag{22}$$

The normalized periodogram with ordinates

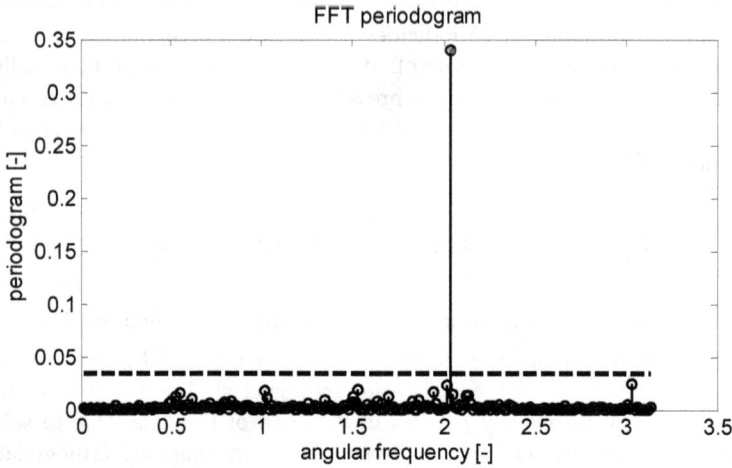

Fig. 25. Periodogram for twill fabric

$$I(\omega_k) / \sum_k I(\omega_k) = A_k / \sum_i (R_i - \bar{R})^2 \tag{23}$$

is then simply interpretable. The k-th ordinate gives the proportion of the total variation due to sinusoidal oscillation at the k-th Fourier frequency, and thus is a partial correlation coefficient R^2. The so called scree plot is in fact dependence of relative contribution to the total variance from individual Fourier frequencies arranged according to their importance. For rough SHV from fig. 4, scree plot is shown in fig. 26.

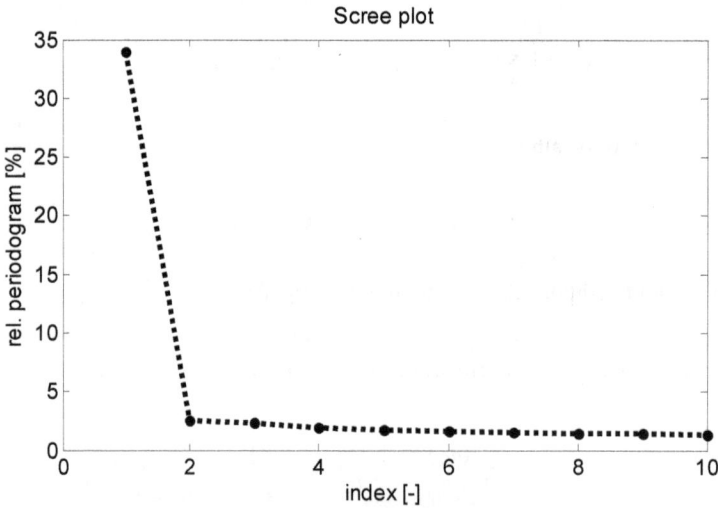

Fig. 26. Scree plot for twill fabric

The well-known trigonometric identity $cos\ (t-s) = (cos\ t)(cos\ s) + (sin\ t)(sin\ s)$ allows to write each paired sinusoid term as

$$a_k\ \cos(\omega_k\ t) + b_k\ \sin(\omega_k\ t) = A_k\ \cos(\omega_k\ t - \phi_k) \qquad (24)$$

with $A_k = \sqrt{a_k^2 + b_k^2}$ and $\tan(\phi_k) = b_k / a_k$. Coefficients A_k creates amplitude spectrum and coefficients ϕ_k creates phase spectrum.

The influence of teeth profile parameters am and ym on the amplitude A_1 of most important Fourier term are shown on the fig. 27.

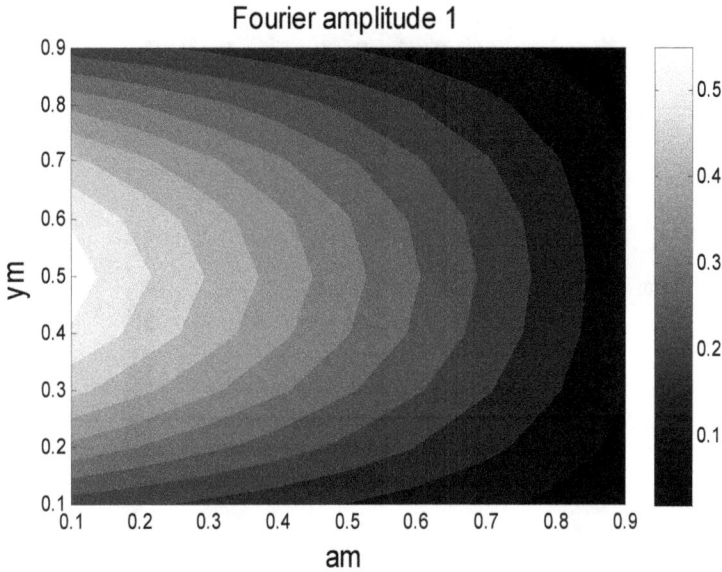

Fig. 27. Influence of am and ym on the A_1

It is visible that the growing of am (i.e. decreasing of tooth thickness) leads to the decrease of the amplitude A_1. Highest values of the amplitude A_1 are around ym equal to the 0.5 (similar behaviour as in the case of MAD). The influence of teeth profile parameters am and ym on the phase ϕ_1 of most important Fourier term are shown on the fig. 28.

It is visible that the growing of am (i.e. decreasing of tooth thickness) have the small influence on the phase ϕ_1 The influence of ym on the phase ϕ_1 is more important with minimum at $ym = 0.5$.

The periodogram is unbiased only in case of Gaussian noise. The variance of periodogram does not decrease with increasing N and has the form

$$D(I(\omega)) \approx I^2(\omega) \left[1 + \left(\frac{\sin(\omega\ N)}{N\ \sin(\omega)} \right)^2 \right] \qquad (25)$$

In some cases, it is useful to express Fourier series in the complex exponential form (Quinn & Hannan, 2001).

$$R_i = \sum_{k=1}^{m} \left(C_k \ \exp(\omega_k \ i \ j) \right) \ i = 0,.. \ N - 1 \tag{26}$$

where j is imaginary unit and complex coefficients C_k have real and imaginary part $C_k = \text{Re}_k + j\text{Im}_k$. The values C_k creates the complex discrete spectrum.

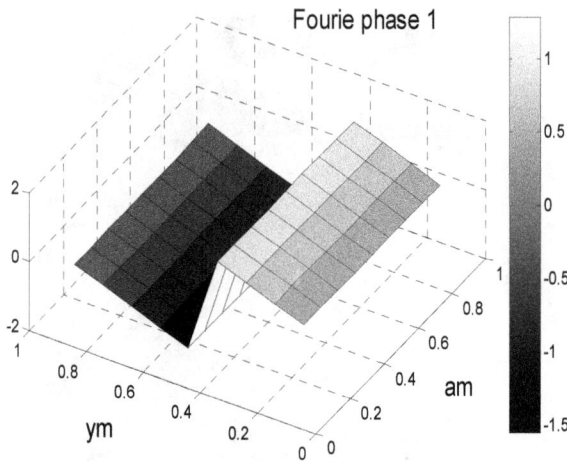

Fig. 28. Influence of *am* and *ym* on the ϕ_1

For discrete data the Fast Fourier Transform (FFT) leads to transformed complex vector *DRF*. The vector *DRF* can be decomposed to the high and low frequency component. After back transformation into original, the SHV part corresponding to noise (high frequencies) and to waviness (low frequencies) can be separated.

For rough SHV from fig. 4 SHV component is corresponding to waviness in fig. 29 and SHV component corresponding to noise is in fig. 30. The number of high frequency components equal to 20 was selected.

Some other techniques for separation of roughness, waviness and form are based on smoothing or digital filtering (Raja et. al, 2002). The smoothing by neural network can be used as well. For estimation of smoothing degree, the minimization of the mean error of prediction is usually applied (Meloun & Militký, 2011).

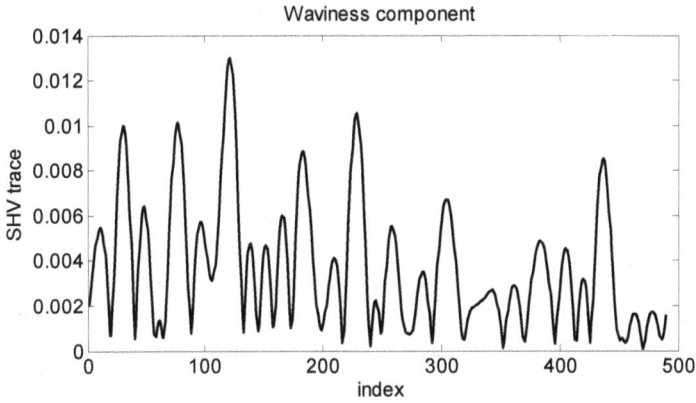

Fig. 29. SHV component corresponding to waviness for twill fabric

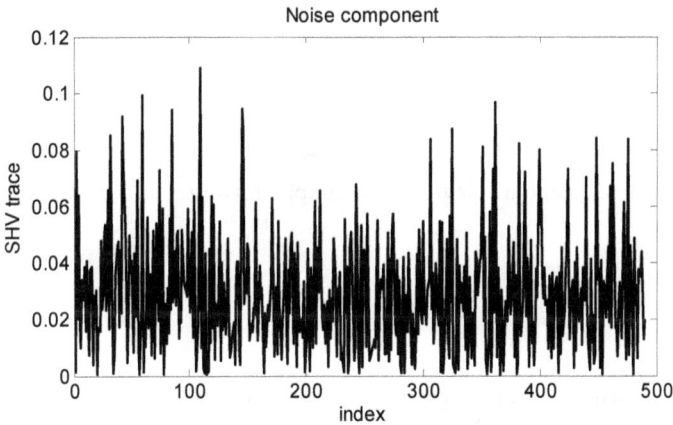

Fig. 30. SHV component corresponding to noise for twill fabric

Vector DRF may be used for creation of power spectral density (PSD)

$$g(\omega) = DRF \ conj(DRF) \ / \ T^2 = abs(DRF)^2 \ / \ T^2 \qquad (27)$$

where conj(.) denotes conjugate vector. The $g(T)$ is estimator of spectral density function and contains values corresponding to contribution of each frequency to the total variance of R.

The periodogram and power spectral density are primary tool for evaluation of periodicities. Frequency of global maximum on the $I(\omega)$ or $g(\omega)$ graphs is corresponding to the length of repeated pattern and height corresponds to the nonuniformity of this pattern.

Spectral density function is therefore generally useful for evaluation of hidden periodicities. The statistical geometry of an isotropic random Gaussian surface could be expressed in the terms of the moment of power spectral function called spectral moments (Zhang & Gopalakrishnan, 1996).

$$m_k = \int_{\omega_H}^{\omega_L} \omega^k g(\omega) d\omega \tag{28}$$

The m_o is equal to the variance oh heights and m_2 is equal to the variance of slopes between bound frequencies. The frequencies ω_H and ω_L are high and low frequency bounds of integration of the spectrogram. These frequency bound can be converted to the wavelength limits. The long wavelength limit is $l_H = 2\pi / \omega_H$ and the short wavelength limit is $l_L = 2\pi / \omega_L$. For rough SHV from fig. 4, the selected spectral moments have the following values:

- zero moment = 0.0006234.
- first moment = 0.2755.
- second moment = 0.0865.
- fourth moment = 6.387e-006.

Corresponding spectral statistical characteristics are:

- spectral variance = 0.010637.
- spectral skewness = -0.0006368.
- spectral kurtoisis = 0.000342.

The roughness Rq = SD (standard deviation) is simply $Rq = \sqrt{m_0}$ and the density of summits is defined as

$$DS = \frac{m_4}{m_2 \ 6 \ \pi \ \sqrt{3}} \tag{29}$$

9. Analysis in spatial domains

A basic statistical feature of $R(d)$ is autocorrelation between distances. Autocorrelation depends on the lag h (i.e. selected distances between places of force evaluation). The main characteristics of autocorrelation is covariance function $C(h)$

$$C(h) = \text{cov}(R(d), R(d+h)) = E((R(d) - E(R(d)) \ (R(d+h) - E(R(d)))) \tag{30}$$

and autocorrelation function ACF(h) defined as normalized version of C(h).

$$ACF(h) = \frac{\text{cov}(R(0) \ R(h))}{v} = \frac{c(h)}{c(0)} \tag{31}$$

ACF is one of main characteristics for detection of short and long-range dependencies in dynamic series. It could be used for preliminary inspection of data.

The computation of sample autocorrelation directly from definition for large data is tedious. The spectral density is the Fourier transform of covariance function $C(h)$

$$g(\omega) = \frac{1}{2\pi} \int_0^\infty C(t) \exp(-i \omega t) dt \qquad (32)$$

The ACF is inverse Fourier transform of spectral density.

$$ACF(h) = \int_0^\infty S(\omega) \exp(i \omega h) d\omega \qquad (33)$$

These relations show that characteristics in the space and frequency domain are interchangeable.

For rough SHV from fig. 4 is ACF till lag 320 component in the fig. 31.

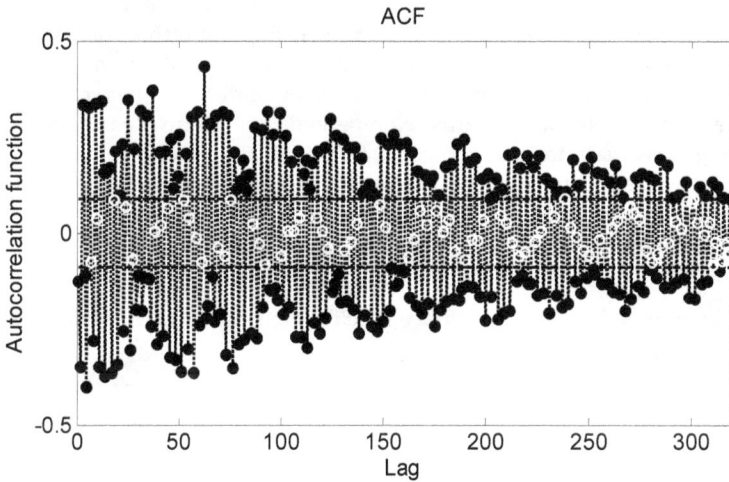

Fig. 31. ACF for twill fabric

The dashed lines in fig. 31 are the approximate 95% confidence limits of the autocorrelation function of an IID process of the same length. Sample autocorrelations lying outside the 95% confidence intervals of an IID process are marked by black circles. The slow decrease of ACF for large lags indicates long-range correlation, which may be due to non-stationarity and/or dynamic non-linearity.

In spatial statistics variogram is more frequent (Kulatilake et. al, 1998) which is defined as one half variance of differences $(R(d) - R(d+h))$

$$\Gamma(h) = 0.5 \ D[R(d) - R(d + h)] \qquad (34)$$

The variogram is relatively simpler to calculate and assumes a weaker model of statistical stationarity, than the power spectrum. Several estimators have been suggested for the variogram. The traditional estimator is

$$G(h) = \frac{1}{2M(h)} \sum_{j=1}^{M(h)} \left(R(d_j) - R(d_{j+h})\right)^2 \tag{35}$$

where $M(h)$ is the number of pairs of observations separated by lag h. Problems of bias in this estimate when the stationarity hypothesis becomes locally invalid have led to the proposal of more robust estimators. It can be summarized that simple statistical characteristics are able to identify the periodicities in data but the reconstruction of "clean" dependence is more complicated.

10. Roughness anisotropy

Above-mentioned roughness characteristics have implicitly assumed that surface roughness is isotropic phenomenon. This assumption can be accepted in the cases when surfaces have the same micro geometric properties no matter what direction they are investigated in. Majority of textiles structures have anisotropic nature. Surface of woven fabric is clearly patterned due to nearly regular arrangements of weft and warp yarns. The special non-random patterns are visible on knitted structures as well. It is well known that anisotropy of mechanical and geometrical properties of textile fabrics are caused by the pattern and non-isotropic arrangement of fibrous mass. Periodic fluctuations of surface heights can be spatially dependent due to arrangements of yarns. Non-periodic complexity spatial dependence is subtler. The roughness characteristics computed from SHV trace are therefore dependent on the direction of measurements i.e. angle of transect line according to fabric cross direction (perpendicular to machine direction). In KES system, it is possible anisotropy treated by averaging of roughness parameters in weft and warp directions only. This approach is generally over simplified and can lead to under or over estimation of surface roughness.

For anisotropic surfaces the so called surface spectral moments $m_{p,q}$ can be used (Longuet-Higgins, 1957)

$$m_{p,q} = \iint \omega_1^p \, \omega_2^q \, S(\omega_1, \omega_2) \, d\omega_1 \, d\omega_2 \tag{36}$$

where $S(\omega_1, \omega_2)$ is bivariate power spectral density of surface. Necessary condition for the case of degenerated spectrum (one dimensional) is

$$2(m_{2,0} \, m_{0,2} - m_{1,1}^2) = 0 \tag{37}$$

For degeneration to more dimensions, similar conditions can be derived (Longuet-Higgins, 1957). The profile spectral moment $m_r(\theta)$ in the direction θ defined by eqn. (27) is connected with surface moments $m_{p,q}$ by relation

$$m_r(\theta) = m_{r,0} \, \cos^r \theta + \binom{n}{1} m_{r-1,1} \cos^{r-1} \theta \, \sin\theta + .. + m_{0,r} \sin^r \theta \tag{38}$$

The second profile spectral moment $m_2(\theta)$, which is equal to the variance of profile slope PS^2, is function of three surface moments $m_{2,0}$, $m_{1,1}$ and $m_{0,2}$ only. This dependence has the simple form derived directly from eqn. (36).

$$m_2(\theta) = m_{2,0} \, \cos^2 \theta + 2 \, m_{1,1} \, \cos\theta \, \sin\theta + m_{0,2} \, \sin^2 \theta \tag{39}$$

The surface moments play central role in description of surfaces topography. The parameters $m_{2,0}$ and $m_{0,2}$, which are the 2nd surface spectral moments, denote the variance of slope in two vertical directions along cross direction and machine direction. The parameter $m_{1,1}$ represents the association-variance of slope in these two directions. These parameters are generally dependent of the selected coordinate system. From the known values of $m_2(\theta_i)$ for selected set of directions θ_i $i = 1,..n$ it is possible to estimate the surface moments $m_{2,0}$, $m_{1,1}$ and $m_{0,2}$ by using of linear regression. The maxima and minima of eqn. (37) are

$$(m_{2\max}, m_{2\min}) = 0{,}5 * [(m_{2,0} + m_{0,2}) \pm \sqrt{\{(m_{2,0} - m_{0,2})^2 + 4m_{1,1}^2\}}] \tag{40}$$

These occur in the angle θ_p called principal direction given by relation

$$\tan \theta_p = \frac{2m_{1,1}}{m_{2,0} - m_{0,2}} \tag{41}$$

As one measure of anisotropy the so-called long-crestedness $1/g$ has been proposed (Longuet-Higgins, 1957), where

$$g = \sqrt{\frac{m_{2\min}}{m_{2\max}}} \tag{42}$$

For an isotropic surface is $g = 1$ and for degenerated one-dimensional spectrum $g = 0$. The better criterion of anisotropy has been proposed in the form (Thomas et. al, 1999)

$$AN = 1 - \frac{2 * \sqrt{m_{2,0} * m_{0,2} - m_{1,1}^2}}{m_{2,0} + m_{0,2}} \tag{43}$$

For $AN = 0$ surface perfectly is isotropic and for $AN = 1$ surface is anisotropic. Lower AN characteristic indicates low degree of anisotropy.

For investigation of surface roughness anisotropy the twill fabric (see fig. 3) and Krull fabric were selected. The $R(d)$ traces have been obtained by means of KES apparatus in the following directions: $\theta_i = 0°$ (weft direction), $30°$, $45°$, $60°$ and $90°$ (warp direction). The Kawabata SMD of individual profiles of twill fabric at chosen directions θ_i is plotted as polar graph in fig. 32 and for Krull fabric in fig. 33.

The surface moments $m_{2,0}$, $m_{0,2}$ and $m_{1,1}$ were computed from eqn. (37) by using of linear least squares regression. Estimated surface moments and AN anisotropy measure are given in the table 1.

The proposed technique is capable to estimate roughness characteristics of anisotropic surfaces typical for textile structures. Beside the anisotropy measure AN the direction θ_p and values of $m_{2\max}$ will be probably necessary for deeper description of textiles surface roughness.

Kawabata SMD

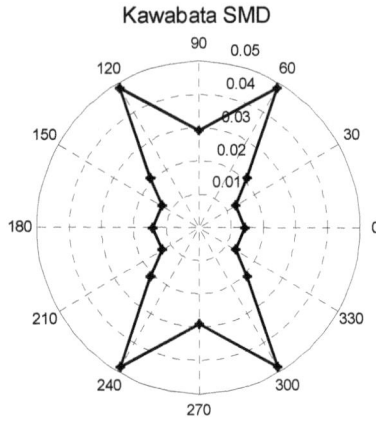

Fig. 32. Kawabata SMD of individual profiles at chosen directions θ_i for twill fabric

Kawabata SMD

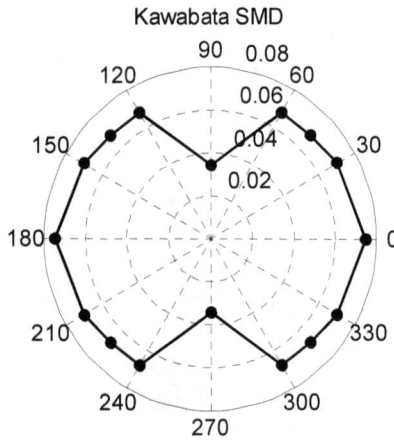

Fig. 33. Kawabata SMD of individual profiles at chosen directions θ_i for Krull fabric

The angular dependence of profile slope variance $m_2(\theta)$ and experimental points for twill are shown on the fig. 34 and for Krull are shown in fig. 35.

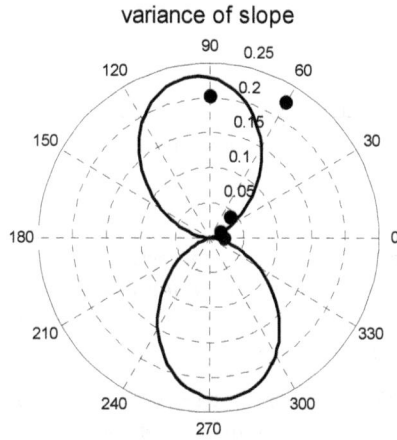

Fig. 34. Angular dependence of profile slope variance for twill fabric

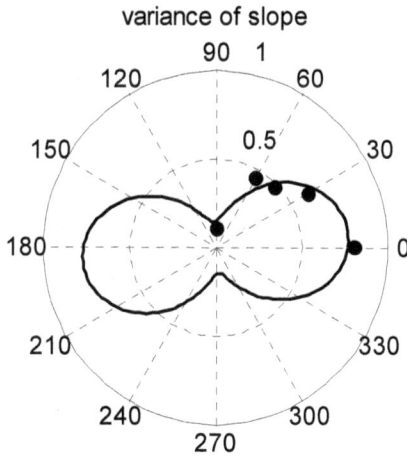

Fig. 35. Angular dependence of profile slope variance for Krull fabric

Pattern	Weave	$m_{2,0}$	$m_{1,1}$	$m_{0,2}$	AN-anisotropy
	Twill	0.0020481	-0.0250585	0.229335	1
	Krull, surface loops	0.7671	0.0613	0.1506	0. 728

Table 1. Surface moments and anisotropy of samples

11. Conclusion

There exists a plenty of other roughness characteristics based on standard statistics or analysis of spatial processes which can be used for separation of noise and waviness (macro roughness). For evaluation of suitability of these characteristics it will be necessary to compare results from sets of textile surfaces.

For deeper analysis of SHV traces from KES device the rough signal registration and digitalization by using of LABVIEW system is beneficial.

The analysis of SHV can be more complex. The other classical roughness characteristics and topothesy can be computed as well and many other techniques of fractal dimension calculation can be included. The analysis can be extended to the chaotic models and autoregressive models. With some modifications it will be possible to use these techniques for characterization of the SHV or surface profiles obtained by other techniques.

The contact less measurement of fabric images by using of RCM device is useful for description of relief in individual slices and in the whole fabric plane.

12. Acknowledgements

This work was supported by the research project 1M4674788501 of Czech Ministry of Education.

13. References

Ajayi, J.O. (1992). Fabric smoothness, Friction and handle, *Text. Res. J.* 62, 87-93

Ajayi, J.O. (1994). An attachment to the constant rate of elongation tester for estimating surface irregularity of fabric, *Text. Res. J.*, 64, 475-476

Anonym, (1997). *ISO 4287: Geometrical product specification, GPS-surface texture, profile method - terms, definitions, and surface texture parameters*, Beuth Verlag, Berlin

Beran, J. (1984).*Statistics for Long -Memory Processes*, Chapman and Hall, New York

Cox, D. R. (1984). *Statistics in Apparaisal*, Iowa State University, 55-74

Davies, S. (1999). Fractal analysis of surface roughness by using of spatial data, *J.R. Stat. Soc.*, 61, 3-37

Eke, A. (2000). Physiological time series: distinguishing fractal noises from motions *Eur. J. Physiol.*, 439, 403-415

Greenwood, J. A. A. (1984). Unified Theory of Surface Roughness, *Proc Roy Soc London*, A393, 133-152

Kawabata, S. (1980). *The standardization and analysis of hand evaluation*, Text. Mach. Soc. Japan

Kulatilake, P. H. S. W. (1998). Requirements for accurate quantification of self affine roughness using the variogram method, *Int. J. Solid Structures* 35, 4167 - 4189

Longuet-Higgins, M. S. (1957). Statistical properties of an isotropic random surface *Phil. Trans. R. Soc. London*, A 250, 157-174

Maisel, L. (1971). *Probability, Statistics and Random Processes*, Simon Schuster, New York

Mandelbrot B.B. & Van Ness J.W. (1968), Fractional Brownian motion, fractional noises and applications, *SIAM Review* 10, 442-460

Meloun, M. & Militký, J. (2011). *Statistical data analysis*, Woodhead Publ. New Delhi

Militký J. & Bajzík V.: (2000) Some open problems of hand prediction, *Journal of Fibers and Textiles*, 7, 141-145

Militký, J. & Bajzík, V. (2001). Characterization of textiles surface roughness, *Proceedings 7th International Asian Textile Conference*, Hong Kong Polytechnic, September

Militký, J. & Bajzík, V. (2002). Surface Roughness and Fractal Dimension, *J. Text. Inst.* 92, Part 3, 1-24

Militký, J. & Bajzík, V. (2004). Characterization of protective clothing surface roughness, *Proc. Int. Textile Congress Technical textiles: World market and future prospects*, Tarrasa, October 251-265

Militký, J. (2007). Evaluation of roughness complexity from KES data, *Proc. 5th International Conference Innovation and Modelling of Clothing Engineering Processes – IMCEP 2007*, Moravske Toplice October

Militký, J. & Klička, V. (2007). Nonwovens Uniformity Spatial Characterization , *Journal of Information and Computing Science* 2, No. 2, 85-92

Militký, J. & Mazal, M. (2007). Image analysis method of surface roughness evaluation, Int. J. Clothing Sci. Technol. 19, 186-193

Militký, J. & Bleša, M. (2008). Evaluation of patterned fabric surface roughness, *Indian Journal of Fibre &Textile Research*. 33, 246-252

Quinn, B.G. & Hannan, E. J. (2001). *The Estimation and Tracking of Frequency*, Cambridge University Press

Raja, J. (2002). Recent advances in separation of roughness, waviness and form, *Journal of the International Societies for Precision Engineering and Nanotechnology* 26, 222–235

Sacerdotti, F. (1996). *Surface Topography in Autobody Manufacture – The state of art*, Brunel University, June

Stockbridge, H. C. (1957). The subjective assessment of the roughness of fabrics, *J. Text. Inst.* 48, T26-34,

Thomas, T. R. (1999). Fractal characterization of the anisotropy of rough surfaces *Wear* 232, 41 – 50

Whitehouse, D. J. (2001). Roughness and fractals dimension facts and fiction, *Wear* 249, 345-346

Zhang, C. & Gopalakrishnan, S. (1996). Fractal geometry applied to on line monitoring of surface finish *Int. J. Mach. Tools Manufact.*, 36, 1137-1150

7

Alkali Treatments of Woven Lyocell Fabrics

Ján Široký, Barbora Široká and Thomas Bechtold
University of Innsbruck, Research Institute for Textile Chemistry and Physics
Austria

1. Introduction

We are surrounded by textile fabrics every second of our daily life. Being produced in enormous amounts, these products were considered simple mechanical constructions for a long time. During the last few decades aspects of functional products and smart textiles directed more scientific attention towards a more profound understanding of the relations between fabric structure and performance properties.

The overall properties of a textile fabric structure are determined by a series of constructive parameters built in at different levels of material design. For example the wetting behaviour and the final water uptake of a fabric will be determined by:

- type of fibre material and fibre properties chosen to produce the fabric e.g. cellulose fiber, polyolefine type fibre
- technical construction of the yarn *e.g.* spinning process, turns per meter, yarn count
- fabric construction, *e.g.* plain weave, single jersey.

As a result an enormous variability of designs can be chosen for a certain application. The best performance of a product will be achieved by appropriate choice of the right technical design parameters for a certain application.

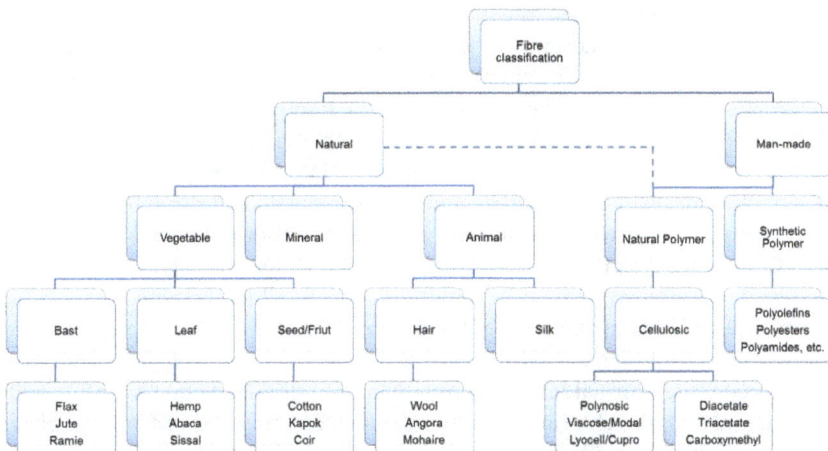

Fig. 1. Classification of textile fibres.

Textile fibres can be divided into two main categories, natural and man-made, as it is depicted in Fig. 1; in addition, there is another category which lies in between, and shares some features of both categories – it is termed 'regenerated fibres' and includes regenerated cellulosic fibres, which are typically wood pulp converted into continuous filaments by dissolving the wood in suitable solvents from which they can be regenerated.

Cellulose fibres exhibit a unique position among the textile fibres, due to their hydrophilicity and their ability to change their dimension by swelling. Swelling occurs in water, polar solvents and in particular in aqueous alkali hydroxide solutions, which are widely applied in textile finishing processes of cotton and regenerated cellulose fibres. Due to their high swelling capability regenerated cellulose fibres are highly sensitive during the alkaline treatment, thus a careful selection of alkalisation conditions for such fabrics is required. This particular behavior of cellulose to swell extensively in alkaline solutions results in a different performance in comparison to synthetic fibres. Herein, the chapter is dedicated to detailed discussion of the fibre behavior and the resulting effects/impacts onto regenerated man-made fabrics, for example lyocell fabric, during alkalisation.

2. Regenerated cellulose

It is difficult to verify when cellulosic materials were used the first time for clothing, because of the biodegradability, but the oldest cotton textiles found go back to 5800 BC and were found in a cave in Tehuacän in ancient Mexico;(Abu-Rous, 2006) other important sources like flax, linen, hemp or wool were also used in early history. For instance, Babylonia was the first country to process and trade in wool (Babylonia = land of wool) and also, the proof of using leather of different skin in clothes and shoes by discovery of the oldest European mummy; Ötzi the Iceman – man who lived about 5,300 years ago, on the border between Austria and Italy. (Hollemeyer et al., 2008; Kutschera & Rom, 2000)

Regenerated cellulose fibres are the first artificial fibres ever made. Processes capable of dissolving the cellulose derived from wood or cotton linters were first discovered by Schnöbein (1845, nitrocellulose soluble in organic solvents), Schweizer (1857, cellulose in cuprammonium solution), Cross, Bevan and Beadle (1885, cellulose sulfidized in sodium hydroxide; 1894, cellulose triacetate in chloroform) followed by commercial production of acetate fibres started in 1919; in 1955 they were joined by triacetate. The introduction of modifiers by Cox in 1950 and the development of high-wet-strength fibres initiated by Tachikawa in 1951 again increased the variety of cellulosic man-mades. In the 1960s, high wet modulus type rayon fibres were developed to improve resistance to alkali and to increase wet fibre mechanical properties and dimensional stability of fabrics. In the 1970s, new kinds of rayon fibres were produced - Avtex Avril III (a multilobal fibre), Rayonier's Prima and Courtaulds' Viloft (a hollow fibre with high water-holding capacity). In the late 1970s due to increasing investment in pollution control, which became cost-determining, several companies (Courtaulds, Lenzing, Enka) started to examine the application of carbon disulfide-free, direct solution systems for cellulose. Several direct solvent systems, such as dimethylsulfoxide/para-formaldehyde, N-methylmorpholine-N-oxide (NMMO) and N,N-dimethylacetamide lithium chloride were investigated, but only NMMO became of practical importance. Lyocell fibre made from an NMMO-solution is today a newest class of man-made cellulosic with very promising properties.

The main raw material from which regenerated cellulosic fibres are manufactured is purified wood pulp; mainly produced from wood and linters, but also from annual plants. For its conversion into textile fibres, it must be dissolved in a suitable solvent from which it can be regenerated as continuous filaments after the solution has been extruded through a fine orifice. At present, the following three methods are mostly used: the viscose process; the lyocell process; and the cuprammonium process.

Most of the world's man-made cellulose fibres are produced via the viscose process; these fibres are called 'rayons' because of the basic fibre-forming process, which involves cellulose regeneration from a cellulose xanthate, a chemical derivative of cellulose and carbon disulfide. The viscose process is characterized by high versatility, which is the result of various modifications that can be made at different stages of the process. The degree of polymerization of cellulose used, additives to modify the viscose solution, coagulation (controlled by coagulation bath additives, exerting an effect on the orientation and alignment of the cellulose molecules in the direction of the fibre axis), and stretching applied during fibre processing, can lead to a huge range of rayon forms and properties. These variations and their consequences on the end products are given in details in Morton and Hearle (1993).

With regard to the macrostructure of viscose fibres, regenerated cellulosics are different in their morphology to cotton as they have no lumen and are non-fibrillar (Fig. 2), which is directly due to the manufacturing process.

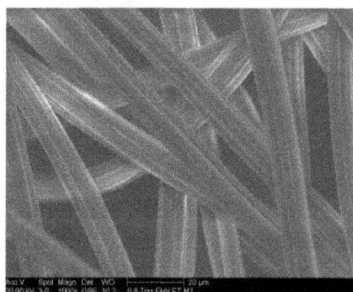

Fig. 2. Scanning electron microscope (SEM) micrographs of viscose fibres.

Lyocell is the generic name of a new generation of regenerated cellulosic fibres made by a solvent spinning process. The development of this fibre was driven by the desire for an environmentally friendly process to produce cellulosic fibres with an improved performance profile and cost compared to viscose rayon, utilizing renewable resources as raw materials.

This cellulosic fibre is derived from wood pulp (typically eucalyptus) produced from sustainably managed forests; the wood pulp (good quality, DP = 400-1000) is mixed at 70-90 °C with approximately 80% (v/v) N-methylmorpholine-N-oxide (NMMO) solution in water with a small quantity of degradation inhibitor. NMMO is capable of physically dissolving cellulose without any derivatisation, complexation or special activation, and it is able to break the inter- and intra-molecular hydrogen-bonds of cellulose.

In the ternary system, cellulose is dissolved in a narrow region and the solution is stabilised using suitable chemicals, *e.g.* isopropyl gallate. The homogenous solution (*dope*) with a minimum of undissolved pulp particles and air bubbles, is put into the evaporator vessel (evaporation of water) operated under vacuum to reduce temperature (ca 90-120 °C), due to the amine oxide solvent in solution degrading if it is overheated. Before spinning, the solution is passed through two stages of filtration. For spinning, the solution is supplied to each jet (a small *air gap* with thousands of tiny holes), and it is then extruded and spun through an air gap (*fibre* or *tow* is obtained) into a spin bath containing dilute amine oxide solution. The fibres are *drawn* or *stretched* in the air gap by the pull of *traction units* or *godets*. Afterwards, the fibres are washed and from the excess liquor and NMMO is recovered by filtration, purification and concentration. Lyocell fibre differs from viscose rayon in the fibre structure and morphology (Fig. 3).

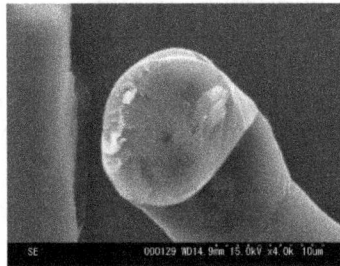

Fig. 3. Lyocell fibre SEM micrograph.(Zhang, 2004)

	Viscose Process	Lyocell Process
Derivatisation	CS_2, NaOH	----
Solvent	NaOH	NMMO
Toxicity	Very toxic (CS_2)	Non-toxic
Spinning Bath	H_2SO_4, $ZnSO_4$	H_2O
Pulps	Small variety	Large Variety
Recovery	Complex	Simple

Table 1. Comparison of the viscose and lyocell processes.(Harvey, 2007)

Production of fibres using this process has very little impact on the environment mainly in terms of chemical used and is a benign technology process. The manufacturing process recovers >99% of the solvent, additionally, the solvent itself is non-toxic and all the effluent produced is non-hazardous. The environmental impact of the viscose and lyocell processes are compared in Table 1. Lyocell is designed as a fully biodegradable cellulosic polymer with beneficial properties, which will be mentioned later. Despite all of these benefits, the production of lyocell was in 2002 only *ca.* 2.5% of the total regenerated cellulosic fibre production (total of 2.76 million tonnes), as shown in Fig. 4.(Abu-Rous, 2006)

The lyocell process is less flexible then the viscose process due to the high orientation of the obtained polymer after the air gap. However, there are other possibilities to influence the structure and properties of lyocell fibres using physical process parameters.

Lyocell fibres have the thinnest and longest crystallites, even the amorphous regions are oriented along the fibre axis, and its crystallinity is of high degree (up to 60-70%). These fibres have a microfibrillar structure because a portion of the molecular chains aggregate to form microcrystals while recrystallizing along the chains, whereas the remaining chains exist in the amorphous phase as links between these two phases.(Okano & Sarko, 1984) In the crystalline regions of cellulose II polymers, the layered structure is very regular, so the length of hydrogen bonds between molecules is the same.

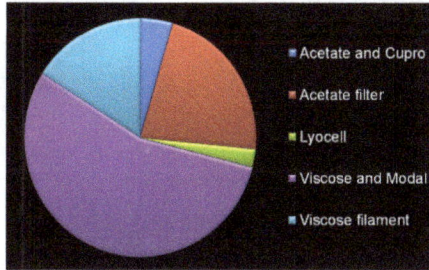

Fig. 4. World production of cellulosic fibres in tonnes.

Although, the physical properties of lyocell are unique among all kinds of rayons (Table 2) remarkably when wet, problems of lyocell properties occur as well.(Woodings, 1995) The weaker lateral links between crystallities are consequence of the highly crystalline lyocell structure and as a result of wet abrasion, at the surface of fibre the separation of fibrous elements known as *fibrillation* (Fig. 5) occurs. Basically, it is the longitudinal splitting of a single fibre filament into microfibers of 1-4 µm in diameter. It can also yield the 'peach skin' touch of fabrics, characteristic surface touch of lyocell fibre, but unwanted and uncontrolled fibrillation can worsen the fabric quality, for example, entanglement of microfibers causes a serious problem of pilling.

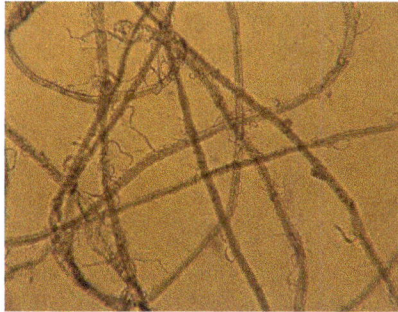

Fig. 5. Fibrillation of lyocell fibre. (Zhang, 2004)

Man-made cellulose fibres, such as viscose or lyocell, morphological structure can be described as a network of elementary fibrils and their more or less random associations; this is called a "fringe fibrillar" structure which is basically one of the macro-conformations of polymer chains depicted in Fig. 6. Schuster et al. (2003) used Ultra Small Angle Neutron Scattering (USANS) to yield information on lyocell fibres and their proposed structure at different dimensional levels is given in Fig. 7. In general, lyocell fibre is distinguished by its high crystallinity, high longitudinal orientation of crystallites, high amorphous orientation,

Property	Regular rayon	Cuprammonium	'Y'-shaped rayon[a]	Modal	Polynosic	Lyocell fibre [b]
Fibre cross-section						
Dry tenacity (cN/tex)	20-24	15-20	18-22	34-36	40-45	40-44
Wet tenacity (cN/tex)	10-15	9-12	9-12	19-21	30-40	34-38
Extension at break (%, dry)	20-25	7-23	17-22	13-15	8-12	14-16
Extension at break (%, wet)	25-30	16-43	23-30	13-15	10-15	16-18
Water imbibitions (%)	90-100	100	100-110	75-80	55-70	65-70
Cellulose DP	250-350	450-550	250-350	300-500	550-700	550-600
Initial wet modulus (at 5 %)	40-50	30-50	35-45	100-120	140-180	250-270

[a]The 'Y'-shaped rayon data are based on Courtaulds' Galaxy fibre;
[b]The solvent-spun rayon data are based on Courtaulds' Tencel fibre

Table 2. Physical properties of selected commercially available rayon fibres.

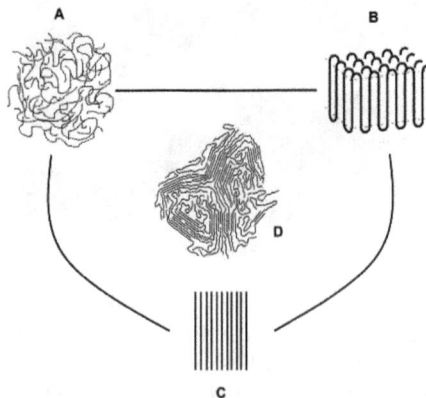

Fig. 6. Illustration of the macro-conformations of polymer chains. A, amorphous; B, regular chain folding; C, chain-extended which represents the limiting cases, and the middle part D, fringe micelle represents the intermediate structure.

low lateral cohesion between fibrils, low extent of clustering and relatively large void (pore) volume in comparison to other cellulosic regenerated fibres.

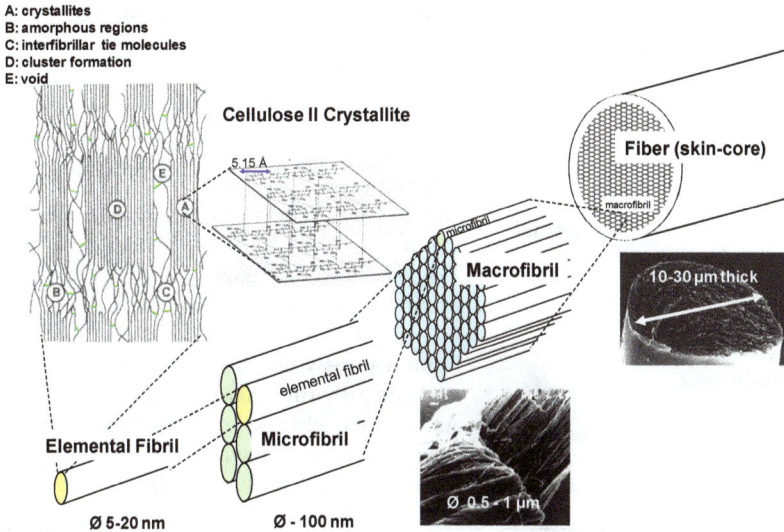

Fringed Micellar Cellulose II
(Elemental Fibril)

A: crystallites
B: amorphous regions
C: interfibrillar tie molecules
D: cluster formation
E: void

Cellulose II Crystallite

5.15 A

Fiber (skin-core)

macrofibril

microfibril

Macrofibril

10-30 µm thick

elemental fibril

Elemental Fibril

Microfibril

Ø 0.5 - 1 µm

Ø 5-20 nm

Ø - 100 nm

Fig. 7. Lyocell fibre structure at different dimensional levels.(based on Schuster et al. (2003))

There is a wide range of possibilities or processes to influence the fiber structure and its properties such as changing the parameters (*e.g.* type of pulp and molecular weight of polymer, dope composition, air gap length, l/d ratio of spinning nozzle, spinning speed, draw ratio, spinning bath composition, *etc.*) during the fibre formation. One of the processes for lyocell fibres mentioned herein is a "softer" precipitation, involving a two-stage precipitation in alcohol and water. By this process a decrease of crystallinity and orientation of lyocell fibers is obtained and, therefore, the fiber structure and core-shell structure is affected.(Klemm et al., 2005) It is also known that the structure and properties of regenerated fibers, like density, crystallite size, orientation, pore number and volume and therefore, skin-core structure as well, can be influenced with applied physical process (spinning) conditions. Recently, a skin-core model for lyocell fibers was proposed by Biganska(Biganska, 2002) where three-component system is presented. A system with compact fiber core, a porous middle zone and a semi-permeable fiber skin. However, Gindl et al. (2006) showed that only two different parts within lyocell fiber do exist, skin and core. They observed that studied fibers have uniform skin-core orientation, in contrast, Kong et al. (2007) obtained non-uniform skin-core orientation by X-ray diffraction as claimed due to the differences of used beam size (5 × 5 µm vs. 500 nm). This non-uniformity resulted in the higher average orientation of the fiber skin than of the core. Additionally, it was shown that the skin-core model of lyocell fibers is influenced by the increased shear forces on the outer region of the fiber during passing the spinning dope through the spinneret, which generates higher crystal orientation at the skin.

3. Activation treatments of cellulose

Activation methods (Table 3) in general are those that only open and widen existing capillaries, voids and interstices, those that are capable of splitting up fibrillar aggregations, *and* those that are able to disrupt crystalline order and/or change the crystal modification in order to increase accessibility and reactivity of fibrous cellulose substrates, *e.g.* fibres or fabrics.(Krässig, 1993b)

Treatment		Action
Degradation	Chemical	Hydrolysis-acid or enzymatic
		Oxidation
	Thermal	Applied at different temperatures in various media
	Radiation	Exposure to UV/VIS or high-energy radiation
Mechanical		Milling – dry and wet, grinding, beating, cutting
		Freezing, ultrasonic agitation
Solvent exchange		From water-swollen state introducing inert media towards providing reactive water-swollen state
Inclusion		As solvent exchange but introducing inert liquids to maintain reactive water-swollen state
Swelling		inter- and intra-fibrillar swelling
		swelling agent/solvent

Table 3. Treatments which have an activating effect onto cellulose fibrous substrates.

Degradation treatment of cellulose substrate can be performed with numerous chemical reaction routs, which of major significance is hydrolysis and oxidation, or with applying different source of energy such as *UV/VIS*, γ-irradiation or thermal treatment. *Mechanical* treatment is commonly employed by grinding, milling, beating or cutting to reduce size as well as increase accessible and reactive internal surface of cellulose with varying of activating time and/or temperature.

Activation of cellulose substrates also occurs in water-free organic environments followed by solvent exchange or by inclusion in a subsequently performed reaction. The *solvent exchange* treatment represents the action from water-swollen state with introduction of media which are unable to swell cellulose itself (inert to cellulose). Very similar is *inclusion* treatment which maintains the reactive water-swollen state by introducing inner liquids (*e.g.* benzene) applying the solvent exchange. Examples of chemicals used which provide highly swelling systems causing *intra-fibrillar* as well as *intra-crystalline swelling* include inorganic acids, various salt solutions, inorganic and organic bases, amines and amine complexes, and metal hydroxide solutions. In general, these treatments penetrate the fibre substrate through existing capillaries and pores, open already present voids between fibrillar elements, disrupt fibrillar associations, and finally enter the more easily accessible regions interlinking the crystallites forming the elementary fibrils. From there, they are able to penetrate at suitable treatment conditions; concentration and temperature, and to some extent tension during treatment, have to be considered from both ends into the lattice structure of elementary crystallites, which break open inter-molecular hydrogen-bonds and van der Waals interactions between lattice layers, leading to a widening of lattice distances or even changes in the crystal lattice.(Krässig, 1993b)

3.1 Accessibility and swelling of cellulose

3.1.1 Cellulose water/moisture interactions

One of the most important features of cellulosic substrates is their propensity to absorb moisture from ambient air, expressed in terms of either moisture regain or moisture content. Water absorption causes swelling of the substrate, which alters the dimensions of the fibre, and this, in turn, will cause both changes in physical properties such as the size, shape, stiffness, and permeability of yarns and fabrics,(Morton & Hearle, 1993) as well as sorption/desorption characteristics (Široká et al., 2008), and in mechanical properties such as tensile modulus and breaking stress,(Kongdee et al., 2004) and, therefore, interaction between cellulose and water plays an important part in the chemistry, physics and technology of cellulose isolation and processing.

Cellulose accessibility largely depends on the available inner surface, supramolecular order, fibrillar architecture, and also fibre pore structure. In most cases, there is interaction with water which consequently destroys weak hydrogen-bonds, but cannot penetrate into the region of high order, in contrast to, for example, aqueous solutions of sodium hydroxide, and therefore cellulose is not dissolvable in water. In Fig. 8, an overview is given about the depth of fibre reorganization/dissolution and also indicates the fields of its application; sorbed water on such a polymer can cause structural changes predominantly in the amorphous or intermediate phases. However, there is also a significant role associated with pores, capillaries, and the network of voids, which do not have uniform size and shape.

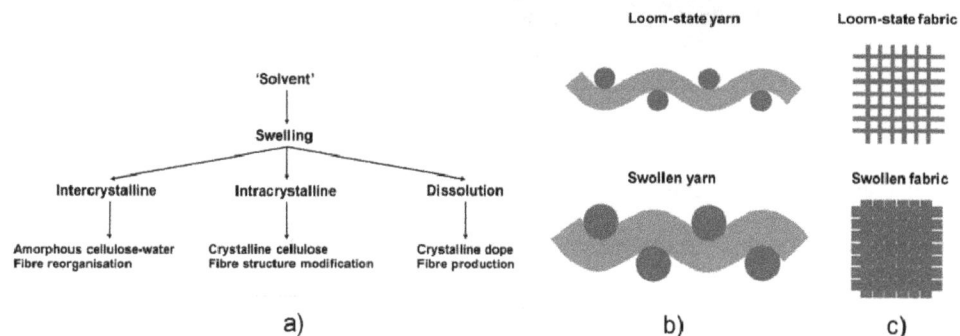

Fig. 8. Different stages of cellulose fibre reorganisation at different level: a) solvent, b) yarn, and c) fabric level.

From the chemistry perspective, cellulose-water interactions are basically a competition of hydrogen-bond formation between hydroxyl groups in the polymer and hydrogen-bond formation between one hydroxyl group of a cellulose chain and a water molecule or a water cluster.(Klemm et al., 1998) Differential scanning calorimetry (DSC) may be used as valuable technique to explore the interaction of either water or moisture with various natural and synthetic polymers with hydrophilic groups, and thermal properties of polymers and water.

Tatsuko and Hyoe Hatakeyama's classified different fractions of water in cellulose structures.(Hatakeyama & Hatakeyama, 1998) The first-order phase transition of water fractions closely associated with the polymer matrix is usually impossible to observe; such fractions are termed *non-freezing water* and it can directly interact with hydroxyl groups of cellulose. Another fraction of water, *freezing bound water*, is less closely associated and in contrast, it exhibits melting/crystallisation; it is able to interact with either hydroxyl groups of cellulose or water already absorbed to cellulose. The sum of the non-freezing and freezing bound water fraction is *the bound water content*. Water, whose melting/crystallisation temperature and enthalpy are not significantly different from those of normal (bulk) water, is designated as *freezing water*. *Bound water* in the water-insoluble hydrophilic polymers, such as cellulose, lignin, *etc.*, breaks hydrogen-bonding between the hydroxyl groups of the polymers, and its content depends on the chemical and supramolecular structure of each polymer. Schematic illustration of all kinds of water within the fibre structure is proposed in Fig. 9, where from it is obvious that the freezing (W_f) and non-freezing (W_{nf}) water provides a dry touch and dry perception respectively. However, with further water absorption within fibre structure, the bulk (W_b) and capillary water (W_c) provides a wet touch and hence, it imparts wet perception respectively.

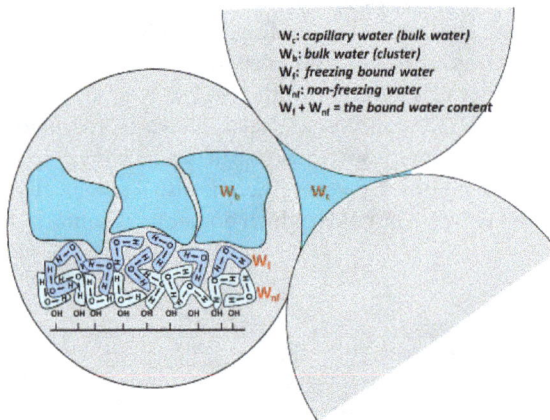

Fig. 9. Water in cellulosic fibre.

Transport of liquid or moisture within the porous solid such as textile fibre or fabric occurs by external forces or capillary forces only. If a liquid is spontaneously transported and it is driven inside porous solid, it is known as wicking. The capillary force increases as the capillaries; gaps between the individual fibres, become thinner – the finer the fibres, the smaller the gaps are, and the better the transport. Wicking is a result of spontaneous wetting in a capillary system due to fact that the capillary forces are caused by wetting. Wetting of fibres or fabric usually involves three phases which may be characterized the best as the displacement from a surface of a fibre-air interface with a fibre-liquid interface. It needs to be noted that the wicking and wetting are not two different processes, but wetting is a prerequisite for wicking.(Kissa, 1996; Rosen, 2004)

Surface wettability (wetting characteristics) of any fabric containing a single fibre type is identical to the wettability of its constituent single fibres.(Hsieh, 1995) Therefore, the fibre surface and the liquid properties are determining factors of fibre wetting properties. The geometric configurations and the pore structure of the porous medium (*e.g.* woven fabric) determine its liquid transport properties and hence, capillary water differs in various types of fabric construction (plain-, twill- or sateen-woven fabrics). Fabric geometry and fabric construction influence properties of macro-pores (pore size, pore size distribution, pore connectivity, total pore volume).(Hsieh et al., 1992) Pores in woven fabrics can be intra-fibre, inter-fibre and inter-yarn;(Hsieh, 1995) intra-fibre pores are smallest and are discontinuous, not being merged with neighboring fibres. Structure and dimensions of inter-fibre and inter-yarn pores are influenced by yarn structure and yarn density in woven fabrics and thus, the observed differences are mainly determined by fabric construction.

When cellulose fibres are dried there is a subsequent reduction in the extent of swelling when the fibres are re-wet; the loss of swelling resulting from a drying and re-wetting cycle is so-called "*hornification*" or "*zip up of voids*".(Crawshaw & Cameron, 2000) Basically, some voids "*zip up*" after drying of swollen substrate by the formation of additional hydrogen-bonds between cellulose molecules, and this may lead to re-crystallization or to spot-wise bridging, which produces non-crystalline, brittle, inaccessible domains on a larger scale. According to this model, during wetting, cellulose chains are hydrated with water molecules, re-arrangement of cellulose chains takes place due to opening of closed pores and joining up of small pores. When drying, the water molecules are partially removed and hydrogen-bonds between cellulose molecules are generated. This interaction induces re-arrangement of fibre structure, loss of fibre pores and collapse of large pores, which is not recovered by re-wetting. Furthermore, amorphous cellulose may become re-crystalline on drying. This change results in lowered pore volume and surface area that is especially found in non-crosslinked type fibres. The limited re-expansions can also be attributed to covalent interactions, such as the formation of lactone bridges between hydroxyl and carboxylic groups in the polymer chains (Diniz et al., 2004). There is also some evidence of drying-induced crystallization in cellulosics, to which is also attributed the phenomenon of hornification (Brancato, 2008). Accessibility and reactivity of cellulosic fibres are reduced due to the already mentioned *hornification* phenomena. In accordance with Crawshaw et al. (2000) findings on drying, the void fraction falls and the void size rises, suggesting that a large number of voids "*zip up*", essentially, some voids merge and creates larger ones. These authors proposed a schematic illustration of water-swollen and dried Tencel fibres. Another widely accepted explanation is so-called fibrillar 'fusion', the formation of hydrogen bonds between surface hydroxyls, which causes the reduction of interfibrillar interstices and hence, the accessible and reactive internal surface is decreased.(Krässig, 1993b) This fusion is enhanced the lower the crystalline order and the higher the degree of swelling prior to drying of cellulose substrate from the water wet state.(Krässig, 1993b) The same reference states that the effect of hornification can be avoided by appropriate swelling treatment (*e.g.* ammonia pre-treatment) or by drying after solvent exchange to more hydrophilic liquids.

An additional effect can be expected due to differences in cross-section type of fibres used. These differences will influence pick-up, sorption properties, and fibre swelling in respect to structural differences and also macroscopic properties like stiffness and handle. Usually, fibres of different cross-section are used in specific applications, *e.g.* hygiene and personal care or technical applications and thus, systematic scientific comparisons on fabric level are rare.

3.1.3 Cellulose-aqueous alkali hydroxide interactions

The process of caustic treatment was devised in 1844 by John Mercer and he was later granted a patent(Mercer, 1851) for work he had done on vegetable fabrics and fibrous material (cotton, flax) and other textures with treatment of caustic soda, potassium hydroxide, dilute sulfuric acid or zinc chloride. The process did not become popular, however, until H. A. Lowe improved it into its modern form in 1890; by holding the cotton during treatment to prevent it from shrinking, Lowe found that the fibre gained a lustrous appearance and good dyeability.(Beaudet, 1999) Afterwards, treatments with aqueous sodium hydroxide solutions found numerous applications as the primary step in many industrial practices leading to acetyl, ether or ester derivatives of cellulose.

One of the most important steps in cellulosic fabric processing is the treatment in aqueous solutions of alkali hydroxides, particularly aqueous sodium hydroxide solution. In general, mercerization and alkalization, respectively of cellulose is a conventional swelling treatment of cellulosics substrates with aqueous alkali solutions, usually carried out in concentrations between 6.8 to 7.6 mol dm^{-3} NaOH, and is a prerequisite to the preparation of many cellulose derivatives, the spinning of viscose ramie and lyocell, or the manufacturing of cellulose-based sponges.(Porro et al., 2007) In its interaction with cellulose, aqueous sodium hydroxide above a certain concentration is able to penetrate the cellulose crystalline lattice to yield a series of more or less well-defined crystalline complexes holding a number of sodium ions and water molecules within their crystalline lattice. From analysis of these complexes by X-ray diffraction it has been found that cellulose and alkali solutions interact strongly with the consequence of swelling of cellulose fibres. This interaction at low alkali concentration with cellulosic fibres causes an increase in their cross-sectional area with simultaneous shrinkage in length; the extent of the dimensional changes caused by the swelling depends on the alkali treatment concentration and temperature.(Krässig, 1993a) This effect is most probably influenced by the degree of swelling observed previously for cotton(Warwicker, 1969), which herein at its maximum (between 3.00 and 3.75 mol dm^{-3} for continuously treated lyocell fabrics) decreases with increasing temperature of the treatment liquor.(Široký et al., 2010) Treatment with aqueous sodium hydroxide solution improves the fabric's mechanical and chemical properties such as dimensional stability, fibrillation tendency, tensile strength, dyeability, reactivity, lustre and smoothness.(Široký et al., 2009) Therefore, the strong swelling tendency of lyocell in alkali solutions substantially influences substrate properties such as the pore structure, crystallinity, fibrillation tendency, and surface characteristics in fibres, and yarn crimp and stiffness in fabrics.(Široký et al., 2010) In addition, during wet processing, lyocell fabrics are sensitive to the generation of permanent crease marks and alkali pre-treatments are recommended to counteract this (Široký et al., 2009).

Recently, Goswami et al.(Goswami et al., 2009) observed that sodium hydroxide treatment causes the density, orientation and crystallinity of lyocell fibres to decrease with increasing sodium hydroxide concentration, and that the greatest change in fibre properties occurs between 3.0 and 5.0 mol dm^{-3} NaOH. This was attributed to the onset of formation of sodium (Na)-cellulose II at 3.0 mol dm^{-3} NaOH; a fully formed Na-cellulose II structure was observed above 6.8 mol dm^{-3} NaOH. In addition, alkalization without tension can increase the moisture regain at a given relative humidity to 1.5 times its previous value;

mercerization under tension does not cause such a large increase,(Morton & Hearle, 1993) and generally causes an increase in fibrillar (crystalline) orientation.(Krässig, 1993a) Moreover, cellulose is susceptible to limited degrees of dissolution in alkali solutions, leading to weight loss in substrates. For instance, lyocell fibre mass reduces by 2-10% with increasing NaOH concentration up to 10 mol dm^{-3}.(Jaturapiree, 2007; Zhang et al., 2005)

The ability of cellulose to adsorb alkali hydroxide ions was originally considered to be a chemical process. In 1907, Vieweg observed two ranges of caustic soda concentrations at which the uptake of sodium hydroxide reached a plateau by applying so-called "change-in-titer" method;(Krässig, 1993a) the principle of this method was to measure the drop in alkali concentration of a given caustic soda solution by a known weight of immersed cellulose, and, at the time, had been use for a decade as a convincing observation for the occurrence of a defined chemical reaction following stoichiometric rules. Nevertheless, Vieweg's findings (and those of other research groups) were subjected to criticism by Leighton (1916) who introduced the centrifugation method; he mainly criticized the assumption made by Vieweg that no water is taken up by solid phase and he showed that "his method" allowed the separation of the uptake of water and the uptake of sodium hydroxide. Moreover, Leighton's method revealed that there is initially a preferential water uptake, which has a maximum at medium concentrations, whereas the uptake of sodium hydroxide steadily increases over the whole concentration range, except at the lowest concentrations. The balance of the two separate uptakes explains the plateau in Vieweg's curve of apparent sodium hydroxide uptake.

From these observations, it was concluded that the uptake of alkali hydroxides by cellulose is an adsorption phenomena based on the Donnan equilibrium theory,(Donnan, 1924; Neale, 1929) rather than a chemical process. It is assumed that cellulose, behaving as a very weak monobasic acid, forms a sodium salt to an extent increasing with the concentration of the alkali. Excess alkali diffuses into the mechanically separable cellulose phase. It results in an unequal distribution of ions what causes an osmosis or movement of water which distends the cellulose until the osmotic pressure is balanced by the forces arising from the cohesion of the gel and therefore, the cellulose swells. In addition, Neale's theoretical osmotic curves showed that as the temperature falls the maximum osmotic pressure rises, and the maximum occurred at a lower concentration of alkali. One of the difficulties with the absorption theory was based on the fact that uptake by native cellulose substrates is not smooth and reversible as might be expected for an adsorption process.(Krässig, 1993a) Another issue with the earlier investigations and results interpretation is that the effect of the morphology, fine structure, and accessibility of the cellulosic fibre substrate was, in most cases, ignored and it was assumed that the fibres were equally accessible throughout.

A comprehensive study of parameters, such as *sodium hydroxide concentration* and *temperature*, which influence the formation of various sodium-celluloses was conducted by Sobue et al. (1939) using an X-ray technique. They proposed a phase diagram of ramie cellulose and its various Na-cellulose crystalline complexes. They also revealed a new highly swollen and poorly crystalline allomorph, namely *Na-cellulose Q*, occurring at NaOH concentrations between 7% to 10% (1.75 to 2.50 mol dm^{-3} NaOH) and temperatures between −10 and 1°C. More recently, it was found that if an adequate pre-treatment of cellulose was applied, within the *Q phase*, total dissolution of cellulose can occur;(Kamide et al., 1984) solutions of 5% (w/v) cellulose could be obtained at a sodium hydroxide concentration of

9% (2.25 mol dm^{-3} NaOH) and temperature of 4°C and below. It is generally accepted that the immersion of ramie or cotton fibres into 12-16% (w/v) NaOH (3.0 to 4.0 mol dm^{-3} NaOH) yields *Na-cellulose I*, whereas the use of more concentrated alkali, typically 20-30% (w/v) NaOH (5.0 to 7.5 mol dm^{-3} NaOH), gives another allomorph: *Na-cellulose II*. The conversion of *Na-cellulose I* into *Na-cellulose III* is obtained by vacuum drying *Na-cellulose I*, whereas washing *Na-cellulose I* and *Na-cellulose II* until neutrality, yields the sodium-free *Na-cellulose IV*, which in its turn will give *cellulose II* upon drying. As Porro *et al.*(2007) stated, the analysis of the X-ray data alone may not be sufficient to give a clear picture of the interaction of the Na$^+$ ions with the cellulose molecules. Therefore, Porro suggested a thorough analysis of ^{13}C solid-state NMR data of the various Na-cellulose complexes due to the low spectral resolution of X-ray which did not lead to a clear assignment of the resonances belonging to the various complexes. These authors proposed a phase diagram, based on the occurrence of *Na-cellulose I* and *Na-cellulose II* together with the *Q region* where cellulose is essentially soluble. It defines six regions, two of them being border zones where two phases are present at the same time.

4. Sodium hydroxide treatment of lyocell woven fabrics in continuous process

The key area in this part will be to examine the effect of aqueous sodium hydroxide treatment (alkalisation) of cellulosic natural-based polymer, in particular lyocell fabric, which are of cellulose II crystalline structure. The alkali treatment or NaOH uptake of cellulose is a long-lasting issue, regarding cotton and other cellulose materials. Sufficient information of alkali pad-batch process (no tension applied) to cellulose substrates is available with its effect onto cellulose fabrics. For example, the maximum NaOH uptake/swelling is between 2.0 – 2.5 mol dm^{-3} for fabric, high shrinkage force is present during treatment causing strong dimensional changes (shrinkage) to treated material, effect on crystallinity and dye sorption was reported. While numerous publications concerning NaOH treatment of cotton have been published, a gap in alkalisation knowledge of lyocell fabrics exists. Limited information is available on the continuous treatment process which constitutes important part of commercial operations. Therefore, a pilot-scale pre-treatment with NaOH was performed with a washer, a special semi-scale apparatus simulating a real alkali process to undertake a comprehensive study of various process parameters on lyocell woven fabric and its physico-chemical and sorption properties. Herein, the influence of NaOH treatment concentration, treatment temperature, tension applied to fabric and their effects to treated substrates will be discussed. Also, an influence of fabric structure/geometry (plain-, twill-, sateen-woven fabrics) on NaOH release (wash-off) from woven lyocell fabrics will be considered.

Before considering the relationship between the structure of cellulose substrates and their physical and chemical properties it is appropriate briefly summarise alkalisation process in textiles. Swelling is a special feature of porous substrate causing its expansion at different levels, *e.g.* fibre, yarn, or fabric. Particularly, this expansion results in close contact of fibres and yarns which in turn, reduces yarns mobility and increases swelling restriction. Therefore, the deformity or irregularity in the fibre cross-sectional area of, for example, lyocell fabric rises with increasing NaOH concentration (0-6 mol dm^{-3}) as observed recently(Goswami et al., 2009) and porosity decreases. Swelling reduction induces restricted effect of chemicals to be swollen and hence, the accessibility of fabric decreases significantly. This effect can be seen in wash-off or alkali release from alkali treated lyocell fabrics which is discussed and shown in recent publication(Široký et al., 2011a). On the level of fabric, the fabric construction has a crucial effect as the swelling and wet pick-up differs for various fabric constructions (*e.g.* plain-, twill- or sateen-woven lyocell fabrics).

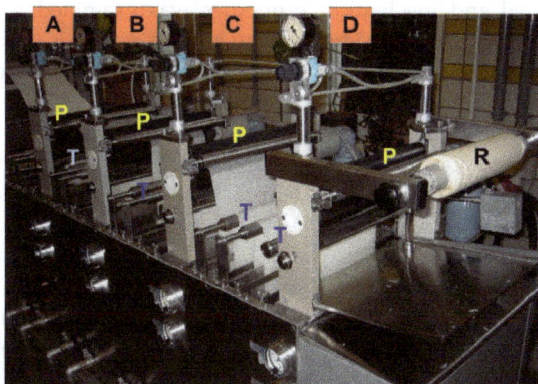

Fig. 10. Picture of apparatus used in continuous alkali pre-treatment process.

The continuing process of alkali pre-treatment was done with a washer simulating real alkali process which is given in Fig. 10. It was divided into four stages (A, B, C, D), each with two sub-compartments (1 and 2) that could be heated independently; the four stages were alkali treatment (A2), stabilisation treatment (B2), washing (C1 and C2), neutralisation (D1), and final washing (D2). The fabric was passed through the apparatus over a series of rollers including tension compensators (T) and pressurised squeeze rollers (P). The fabric after passing through the last compartment (D) was wound on a take-up roller (R). In treatment stage, the NaOH concentration, applied tension and temperature varied. Due to changes in the concentration in the treatment stage (caused by process), the "effective" concentration of alkali was established at 0.0, 2.53, 3.33, 4.48, 4.65 and 7.15 mol dm^{-3} for treatment bath and 0.0, 0.73, 1.08, 1.18, 1.48 and 2.15 mol dm^{-3} for stabilization bath. Residence time was adjusted at 75±2 s and 71±3 s in treatment bath and stabilization bath respectively. Tension in treatment compartments was applied either at 49 N m^{-1} or 147 N m^{-1}, tension in stabilisation compartment was applied at 147 N m^{-1}, and tension in washing and neutralization compartments was applied at 49 N m^{-1}. During the alkali treatment stage the temperature of the solution was set at 25 °C or 40 °C, and in the stabilization stage the temperature was 60°C. The speed of passage of fabric through the system was set at 2 m min^{-1}. Entire process and its all adjustments during the process are described in detail in our previous work.(Široký et al., 2009)

4.1 Influence of physical properties

Alkali treatment has a substantial influence on dimensional and mechanical properties as well as on morphological, molecular and supramolecular properties of cellulose II fibres (*e.g.* lyocell type) causing changes in their structure and performance. These physical changes were observed and examined by dimensional change, flexural rigidity (in dry and wet state), mass per area, water retention and pick-up, abrasion resistance, tensile strength, elongation at break, and crease recovery following continuous treatment of the lyocell plain woven fabrics with aqueous sodium hydroxide solution under varying condition parameters.

The influence of process parameters in alkali treatment of lyocell fabrics by a continuous process on dimensional and mechanical properties was examined with focus on: alkali concentration, treatment temperature, and tension applied on fabric.(Široký et al., 2009) The alkali concentration and treatment temperature exerted significant influence on these fabric properties, but no significant influence of tension was observed in the results. It was proposed herein, that at the maxima of fibre swelling in alkali influencing fabric macroscopic structure and properties at certain NaOH concentrations and temperatures (observed peaks) the greatest NaOH effect occurs. Peaks of fabric shrinkage were observed in samples treated with 3.33 mol dm^{-3} NaOH at 25°C, and with 4.48 mol dm^{-3} NaOH at 40°C. The shrinkage peaks corresponded to peaks in flexural rigidity, and minima in water retention, crease recovery, and breaking force. The shrinkage in fabrics and corresponding changes in their properties are attributed to fibre swelling in alkali influencing fabric macroscopic structure and properties. In the range of concentrations examined in this particular work(Široký et al., 2009), the best combination of properties was observed in fabrics treated with 2.53 mol dm^{-3} NaOH. These fabrics exhibited low wet flexural rigidity, high water retention, high crease recovery, high abrasion resistance, and high breaking strength as compared to the untreated fabric as well as among alkali treated fabrics. Also, it was found that the NaOH concentrations in treatment baths decreased during the course of fabric passage through liquors while, the NaOH concentration in the stabilisation baths increased into extents that increased with rise in liquor alkali concentrations. The NaOH depletion was attributed to sorption of alkali by substrates, and the rise in the stabilisation bath was attributed to the transport of NaOH from the treatment to the stabilisation bath by the fabric.

4.2 Attenuated total reflectance fourier-transform infrared spectroscopy to characterize crystallinity changes

In the structure of cellulose, a linear 1,4-β-glucan polymer, there are three hydroxyl groups able to interact through intra- and inter-molecular hydrogen bonding, which form highly ordered structures. Alkali treatment of all allomorphs of cellulose has a substantial influence on morphological, molecular and supramolecular levels, causing changes in crystallinity as a result of the degree of swelling, the rate of fibre swelling, changes in the voids and pore networks, and further, on supramolecular level, the diffusion of alkali solution into the less ordered amorphous and quasi-crystalline phases of the cellulose II polymer. Fourier-Transform Infrared spectroscopy (FT-IR) is currently one of the best available techniques in the textile science to obtain structural information (Široká et al., 2011), which has been universal accepted for many years as a powerful tool for analytical and chemical characterisation in wide range of scientific disciplines and widely used in polymer characterisation, providing information about chemical nature, isomerisation, conformational order, state of order, and orientation. Using FT-IR, a clear relationship between interaction of hydroxyl groups and crystallinity in cellulose has been established, and numerous papers have been published in relation to infrared spectroscopy analysis of native cotton and mercerised cotton as well as regenerated fibres (Hurtubise & Krassig, 1960; Nelson & O'Connor, 1964; Široký et al., 2010) in order to establish or develop the ratio indices of absorption bands, crystallinity indices, the so-called 'total crystallinity index' (TCI) and 'lateral order index' (LOI). TCI represents the overall degree of order in cellulose given by the ratio of absorption at 1372 cm^{-1} (C–H deformation in cellulose II) and 2892 cm^{-1} (C–H stretch in cellulose II). LOI reflects the ordered regions perpendicular to the chain direction, which is greatly influenced by chemical processing of cellulose. This index is calculated as the ratio of two signals, at 1418 cm^{-1} (CH$_2$ scissoring at C(6) in cellulose II) and at 894 cm^{-1} (C–O–C valence vibration of β-glycosidic linkage). First band decreases and second one increases as crystallinity decreases. Attenuated Total Reflectance Fourier-Transform Infrared (ATR-FTIR) spectroscopy analysis is appropriate methods for gathering spectroscopic data, IR absorbance bands are particularly sensitive to the vibrations of polar groups such as OH groups as well as quite reasonable to the non-polar skeletal bonds such as C–C and C–O, and C–H. Therefore, it was used to observe the qualitative crystallinity changes in lyocell fabrics following continuous treatment with sodium hydroxide.

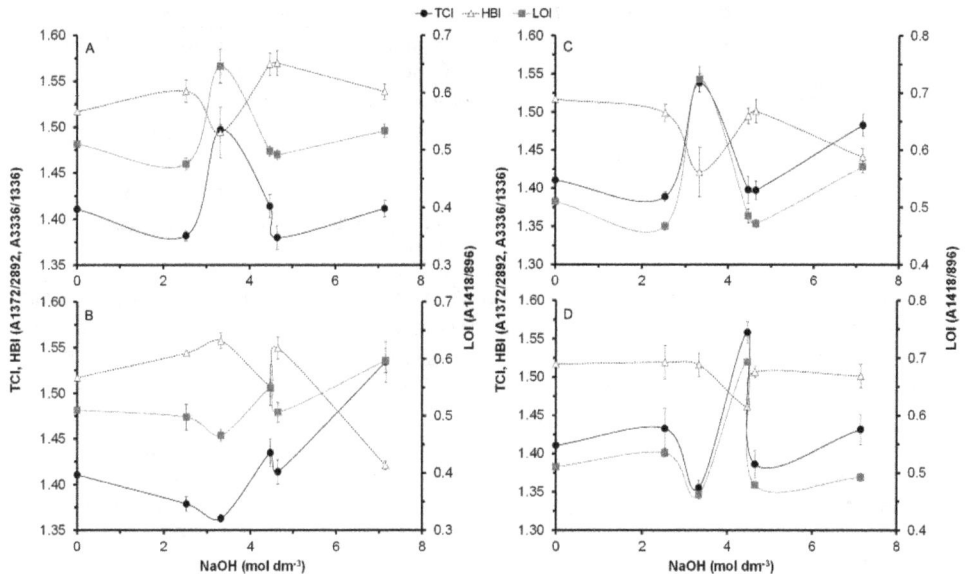

Fig. 11. Total crystallinity index (TCI; ◆), lateral order index (LOI; ●) and hydrogen bonding intensity (HBI; O) of plain woven lyocell fabrics with increasing sodium hydroxide concentration in treatment stage, under varying (49 or 147 N m^{-1}) and temperature (25 or 40 °C) in treatment stage of continuous alkali treatment: A) 49/25; B) 49/40; C) 147/25; D) 147/40.

The results and comparisons of crystallinity indices (Fig. 11), TCI and LOI, and hydrogen-bond intensity (HBI) against NaOH treatment concentration shown parallel behavior over the whole range of concentrations and process conditions, although the absolute values differ. There were observed maxima for TCI and LOI, and minima for HBI for all examined lyocell fabrics at effective concentrations of 3.33 and 4.48 mol dm^{-3} NaOH, when treated at

25°C and 40°C, respectively. Under these treatment conditions, it was proposed therein that maximum molecular reorganisation occurs in the amorphous and quasi-crystalline phases of the cellulose II polymer, which are corresponding to the changes evidenced from evaluation of physico-chemical properties (Široký et al., 2009). Moreover, it was demonstrated that even subtle changes in ATR-FTIR spectra can substantially affect crystallinity indices.

4.3 Sorption properties

It is well known that the regenerated cellulosic fibres absorb significant quantities of water by the expansion of void spaces within the semi-crystalline morphology, forming a water-cellulose two-phase structure. Dyestuffs used for cellulosic fibres are highly water soluble, with molecular structures which are designed to interact at the internal interface between cellulose and water. The uptake of dyes is often used to monitor changes in fibre properties brought about by variations in processing, hence, dyes can be considered as coloured probe molecules that provide information on the detailed internal pore structure of fibres (Ibbett et al., 2006). The adsorption study of *hydrolyzed reactive dye*, namely C.I. Reactive Red 120 (RR120), on lyocell fabric was performed (Široký et al., 2011b) to gather information of the alkali treatment effect on the thermodynamics of the dye sorption and on changes in the fabric/fibre structure. Hydrolyzed reactive dye behaves essentially like a direct dye, which provides particular advantages of assurance that covalent bonding between the dye and fibre did not interfere with sorption, the substantivity (physical adsorption) is high and therefore, relatively short dyeing times can be employed as well as being used as molecular sensors to characterize cellulose substrate properties such as pore structure.(Inglesby & Zeronian, 2002; Luo et al., 2003)

Thorough study of dye adsorption revealed that sorption observed herein may occur *via* a combination of Langmuir, typical of limited side-specific adsorption, and Freundlich, indicating unlimited adsorption at non-specific sites, isotherms, although, sorption theory teaches that the interaction between dye molecules and cellulosic fibres is typically of Freundlich isotherm and based on hydrogen bonding and van der Waals interactions. On other site, carboxylate function groups, which are formed through oxidation of cellulosic fibres following mercerisation, are capable to provide anionic side-specific sides for adsorption. Interesting relation was obtained when the observations for equilibrium dye sorption (q_e) from aqueous solution onto a lyocell was compared with the *reacting structural fraction* (RSF) concept developed and established by Fink *et al.* (1986). The q_e of hydrolyzed RR120 by lyocell fabrics treated with varying concentrations of NaOH showed very similar shape of the plots with a maximum at *ca.* 3.50 mol dm^{-3} NaOH for lyocell, assigned to transition of cellulose II to its Na-cellulose II form following break of inter-molecular hydrogen bonds. Subsequent decrease within the respective transition ranges occurred to *ca.* 4.50 mol dm^{-3} NaOH for cellulose II. A second maximum occurred at *ca.* 4.75 mol dm^{-3} for cellulose II which is caused by an additional increase in the RSF of the polymer, relative to disruption of the intra-molecular hydrogen bonds in the Na-cellulose II form. Increasing treatment concentration of NaOH causes strong lateral fibre swelling and subsequently increases accessible internal volume within the structure of the fibre. Therefore, the observed differences in sorption may be explained by the accessible pore volume (APV) of lyocell fibres to probes as a function of NaOH concentration previously studied by Özturk *et al.* (2009) using inversion size exclusion chromatography (ISEC). It was demonstrated that for pore diameters of 19 Å and below that APV increased with increasing NaOH concentration to a maximum around 2.5 mol dm^{-3} NaOH, which was ca. 10% higher than for untreated lyocell. Applying molecular dynamics, the geometry and electronic properties of hydrolyzed RR120 were explored, and it was found that the dimensions of employed dye are about 30 Å (length) × 14 Å (wide) × 3.5 Å (depth) in one of the minimum energy states, providing a molecular diameter of 14 Å (end-on). It is less than 19 Å and therefore, it is likely that the maximum APV at around 2.5 mol dm^{-3} NaOH observed by (Özturk et al., 2009) influences the accompanying maxima in the theoretical monolayer capacity (q_0) and adsorption energy (ΔG^0) also known as a standard free energy of dyeing or standard affinity of the dye for the substrate, which was indeed observed from the experimental results.

The measurement of *iodine sorption* has been frequently applied in order to study an accessibility of cellulosic fibers.(Nelson et al., 1970; Schwertassek K., 1956; Široká et al., 2008; Strnad et al., 2001) This method has also been employed as a measure of fiber crystallinity.(Elesini U. S. & Cuden A. P., 2002; Hessler & Power, 1954; Zemljic et al., 2008) The computation of substrate crystallinity from iodine sorption value is based on the principle that iodine sorption in substrates is limited only to amorphous regions. However, earlier studies have shown that iodine penetrates into crystalline regions when the adsorption exceeds 11-12%, and the potassium iodide, in which the iodine is dissolved, acts as a swelling agent for cellulose.(Doppert H. L., 1967) These factors will influence results of iodine sorption. Hence, the iodine sorption may be better regarded as a general measure of overall accessibility in substrates. From the iodine sorption experiments performed on alkali treated lyocell fabrics, it was observed that the results do not follow the crystallinity results obtained by ATR-FTIR method. It seems that not only crystallinity of alkali treated lyocell fabrics swelling which is generally considered to increase accessibility of substrates, but also the construction of substrate can significantly influence the accessibility to iodine.

4.4 Woven lyocell structure effect on sodium hydroxide release (wash-off)

While in previous sections the discussion was oriented to the extended study of lyocell plain woven fabrics with effect of alkali concentration, treatment temperature and applied tension based on the previous work (Široký et al., 2010, 2011b; Široký et al., 2009), in this chapter, the influence of fabric structure (plain-, twill-, or sateen-woven fabrics) on NaOH release from lyocell after pad-batch pre-treatment (Široký et al., 2011a) will be explored by conductivity measurements in the system of deionized water-NaOH impregnated assemblies-the wash-off bath. Weaving of the fabric is distinguished according to the manner the yarns or threads, longitudal-warp and lateral-weft, are interlaced to form fabric or cloth. Three basic weaves are used the most, plain-, twill- or sateen-weaving. Differences in fabric weaving or construction provide also differences in terms of accessibility, substrate surface and compactness, fabric diffusion, swelling, and fabric bulk density and porosity.

Different fabric construction of the same fibre and yarn, herein weaving, plays a crucial role in liquid-fabric interactions. For example, wet pick-up (*WPU*) differs, which in turn effects the swelling, bulk density, or porosity of such a porous substrate as it shown in Fig. 12. Due to mathematical complexicity, only the changes on the macro-level were further investigated by measuring the conductivity in the wash bath and applying Crank's approximation, the Crank's equation for the flow through a membrane(Crank, 1975):

$$\frac{M_t}{M_\infty} = 1 - \frac{8}{\pi^2}\exp\left(\frac{-\pi^2 D\,t}{1^2}\right) \tag{1}$$

where, M_t is conductivity at a given time (mS cm⁻¹), M_∞ is conductivity at wash-off equilibrium or the corresponding conductivity during infinite time (mS cm⁻¹), D is the diffusion coefficient (m² s), l is half of the sheet (layer) thickness, because diffusion takes place from both its sides (m), and t is time (s). Herein, the factor ($-\pi^2 D/l^2$) was used as "alkali transport coefficient" K and was determined from the slope of $\ln(1 - M_t/M_\infty) = f(t)$.

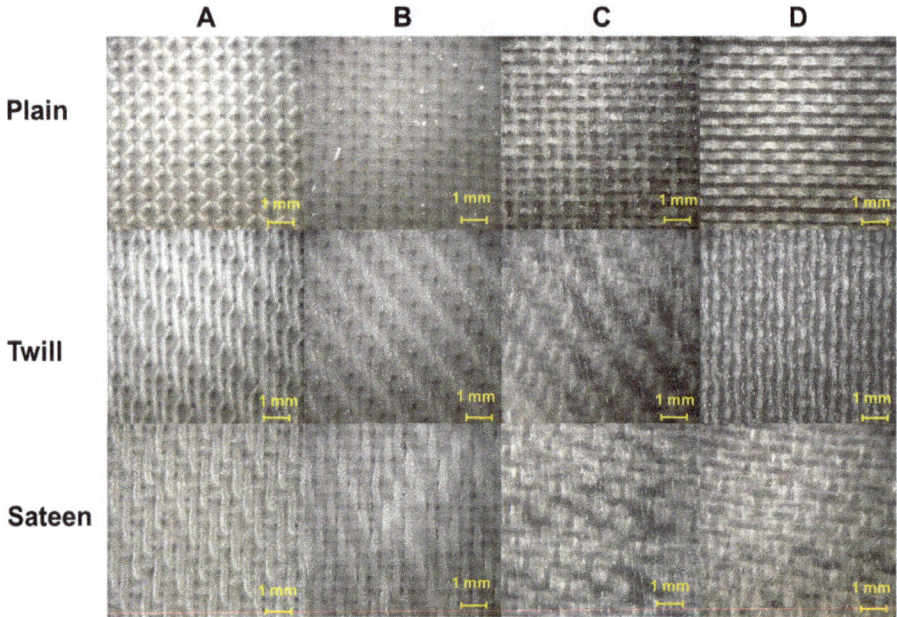

Fig. 12. Changes in wet pick-up and porosity of loom-state fabric (A-untreated) and swollen fabric after dropping NaOH solution of different concentration: B-0.7, C-3.0, or D-5.0 mol dm⁻³ on fabric of different construction (plain-, twill-, or sateen-woven fabric). An optical microscope (Krüss, Germany) combined with a digital camera (Canon Power Shot S40) was used for taking microphotographs. A sample of 1 x 1 cm size was placed on a glass slide and NaOH solution was dropped on the substrate. After 10 s, the picture was taken.

The kinetics of the system is a complex system with many parameters/components involved. Briefly, cellulose accessibility is dependent on the available inner surface, supramolecular order (range of degrees of order), fibrillar architecture, and pore structure. Interactions of cellulose with water which consequently destroys weak hydrogen-bonds, but cannot penetrate into regions of high order, in contrast to aqueous solutions of sodium hydroxide. Additionally, a significant role associated with pores, capillaries, and the network of voids, which do not have uniform size and shape need to be considered, and also NaOH treatment causes changes in pore structure (shape and size).(Bredereck & Hermanutz, 2005) Liquid absorption leads to swelling of the substrate (see Široký et al.

(2011a)), which alters the dimensions of the fibre, and this, in turn, causes changes in both physical properties (size, shape, stiffness, and permeability of yarns and fabrics, sorption/desorption characteristics)(Morton & Hearle, 1993; Široká et al., 2008) and mechanical properties (tensile modulus, breaking stress).(Kongdee et al., 2004) Moreover, the "skin-core" effect of lyocell fibres exists, wherein a fibre structure consists of different regions, which may lead to a diffusional boundary layer and hence, cause significant changes in concentration gradient within the substrate.

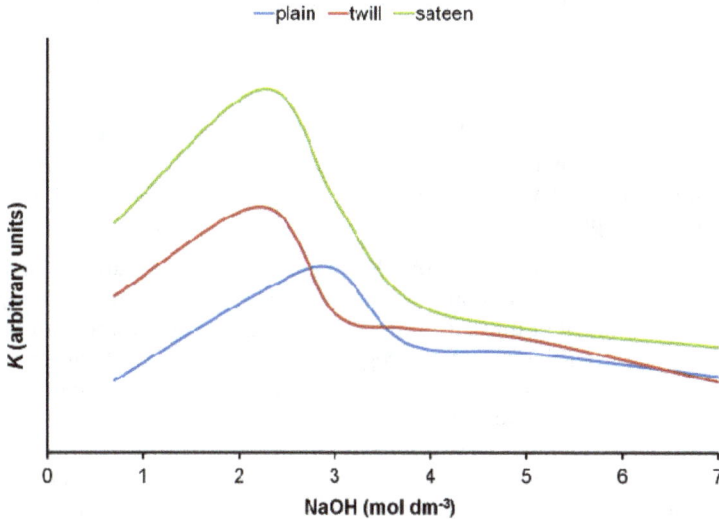

Fig. 13. Trend lines based on calculated liquid transport coefficients (K) at different NaOH treatment concentrations.

An alkali transport coefficient (K) was established to quantify alkali release, which represents the liquid-side mass transfer of alkali release after a pad-batch process into the wash-bath.(Široký et al., 2011a) Three regions of NaOH-release behavior were observed (Fig. 13) and two competitive phenomena related to swelling, an increase of the substrate surface, and the substrate compactness, have been recognized. Firstly (up to 2.25 mol dm^{-3}), the influence of fabric-alkali interactions and non-uniform access of NaOH determines alkali release, and also the absolute K value is mainly determined by the bulk densities (plain 0.4998, twill 0.4622 and sateen 0.4096 g cm^{-3}) and porosities (plain 0.671, twill 0.696 and sateen 0.731) and thus, it increases continuously as the material is still not fully swollen. The maximum swelling of cellulose fibers occurs between 2.25-3.00 mol dm^{-3} NaOH; here, maximum K is observed as well as dependency on fabric construction is still determining factor. Above 3.75 mol dm^{-3} NaOH, the treated substrates are highly swollen sheets and the K does not show dependency on fabric constructions, thus, the fabric construction is not relevant at this range anymore for the alkali-diffusion and the approximation of plain sheet diffusion corresponds to the real situation. As shown in this study, alkalization and release during wash-off are governed by fabric structure and alkali concentration. This finding is of particular relevance for an optimized processing of fabrics from regenerated cellulose fibers.

5. Conclusion

The unique feature of cellulosic substrates to absorb moisture from the air or absorption of water or other solution, for example sodium hydroxide, causes swelling of this substrate. Then expansion of the fibre, yarn, or fabric occurs due to the swelling in liquid media. By continuous growth of fibre, yarn, or fabric, the accessibility becomes restricted as the porosity or the inter- and intra-yarn pores and/or inter-spaces diminish. Hence, NaOH treatment influences physical properties such as stiffness, shrinkage, water retention value and wet pick-up, mechanical properties such as tension, break force, elongation, crease recovery angle and mass loss.

Also, it was shown and discussed herein that alkalization and release during wash-off are governed by fabric structure and alkali concentration. This finding is of particular relevance for an optimized processing of fabrics from regenerated cellulose fibres. While in technical processing, materials consisting from the same type of fibre and with similar mass per area are considered to behave identical during alkalization the present results show the need for individual process adaptation.

It is obvious that the cellulose is very complex material and there is still limited knowledge available in this field rather containing more questions and uncertainties. A representative example is alkali treatment of this substrate with ongoing research more than 160 years and always some gaps within this area are recognised. In recent years, it can be seen that the field of cellulose, cellulose derivatives, or polysaccharides is expanding largely due its recognised wide potential and also its environmental benefit. Therefore, the long-term task for present and future research in cellulose is the development of novel processes which yield no or minimal ecologically harmful by-products. If these efforts are successful, cellulose will maintain and strengthen its position as a renewable and environmentally beneficial, industrially important raw material competing with synthetically produced polymers.

6. Acknowledgment

The authors are grateful to the Amt der Vorarlberger Landesregierung, Europäischer Fonds für Regionale Entwicklung (EFRE), and COMET K-project "Sports Textiles" project number 820494 funded by Die Österreichische Forschungsförderungsgesellschaft (FFG), for their provision of the financial support.

7. References

Abu-Rous, M. (2006). *Characterisation of the wet-state pore structure of lyocell and other man-made cellulosic fibres by fluorescence and electron microscopy.* PhD Thesis, University of Innsbruck, Dornbirn.

Beaudet, T. (1999). In: *What is Mercerized cotton?*, Available from:
   http://fiberarts.org/design/articles/mercerized.html

Biganska, O. (2002). *Etude physico-chemique des solutions de cellulose dans la N-methylmorpholine-Noxyde.* PhD Thesis, Centre de Mise en Forme des Materiaux (CEMEF), Sophia Antipolis, France.

Brancato, A. A. (2008). In: *Effect of progressive recycling on cellulose fiber surface properties.*, Retrieved 23.11.2009, Available from: http://etd.gatech.edu/theses/available/etd-07292008-213453/

Bredereck, K., & Hermanutz, F., (2005). Man-made cellulosics. *Review of Progress in Coloration and Other topics,* Vol. 35, No. 1, pp. 59-75,

Crank, J. (1975). Diffusion in a plane sheet, In: *The mathematics of diffusion,* pp. 44-68, Clarendon Press, ISBN: 0-19-853411-6, Oxford

Crawshaw, J., & Cameron, R. E., (2000). A small angle X-ray scattering study of pore structure in Tencel cellulose fibres and the effects of physical treatments. *Polymer,* Vol. 41, No. 12, pp. 4691-4698,

Crawshaw, J., Vickers, M. E., Briggs, N. P., Heenan, R. K., & Cameron, R. E., (2000). The hydration of TENCEL cellulose fibres studied using contrast variation in small angle neutron scattering. *Polymer,* Vol. 41, No. 5, pp. 1873-1881,

Diniz, J. M. B. F., Gil, M. H., & Castro, J. A. A. M., (2004). Hornification - its origin and interpretation in wood pulps. *Wood Science and Technology,* Vol. 37, No. 6, pp. 489-494, ISSN: 0043-7719

Donnan, F. G., (1924). The theory of membrane equilibria. *Chemical Reviews,* Vol. 1, No. 1, pp. 73-90, ISSN: 0009-2665

Doppert H. L., (1967). Adsorption of iodine from aqueous solutions by samples of tire yarn from regenerated cellulose. *Journal of Polymer Science. Part A-2: Polymer Physics,* Vol. 5, No. 2, pp. 263-270, ISSN: 0449-2978

Elesini U. S., & Cuden A. P., (2002). Study of the green cotton fibres. *Acta Chimica Slovenica,* Vol. 49, No. 4, pp. 815-833,

Fink, H. P., Dautzenberg, H., Kunze, J., & Philipp, B., (1986). The composition of alkali celluloce - a new concept. *Polymer,* Vol. 27, No. 6, pp. 944-948, ISSN: 0032-3861

Gindl, W., Martinschitz, K., Boesecke, P., & Keckes, J., (2006). Orientation of cellulose crystallites in regenerated cellulose fibres under tensile and bending loads. *Cellulose,* Vol. 13, No. 6, pp. 621-627, ISSN: 0969-0239

Goswami, P., Blackburn, R. S., El-Dessouky, H. M., Taylor, J., & White, P., (2009). Effect of sodium hydroxide pre-treatment on the optical and structural properties of lyocell. *European Polymer Journal,* Vol. 45, No. 2, pp. 455-465, ISSN: 0014-3057

Harvey, A. (2007). *Physicochemical mechanisms involved in the binding of antimicrobial agents to textile fibres.* PhD Thesis, University of Leeds, Leeds, UK.

Hatakeyama, H., & Hatakeyama, T., (1998). Interaction between water and hydrophilic polymers. *Thermochimica Acta,* Vol. 308, No. 1-2, pp. 3-22, ISSN: 0040-6031

Hessler, L. E., & Power, R. E., (1954). The Use of Iodine Adsorption as a Measure of Cellulose Fiber Crystallinity. *Textile Research Journal,* Vol. 24, No. 9, pp. 822-827, ISSN: 0040-5175

Hollemeyer, K., Altmeyer, W., Heinzle, E., & Pitra, C., (2008). Species identification of Oetzi's clothing with matrix-assisted laser desorption/ionization time-of-flight mass spectrometry based on peptide pattern similarities of hair digests. *Rapid Communications in Mass Spectrometry,* Vol. 22, No. 18, pp. 2751-2767, ISSN: 1097-0231

Hsieh, Y. L., (1995). Liquid Transport in Fabric Structures. *Textile Research Journal,* Vol. 65, No. 5, pp. 299-307, ISSN: 0040-5175

Hsieh, Y. L., Yu, B. L., & Hartzell, M. M., (1992). Liquid Wetting, Transport, and Retention Properties of Fibrous Assemblies .2. Water Wetting and Retention of 100-Percent and Blended Woven Fabrics. *Textile Research Journal*, Vol. 62, No. 12, pp. 697-704, ISSN: 0040-5175

Hurtubise, F. G., & Krassig, H., (1960). Classification of Fine Structural Characteristics in Cellulose by Infrared Spectroscopy - Use of Potassium Bromide Pellet Technique. *Analytical Chemistry*, Vol. 32, No. 2, pp. 177-181, ISSN: 0003-2700

Ibbett, R. N., Kaenthong, S., Phillips, D. A. S., & Wilding, M. A., (2006). Characterisation of the porosity of regenerated cellulosic fibres using classical dye adsorption techniques. *Lenzinger Berichte*, Vol. 85, No., pp. 77-86,

Inglesby, M. K., & Zeronian, S. H., (2002). Direct dyes as molecular sensors to characterize cellulose substrates. *Cellulose*, Vol. 9, No. 1, pp. 19-29, ISSN: 0969-0239

Jaturapiree, A. (2007). *Porosity and transport properties in the system cellulose fibre/aqueous medium-determination and modification of porosity* PhD Thesis, University of Innsbruck, Dornbirn, Austria.

Kamide, K., Okajima, K., Matsui, T., & Kowsaka, K., (1984). Study on the Solubility of Cellulose in Aqueous Alkali Solution by Deuteration Ir and C-13 Nmr. *Polymer Journal*, Vol. 16, No. 12, pp. 857-866, ISSN: 0032-3896

Kissa, E., (1996). Wetting and wicking. *Textile Research Journal*, Vol. 66, No. 10, pp. 660-668, ISSN: 0040-5175

Klemm, D., Heublein, B., Fink, H. P., & Bohn, A., (2005). Cellulose: Fascinating biopolymer and sustainable raw material. *Angewandte Chemie-International Edition*, Vol. 44, No. 22, pp. 3358-3393, ISSN: 1433-7851

Klemm, D., Philipp, B., Heinze, T., Heinze, U., & Wagenknecht, W. (1998). Fundamentals and analytical methods, In: *Comprehensive Cellulose Chemistry*, Wiley - VCH, Weinheim, Germany

Kong, K., Davies, R. J., McDonald, M. A., Young, R. J., Wilding, M. A., Ibbett, R. N., & Eichhorn, S. J., (2007). Influence of domain orientation on the mechanical properties of regenerated cellulose fibers. *Biomacromolecules*, Vol. 8, No. 2, pp. 624-630, ISSN: 1525-7797

Kongdee, A., Bechtold, T., Burtscher, E., & Scheinecker, M., (2004). The influence of wet/dry treatment on pore structure-the correlation of pore parameters, water retention and moisture regain values. *Carbohydrate Polymers*, Vol. 57, No. 1, pp. 39-44,

Krässig, H. A. (1993a). *Cellulose: Structure, Accessibility and Reactivity*, Gordon and Breach Science, ISBN: 2881247989, Yverdon, Switzerland

Krässig, H. A. (1993b). Effect of structure and morphology on accessibility and reactivity, In: *Cellulose: Structure, Accessibility and Reactivity*, pp. 215-277, Gordon and Breach Science, ISBN: 2881247989, Yverdon, Switzerland

Kutschera, W., & Rom, W., (2000). Ötzi, the prehistoric Iceman. *Nuclear Instruments and Methods in Physics Research Section B: Beam Interactions with Materials and Atoms*, Vol. 164-165, No. 0, pp. 12-22, ISSN: 0168-583X

Leighton, A., (1916). The adsorption of caustic soda by cellulose. *Journal of Physical Chemistry*, Vol. 20, No. 1, pp. 32-50, ISSN: 0022-3654

Luo, M., Zhang, X. L., & Chen, S. L., (2003). Enhancing the wash fastness of dyeings by a sol-gel process. Part 1: Direct dyes on cotton. *Coloration Technology*, Vol. 119, No. 5, pp. 297-300, ISSN: 1472-3581

Mercer, J. (1851). Improvement in chemical processes for fulling vegetable and other textures, England, Patent No. 8,303, US Patent.

Morton, W. E., & Hearle, J. W. S. (1993). An Introduction to fibre structure, In: *Physical Properties of Textile Fibres*, pp. 1-73, The Textile Institute, ISBN: 1-870812-41-7, Midsomer Norton, Bath, UK

Neale, S. M., (1929). The swelling of cellulose, and its affinity relations with aqueous solutions. Part I - Experiments on the behaviour of cotton cellulose and regenerated cellulose in sodium hydroxide solution, and their theoretical interpretation. *Journal of the Textile Institute Transactions*, Vol. 20, No. 12, pp. T373-T400,

Nelson, M. L., & O'Connor, R. T., (1964). Relation of Certain Infrared Bands to Cellulose Crystallinity and Crystal Lattice Type. Part II. A New Infrared Ratio for Estimation of Crystallinity in Celluloses I and II. *Journal of Applied Polymer Science*, Vol. 8, No. 3, pp. 1325-1341,

Nelson, M. L., Rousselle, M.-A., Cangemi, S. J., & Trouard, P., (1970). The Iodine Sorption Test. Factors Affecting Reproducibility and a Semimicro Adaptation. *Textile Reseach Journal*, Vol. 40, No. 10, pp. 872-880,

Okano, T., & Sarko, A., (1984). Mercerization of Cellulose .1. X-Ray-Diffraction Evidence for Intermediate Structures. *Journal of Applied Polymer Science*, Vol. 29, No. 12, pp. 4175-4182, ISSN: 0021-8995

Özturk, H. B., Potthast, A., Rosenau, T., Abu-Rous, M., MacNaughtan, B., Schuster, K. C., . . . Bechtold, T., (2009). Changes in the intra- and inter-fibrillar structure of lyocell (TENCEL^R) fibers caused by NaOH treatment. *Cellulose*, Vol. 16, No. 1, pp. 37-52, ISSN: 0969-0239

Porro, F., Bedue, O., Chanzy, H., & Heux, L., (2007). Solid-state C-13 NMR study of Na-cellulose complexes. *Biomacromolecules*, Vol. 8, No. 8, pp. 2586-2593, ISSN: 1525-7797

Rosen, M. J. (2004). Wetting and its modification by surfactants, In: *Surfactants and interfacial phenomena*, M. J. Rosen (Ed.), Wiley-Interscience, Hoboken, New Jersey

Schuster, K. C., Aldred, P., Villa, M., Baron, M., Loidl, R., Biganska, O., . . . Jericha, E., (2003). Characterising the emerging lyocell fibres structures by ultra small angle neutron scattering. *Lenzinger Berichte*, Vol. 82, No., pp. 107-117,

Schwertassek K., (1956). Praktische Methoden der Strukturforschung an Faserstoffen. *Textilpraxis*, Vol. 11, No., pp. 762-767,

Široká, B., Noisternig, M., Griesser, U. J., & Bechtold, a. T., (2008). Characterization of cellulosic fibers and fabrics by sorption/desorption. *Carbohydrate Research*, Vol. 343, No. 12, pp. 2194-2199,

Široká, B., Široký, J., & Bechtold, T., (2011). Application of ATR-FT-IR Single-Fiber Analysis for the Identification of a Foreign Polymer in Textile Matrix. *International Journal of Polymer Analysis and Characterization*, Vol. 16, No. 4, pp. 259 – 268,

Široký, J., Blackburn, R. S., & Bechtold, T., (2011a). Influence of fabric structure on NaOH release from woven lyocell type material. *Textile Research Journal*, Vol. 81, No. 16, pp. 1627-1637,

Široký, J., Blackburn, R. S., Bechtold, T., Taylor, J., & White, P., (2010). Attenuated total reflectance Fourier-transform Infrared spectroscopy analysis of crystallinity changes in lyocell following continuous treatment with sodium hydroxide. *Cellulose*, Vol. 17, No. 1, pp. 103-115, ISSN: 0969-0239

Široký, J., Blackburn, R. S., Bechtold, T., Taylor, J., & White, P., (2011b). Alkali treatment of cellulose II fibres and effect on dye sorption. *Carbohydrate Polymers*, Vol. 84, No. 1, pp. 299-307, ISSN: 0144-8617

Široký, J., Manian, A. P., Široká, B., Abu-Rous, M., Schlangen, J., Blackburn, R. S., & Bechtold, T., (2009). Alkali Treatments of Lyocell in Continuous Processes. I. Effects of Temperature and Alkali Concentration on the Treatments of Plain Woven Fabrics. *Journal of Applied Polymer Science*, Vol. 113, No. 6, pp. 3646-3655, ISSN: 0021-8995

Sobue, H., Kiessig, H., & Hess, K., (1939). The cellulose-sodium hydroxide-water system subject to the temperature. *Zeitschrift Fur Physikalische Chemie-Abteilung B-Chemie Der Elementarprozesse Aufbau Der Materie*, Vol. 43, No. 5, pp. 309-328, ISSN: 0372-9664

Strnad, S., Kreze, T., Stana-Kleinschek, K., & Ribitsch, V., (2001). Correlation between structure and adsorption characteristics of oriented polymers. *Materials Research Innovations*, Vol. 4, No. 2-3, pp. 197-203, ISSN: 1432-8917

Warwicker, J., (1969). Swelling of Cotton in Alkalis and Acids. *Journal of Applied Polymer Science*, Vol. 13, No. 1, pp. 41-54,

Woodings, C. R., (1995). The development of advanced cellulosic fibres. *International Journal of Biological Macromolecules*, Vol. 17, No. 6, pp. 305-309, ISSN: 0141-8130

Zemljic, L. F., Persin, Z., Stenius, P., & Kleinschek, K. S., (2008). Carboxyl groups in pre-treated regenerated cellulose fibres. *Cellulose*, Vol. 15, No. 5, pp. 681-690, ISSN: 0969-0239

Zhang, W. (2004). *Characterisation of cellulosic fibres: The structure of cellulose fiber and changes of fibre properties in textile processing*. PhD Thesis, University of Innsbruck, Innsbruck, Austria.

Zhang, W. S., Okutbayashi, S., & Bechtold, T., (2005). Fibrillation tendency of cellulosic fibers - Part 4. Effects of alkali pretreatment of various cellulosic fibers. *Carbohydrate Polymers*, Vol. 61, No. 4, pp. 427-433, ISSN: 0144-8617

8

Sensorial Comfort of Textile Materials

Gonca Özçelik Kayseri[1], Nilgün Özdil[2] and Gamze Süpüren Mengüç[1]
[1]Ege University, Emel Akın Vocational Training School, Izmir,
[2]Ege University, Textile Engineering Department, Izmir,
Turkey

1. Introduction

Sensorial evaluation plays an important role for identification of materials in addition to technical specification. Textiles differs from each other with their technical structures in that it must have sufficient strength, performance characteristics and at the same time it has to be flexible, elastic, easy to pleat and shape, comfortable in aesthetic and sensorial aspects.

In order to find a method for the sensational evaluation of textiles, the concept of "fabric hand" is commonly used (Mäkinen et al., 2005). Understanding and measuring consumer preferences have opened up an important field of interest in recent researches in textile industry known as "handle" or in a broader sense "skin sensational wear comfort" or "tactile comfort" (Bensaid et al., 2006; Das & Ishtiaque, 2004) which refers to the total sensations experienced when a fabric is touched or manipulated in the fingers. Since fabric handle is based on people's subjective preferences, obviously it can mean different things to different people and consumers having different backgrounds. The preferences for certain fabric types are diverse and, in extreme cases, even opposite (Pan et al., 1988). Hand influences consumers' preferences and their perception of the usefulness of the product, and consequently retailer's saleability of the apparel. This fabric property is critical to manufacturers, garment designers, and merchandisers in developing and selecting textile materials, especially the textiles intended for use in apparel (Kim & Slaten, 1999; as cited in Pan & Yen, 1992; as cited in Pan et al., 1993).

In the literature, handle is defined in several ways as given in the following:

- The judgement of the buyer which depends on time, place, seasons, fashions and personal preferences (Bakar, 2004; as cited in Peirce, 1930)
- The quality of fabric or yarn assessed by the reaction obtained from the sense of touch (Bishop, 1996; as cited in The Textile Institute, 1995)
- A person's estimation when feeling fabrics between fingers and thumb (Bishop, 1996; as cited in Thorndike & Varley, 1961)
- The total of the sensations expressed when a textile fabric is handled by touching, flexing of the fingers, smoothing and so on (Bishop, 1996; as cited in Dawes & Owen, 1971)
- A perception of clothing comfort which is sensory responses of nerves ending to external stimuli including thermal, pressure, pain, etc producing neuro physiological impulses which are sent to the brain (Dhinakaran et al., 2007).

It is no doubtful that the hand judgement of fabrics is one of the important fabric tests and has been used widely by many people that can be classified as experts in textile factories and general consumers. The experts in factories especially in finishing departments carry out this judgement to control the property of their products every day. On the other hand each of the consumers also examines the property of the fabric by his "hand" to select a good clothing material according to his feeling and experience during purchasing (Kawabata, 1980). In both cases, fabric hand is examined mainly by the sense of touch and the sensory signals sent to the brain are formulated and clustered as subjective perception of sensations as follows:

- Tactile sensations: prickly, tickling, rough, smooth, craggy, scratchy, itchy, picky, sticky.
- Moisture sensations: clammy, damp, wet, sticky, sultry, non-absorbent, clingy.
- Pressure (body fit) sensations: snug, loose, lightweight, heavy, soft, stiff.
- Thermal sensations: cold, chill, cool, warm, hot (Bishop, 1996; Dhinakaran et al., 2007).

Fabric handle is a complex parameter and is related to the several fabric properties such as flexibility, compressibility, elasticity, resilience, density, roughness, smoothness, surface friction and thermal characteristics (Ozcelik et al., 2007, 2008). Fabric properties mostly influencing fabric handle are generally listed as fabric smoothness (28%), softness (22%), stiffness (8%), roughness (7%), thickness (5%) and weight (5%). Such fabric properties like warmth, hardness, elasticity, creasing propensity, drape and other properties less influencing the textile hand enter the residual part of 25% (Bishop, 1996; Grinevičiūtė et al., 2005).

This chapter presents one of the most challenging properties of the textile materials, "the sensorial comfort" in detail. The first part deals with the factors influencing the fabric handle, such as fiber, yarn and fabric properties and finishing treatments. The second part of the chapter summarizes the mechanical properties related with the sensorial properties such as bending, shear, tensile, thickness and compression, drape, friction and roughness as well as other fabric sensory properties related to the assessment of fabric handle and quality such as fabric thermal properties, surface appearance, prickle, noise and odour. In the third part, subjective evaluation methods, systems and devices for the objective measurement of fabric handle are examined. Finally, the relationship between the subjective evaluation and objective measurement of fabric handle is given.

2. Factors influencing the fabric sensorial comfort

In textile products the basic elements that can fundamentally affect fabric handle are given below. All these characteristics have an interacted relation in terms of the mechanism of influencing sensorial comfort of the end product. Namely, yarn handle characteristics are the results of the fiber properties and similar relationship can be observed between yarn and fabric features.

- Fiber characteristics: Material type, morphological structure, fineness, length, friction property, resilience, compressibility etc.
- Yarn characteristics: Yarn type (staple fiber, continuous filament, textured), linear density, twist etc.
- Fabric characteristics: Production method (woven, knitted, non-woven), fabric construction, weight, thickness, surface roughness, structure, yarn density etc.

- Method and type of dyeing and finishing processes (heat treatment, brushing, calendaring, softening, etc.) (Dillon et al., 2001; Behery, 2005; Shanmugasundaram, 2008).

2.1 Effect of fibers

The fiber type (natural/man made, staple/filament) is the first criterion for obtaining various fabrics having diffferent sensorial comfort characteristics.

Linen is a fiber that gives fabrics comparatively higher toughness. 100% linen fabrics offer the highest tensile resilience, bending rigidity and bending hysteresis, whereas the lowest values are obtained in terms of the shear rigidity and shear hysteresis values. Cotton, viscose and cotton/linen, viscose/linen blended fabrics give comparatively lower surface friction, surface roughness and bending rigidity compared to 100% linen fabrics. (Behera, 2007). The surface properties of the yarns and fabrics are affected by the morphological properties of the animal fibers (Supuren, 2010). As the fiber diameter increase, the prickliness of the fabrics increases for this type of fabrics (Behera, 2007). Linen is a fiber that gives the fabrics comparatively higher toughness; the toughness is reduced by blending the fiber with viscose. 100% linen fabrics offer the highest tensile resilience, bending rigidity and bending hysteresis, whereas the lowest values are obtained in terms of the shear rigidity and shear hysteresis values. Cotton, viscose and cotton/linen, viscose/linen blended fabrics give comparatively lower surface friction, surface roughness and bending rigidity compared to 100% linen fabrics (Behera, 2007). The surface properties of the yarns and fabrics are affected by the morphological properties of the animal fibers (Supuren, 2010). As the fiber diameter increase, the prickliness of the fabrics increases for this type of fabrics (Behera, 2007).

The finer the fibers are, the smoother and more flexible the yarn is and the fabric drape gets better. Longer fibers and smaller variation in the fiber length distribution result in smoother yarn and fabric surfaces. Micro denier filament fabrics give a better drape and handle properties compared to the normal denier filament fabrics (Behera et al., 1998).

The cross-sectional shape of the fiber affects the smoothness and bending of the yarn (Shanmugasundaram, 2008; Behery, 2005). It also determines how light interacts with the fiber. For example, a round fiber will appear more lustrous than a trilobal fiber made of the same polymer (Behery, 2005).

Another property that is important for fabric handle is the fiber friction. The fiber–fiber friction influences the way that the fibers interact with each other. The friction properties affect the flexibility of the yarns. As the fiber–fiber friction increases, the ability of the fibers to slide from each other during yarn and fabric deformation decreases (Behery, 2005).

Besides the typical mechanical methods to alter the handle of fabrics/fibers, handle can be improved by by chemical treatments. In treating the fiber surface, chemicals called 'softeners' are usually used. Softeners work by lubricating the surface of the fiber. This reduces the fiber–fiber friction, which makes the fabric move and flow more easily. Another method of changing fiber hand is to alter the chemical nature of the fibers themselves. A very common method is the mercerization of cellulose. As cellulose is mercerized, the overall shape of the fiber becomes more circular and more uniform than its irregular form thus becoming stronger and smoother to touch. With its round shape, mercerized cellulose becomes more lustrous (Behery, 2005).

Crystallinity also affects the handle of the fabrics by influencing the way that the fibers move and respond to bending. If the molecules in a fiber are aligned along the fiber axis, the fiber will be strong in uniaxial tension along the fiber axis. A more crystalline fiber is more resistant to bending (Behery, 2005).

2.2 Effect of yarns

The twist of the yarns which the fabrics are made of, is one of the main parameter affecting the fabric behaviour including bending, stiffness and shearing property (Shanmugasundaram, 2008). The amount of twist, together with the characteristics of the fibers (luster, hand, cross-sectional shape, etc.), determines the appearance and feel of the yarn. Fabrics composed of yarns with higher levels of twist are known to have higher bending stiffness, less compressibility, less fiber mobility, lower surface friction, less bulkiness than similar fabrics composed of yarns with less twist. Increased yarn twist leads to greater internal (fiber-to-fiber) friction within the yarn structure and reduce softness and bulkiness, in general, and hairiness in the case of spun yarns (Behery, 2005).

Another factor affecting handle is the number of the yarns folded. In plied yarns, i.e. two or more single yarns twisted together, the stiffness is increased compared to single yarns (Shanmugasundaram, 2008).

Filament yarns are sometimes put through an additional process known as texturizing. The process modifies the handle of the filament yarns by adding bulkiness and/or stretchiness to the filaments and therefore changes the smooth surface feel of fabrics (Behery, 2005). The feel of textured-yarn fabrics against the skin is considerably different than that of flat-yarn fabric. Textured yarns give a fabric more pleasant hand, fabric becomes warmer and softer and it has less synthetic feeling (Shanmugasundaram, 2008; as cited in Mahar & Pestle, 1983).

Fiber linearity and fiber-packing density in yarn structures are also important for the tactile qualities of a fabric, when not masked by twist (Behery, 2005). Comparing multifilament yarns of the same size and fiber composition, yarns containing more filaments (finer) are much less stiff than multifilament yarns containing less filaments (coarser). Fabrics produced from coarser filament textured yarns have higher thickness, compression characteristics than that of the fabrics made of finer filament textured yarns but there is no significant difference in fabric recovery and resiliency (Mukhopadyhay et al., 2002).

Yarn type has a significant effect on the handle properties of the fabrics. The static friction coefficient of cotton fabrics made of combed ring spun yarn is much lower than that for fabrics made of rotor yarn due to the more regular, denser and smoother structure of the ring spun yarn compared to the rotor yarn, on the surface of which there are visible wrapped fibers. The plain weft knitted fabric produced from the sirospun yarn is thicker, softer, of poorer recovery from compression, less rough and less stiff. For finer sirospun yarn, the shape retention is worse as the hysteresis of bending and shear is greater. (Sun & Cheng, 2000). Air-jet spun yarns produce thicker, less compressible, more extensible and stiffer fabrics than fabrics made from ring spun yarns (Behery, 2005; as cited in Vohs et al., 1985).

The fancy yarn structure also influences the subjective handle properties of the fabrics. In the production of fancy yarns, as the feeding rate increases; the fabric thickness, fabric weight, bending rigidity and friction coefficient of the fabric increase (Özdil et al., 2008).

2.3 Effect of fabrics

The handle of the fabrics is affected by mainly fabric structure and fabric geometry. Fabric construction and yarn densities play major role in determining fabric handle (Shanmugasundaram, 2008; Na & Kim, 2001). Variations in warp and weft densities and in the number of warp and weft yarns have significant effects on the handle characteristics of the fabrics (Shanmugasundaram, 2008). Sensory analysis shows that fabric hand can be influenced more by fabric weave than by the component yarn. Weaves that use fewer yarn interlacing improve the handle characteristics of the fabrics (Behery, 2005; as cited in Vohs et al., 1985).

Fabrics for which the warp and weft twist are unidirectional (Z & Z twist direction) perform higher shearing rigidity, (smooth rounded curvature) in comparison to those for which the warp and weft twist directions are opposite to each other (Yazdi & Özçelik, 2007).

The tightness factor of the knitted fabrics is another factor affecting the handle, with the increment of this factor, stiffness, which is related to bending rigidity, increase. Fullness, softness and smoothness decrease with the higher tightness factor (Park & Hwang, 2002).

Nonwoven fabrics differ from knitted or woven fabrics, because they are not based on webs of individual fibres, which can be bonded to each other by several means that changes the texture ranges from soft to harsh. Fiber composition influences performance far more for nonwoven fabrics than for fabrics containing yarns. High strength combined with softness is one of the most difficult property combinations to be achieved in nonwoven fabrics because the geometrical factors that permit high strength also lead to increased stiffness (Shanmugasundaram, 2008).

2.4 Effect of finishing

There are many researches in the literature related with the effects of finishing process on sensational properties of the fabrics. The diversity of fabric types with finishes available for any end-use continues to increase, making the selection of the most appropriate fabric an increasingly difficult task (Shanmugasundaram, 2008).

Different kind of end products can be produced from the same unfinished woven or knitted fabric by using various finishing treatments. High-speed scouring and milling create a fibrous surface as well as modifying other properties of the fabric, notably shear rigidity and specific volume. Heat treatments may cause fibres to crimp, increasing the bulkiness of the fabric. Soft handle and in some cases the 'peach skin' effect can be obtained by enzymatic treatment. By using cellulose enzyme, which acts on the fibril ends and causes their shortening, a slight pick-out surface can be obtained. Light brushing is another mechanical treatment that gives peach-skin effect to the fabrics. Raising and teaseling are the mechanical processes which draw fibres to the surface of a fabric to create a pile. Calendaring and also many chemical treatments (softening compounds, resins) give flatter surface that affect the fabric hand. Pressing and decatising are designed to flatten the fabric and create a smooth surface. Cropping and singeing are the processes which are designed to remove fibrous protrusions from the body of the fabric and thereby create a smooth surface. After the process of flame-retardant finishing, the stiffness parameters, especially bending and shear properties, increase significantly (Le et al., n.d., as cited in Stewart & Postle, 1974; as cited in Boos, 1988; Shanmugasundaram, 2008; Mamalis et al., 2001; Frydrych et al., 2002).

Pigment printing is the most effective finishing process on the stiffness of the fabrics and mercerization before dyeing and printing results in improvements in various fabric properties; however cause negative effect in the fabric softness values (Özgüney et al., 2009).

The effects of the stiffness and softening agents on the friction and stiffness properties of the fabrics are apparent. When compared to the untreated fabric, all the softening and stiffening treatments result in decrease in the both static and dynamic friction coefficient values. Especially for softeners, this decrease is obvious. Shirley stiffness values of the fabrics treated with stiffeners are quite higher compared to untreated one whereas all the fabrics treated with softeners have lower values. The most stiffening effect is obtained with starch (Namligöz et al, 2008).

3. Fabric basic properties affecting sensorial comfort

Fabric handle is related to the basic mechanical properties of fabrics, especially initial low-stress region of those properties (Mitsuo, 2006). Since the stresses involved in fabric handling are low compared with those applied in other types of textile performance testing (e.g. for ultimate tensile strength, tear strength, seam strength, etc), the methodology is sometimes referred to as measurement of "low-stress fabric mechanical and surface properties" (Bishop, 1996).

The stimulation of the feeling sensors greatly depend on the mechanical properties of the textile products, for instance a lower value of bending rigidity supports the positive impression of sensorial comfort (Ozcelik et al., 2005). The main mechanical and surface properties of fabrics that influence the sensorial properties of fabrics are tensile, bending, shearing and thickness (Figure 1).

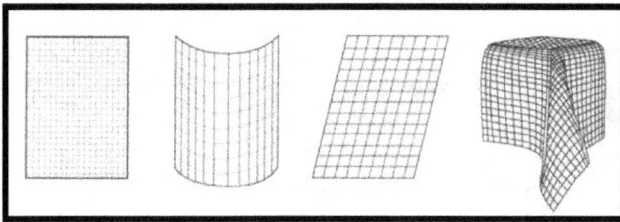

Fig. 1. The types of fabric deformation (Hu, 2004)

3.1 Tensile properties of woven fabrics

Tensile properties are one of the most important properties governing the fabric performance during usage. Each pieces of fabric consist of large quantity of fibers and yarns, and hence any slight deformation of the fabric will lead to a chain of complex movements among these constituent fibers and yarns (Hu, 2004).

There are three stages for the extension mechanism. The first part is dominated by inter fiber friction that is the frictional resistance due to the yarn bending. Second part, a region of lower modulus, is the decrimping region resulting from the straightening of the yarn set in the direction of application of load, with the associated increase in crimp in the direction perpendicular to the yarn direction. This is commonly referred to as "crimp interchange".

The last part of the load extension curve, is due to the yarn extension. As the crimp is decreased, the magnitude of the loading force rises very steeply, and as a result, the fibers themselves begin to be extended. This is clearly a region of higher modulus (Figure 2a).

If the fabric undergoes in a cycling loading process, the fabric is first stretched from zero stress to a maximum and the stress is fully released, then an unloading process follows the loading process. As a result, a residual strain ε_0 is observed, since textile materials are viscoelastic in nature. Due to the existence of residual strain, the recovery curve never return to the origin, as shown in Figure 2b. This is the hysteresis effect which denotes the energy lost during the loading and unloading cycle. Due to the existence of hysteresis, a deformed fabric cannot resume its original geometrical state (Hu, 2004; Schwartz, 2008).

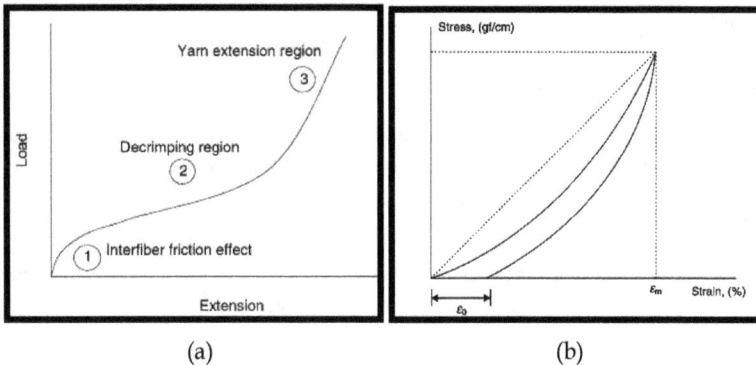

(a)   (b)

Fig. 2. (a) Schematic of a typical load- extension curve for a woven fabric (Schwartz, 2008), (b) Loading and unloading cycle in the tensile stress-strain curve (Hu, 2004)

3.2 Bending properties of woven fabrics

Bending properties of fabrics govern much of their performance, such as hang and drape, and are an essential parts of complex fabric deformation analysis. The bending properties of a fabrics are determined by yarn bending behavior, the weave of the fabric and the finishing treatment of the fabric, the relationship among them are highly complex (Schwartz, 2008). Two parameters that characterize the fabric bending behavior are its bending rigidity and bending hysteresis. Bending rigidity can be defined as the resistance of textile against flexion by its specific weight and external force. Bending hysteresis can be considered as a measure of fabric's ability to recover (Pavlinić & Geršak, 2003).

Figure 3 illustrates the typical bending curve for a woven fabric. For this curve, there are two stages with a hysteresis loop under low stress deformation. Firstly, there is an initial, higher stiffness non linear region (OA); within this region the curve shows that effective stiffness of the fabric decreases with increasing curvature from the zero motion position, as more and more of the constituent fibers are set in motion at the contact points. Secondly, a close to linear region (AB), since all the contact points are set in motion, the stiffness of the fabric seems to be close–to-constant (Schwartz, 2008).

In applications, where the fabrics are subjected to low curvature bending, such as in drapes, the hysteresis is attributed to the energy loss in overcoming the frictional forces. Under high curvature bending, the viscoelastic properties of the fibers must be considered (Schwartz, 2008).

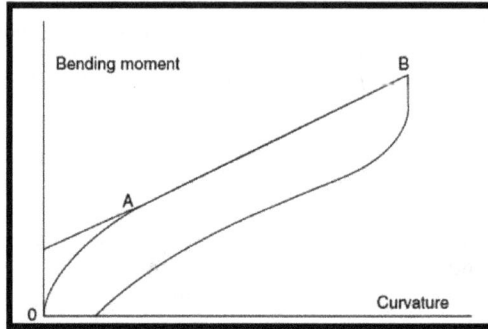

Fig. 3. Typical bending curve of woven fabrics (Schwartz, 2008)

3.3 Shear properties of woven fabrics

The shearing behaviour of a fabric determines its performance properties when subjected to a wide variety of complex deformations in use (Hu, 2004). The shear mechanism is one of the important properties influencing drapeness, pliability and handling of woven fabrics (Schwartz, 2008).

Shear deformation is very common during the wearing process since the fabric needs to be stretched or sheared to a greater or less degree as the body moves (Hu, 2004). This property enables fabric to undergo complex deformations and to conform to the shape of the body. The ability of a woven fabric to accept shear deformation is a necessary condition for a conformable fitting to a general three dimensional surface and is the bases for the success of woven textiles as clothing materials. This property is affected by the yarn characteristic and fabric structure (Yazdi & Ozcelik, 2007).

Shear hysteresis is defined as the force of friction occurring among interlacing points of the warp and weft yarns, when they are moving over each other, having their origin in the forces of stretching/shrinking, since the system of warp and weft yarns stretches/shrinks under strain (Pavlinić & Gerśak, 2003). Shear angle is one of the main criteria for characterizing the formability of fabrics. As the fabric is fitted onto a spatial surface, shearing occurs increasingly until the critical shearing angle is reached. When this angle exceeds a strict value, the specimen starts to buckle, i.e. wrinkling is observed (Domskienė & Strazdienė, 2005).

If a fabric is deformed at low levels of strain, the shear resistance is initial large and decreases with increasing strain (Figure 4). In this region the shear behaviour is dominated by frictional mechanism and the decreasing incremental stiffness is generally attributed to the sequential movement of frictional elements. As soon as the stress is large enough, to overcome the smallest of the frictional restraints acting at the intersection regions, the system starts to slip and the incremental stiffness falls (AB). At point B, the incremental stiffness reaches to the minimum level and remains almost linear over a range of amplitudes with slope that are called elastic elements in the fabric. At amplitudes greater than a certain amount C, the incremental stiffness again begins to rise and closed curves increase in width with increasing amplitudes of shear angle (Schwartz, 2008).

In the case of closely woven fabrics, there is not much slippage between warp and weft yarns under shearing strain, the result being just a higher friction between individual yarns. More loosely woven fabrics, with lower cover factor, exhibit lower friction between warp and weft yarns (Pavlinić & Gerśak, 2003). Shear deformation of woven fabrics also affects the bending and tensile properties of woven fabrics in various directions rather than in the warp and weft directions only (Hu 2004; as cited in Chapman, 1980; as cited in Skelton, 1976).

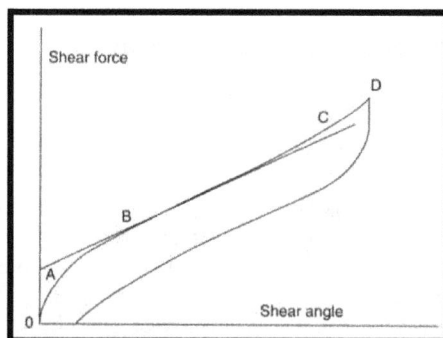

Fig. 4. Stress-strain curve of woven fabric during shear deformation (Schwartz, 2008)

3.4 Thickness and compression properties of woven fabrics

Thickness and compressional properties of the fabric are very important characteristics in terms of fabric handle, especially for the fabrics used in garment manufacture. Fabric compressional characteristics depend on several factors like the compressional properties of the constituent warp and weft threads and the structure of the fabric (Mukhopadyhay et al., 2002).

The thickness of a fabric is one of its basic properties giving information on its warmth, heaviness or stiffness in use. In practice thickness measurements are rarely used as they are very sensitive to the pressure used in the measurement (Saville, 1999).

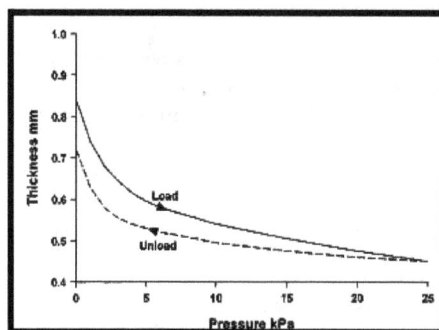

Fig. 5. The change in thickness with pressure (Saville, 1999)

Fabric thickness is generally evaluated by measuring the distance between two parallel plates separated by a fabric sample, with a known arbitrary pressure applied and maintained between the plates (Majumdar & Saha, 2008; as cited in BS Handbook, 1974).

Figure 5 shows the change in thickness with pressure for a soft fabric together with the recovery in thickness as the pressure is removed. The steep initial slope of the curve makes it very difficult to measure thickness with any accuracy as a small change in pressure in this region causes a large change in measured thickness. Thickness at zero pressure always has to be obtained by extrapolation of the curve, as a positive pressure is needed to bring any measuring instrument into contact with the fabric surface (Saville, 1999).

3.5 Other fabric properties related to sensorial comfort

It is evident from studies on the subjective evaluation of fabric handle that warm-cool feeling of a fabric very often makes an important contribution to the perception of its over-all handle and quality in the context of a given end use (Bishop, 1996). Alambeta instrument developed by Hes can be used for measuring the warm-cool feeling (Bishop, 1996; Özçelik et al., 2008; Mazzuchetti et al., 2008; Hes, 2004). This feeling, which is generated when fabric initially contacts the skin, is related to the heat flow between the skin and the contacted object. Rough fabric surface reduces the area of contact appreciably, and a smoother surface increases the area of contact and the heat flow, thereby creates a cooler feeling. The correlation between transient heat flux and the warm/cool feeling was investigated by several researchers (Barker, 2002; as cited in Rees, 1941; as cited in Hollies et al., 1953; as cited in Kawabata et al. ,1985). It has been reported that transient heat flux significantly affects clothing comfort in next-to-skin fabrics (Barker, 2002; as cited in Kawabata et al., 1985).

Prickle is a rather negative attribute associated particularly with fabrics containing a proportion of coarse animal fibres, having diameters of about 30 μm or more. When evaluating the handle and quality of wool containing fabrics, there has been obtained good correlation and therefore prickle characteristics of the fabrics should also be taken into consideration in objective handle measurements (Bishop, 1996). Although the luster and fuzzes of the fabric surface are not generally included in objective handle measurements, since the appearance of fabrics make an important contribution to consumers` garment preferences, the measurement of these properties should also be of interest in handle evaluation. The fabric noise and odour are not redundant in the subjective evaluation of textiles but these properties can be useful to evaluate fabric handle (Mitsuo, 2006).

Fabric handle according to fabric sound has been investigated in recent days. Fabric sound in the forms of sound spectra through Fast Fourier Transform analysis was analyzed (Mitsuo, 2006; as cited in Yi & Cho, 2000). Level pressure of total sound, level range and frequency differences of fabric sound were compared to mechanical parameters measured by KES system (Mitsuo, 2006; as cited Cho et al., 2001, 2005). The physiological responses evoked by friction sounds of warp knitted fabrics to that of electroencephalog, the ratio of high to low frequency were also studied (Mitsuo, 2006).

4. Sensorial comfort of fabrics

4.1 Subjective evaluation of fabric handle

Subjective assessment treats fabric hand as a psychological reaction obtained from the sense of touch (Bakar, 2004). Traditionally, in the textile and clothing industries, the assessment of the fabric handle is carried out subjectively by individual judges. The judgements strongly rely on personnel criteria (Yick et al., 1995).

When a person runs their finger across the surface of a fabric, a complex multi-sensory, emotional and cognitive experience takes place. A memory is stirred, an emotion, feeling and association is evoked and a decision is made, an impression becomes embossed in the mind. Decisions and motivations are based on anticipated reality of preference, personality, emotion and moods, for audience or non-audience participation (Moody et al., 2001).

The subjective hand is the result of touch sensation and therefore is dependent on the mechanisms of human tactile sensations. The somatic senses are those nervous system mechanisms by which sensory information is collected from within body. The somatic senses are classified to the three groups:

1. Mechanoreceptors – stimulated by mechanical displacement of various tissues at the body
2. Thermo receptors – stimulated by temperature changes
3. Nocio receptors – representing the human pain sense

It is then clear that subjective hand sensing is the combination of various receptors responsible for feeling of texture, pressure, stretching, thermal feedback, dynamic deformation and vibration and from the sum of their complex responses humans can perceive and discriminate between the textiles (Militky& Bajzík, 1997).

In the subjective assessment process of textiles, fabric hand is understood as a result of psychological reaction through the sense of touch. There are variations in how individuals actually feel textiles because people do not have the same sensory perception of identical occurrences. Affecting aspects can be grouped in sociological factors and the physiological factors.

The other main factors affecting the subjective handle evaluation can be defined as; the judges, the criteria of judgement, assessment conditions, assessment technique, the method of ranking and scaling the assessment, analysis of the results (Mahar et al., 1990; as cited in Brand, 1964; Bishop, 1996). Gender, age, education and cultural backgrounds are potential influencing factors. Female individuals in general respond more delicately and sensitively than male individuals and therefore have a finer assessment of a specific parameter (Mäkinen et al., 2005; as cited in Kweon et al., 2004; as cited in Dillon et al., 2000).

Since the services of expert judges are not widely available for research activities, generally students, laboratory assistants and other consumer groups work as panelists. Such panels should be capable of making consistent judgements of textile attributes and due to the high variability of these panelists compared to the expert judges, larger panel sizes should be preferred, at around 25-30 persons (Bishop, 1996).

In order to ensure the reliability of subjective assessments, it is critical to choose the right expressions for the description of a fabric handle parameter (Mäkinen et al., 2005). There are different sensory attributes identified by numerous authors. These are grouped as given in Table 1 (Hu, 2008). The words "thickness," "thinness," "softness," "stiffness," "slippery," "roughness," "tightness," "fullness" and "pliable" are mostly used expressions to describe the feel of a fabric (Sülar & Okur, 2007).

Generally, in order to evaluate the handle of the fabric, fingers are slid on the surface of the fabric, compressed between the thumb and sign finger. The fingers containing more than

250 sensors per cm² are the crucial important organs determining the fabric quality (Bensaid et al., 2006). Tightening of the fabric between fingers gives idea about thickness, bulkiness, compressibility, thermal absorptivity and surface properties of the fabrics, whereas slipping of the fingers on the surface of the fabrics with a pressure renders about structure and elongation of the fabrics (Aliouche & Viallier, 2000).

Stiffness/crispness/pliability/flexibility/limpness	Anti-drape/spread/fullness
Softness/harshness/hardness	Tensile deformation/ bending/surface friction/sheer
Thickness/bulkiness/sheerness/thinness	Compressibility
Weight/heaviness/lightness	Snugness/loosenes
Warmth/coolness/coldness (thermal characteristics)	Clinginess/flowing
Dampness/dryness/wetness/clamminess	Quietness/noisiness
Prickliness/scratchiness/roughness/coarseness/itchiness/tickliness/stickiness/	Smoothness/fineness/silkiness
Looseness/tightness	

Table 1. Sensory attributes of fabrics

There are generally four handle methods for the evaluation of fabric handle, as shown in Figure 6. The multiple finger pinch and the touch-stroke are the most relevant ones.

However, using just the index finger has also proved acceptable. Evaluated properties of the fabrics by these handle techniques are given in Table 2 (Moody et al., 2001; as cited in Dillon & Moody, 2000).

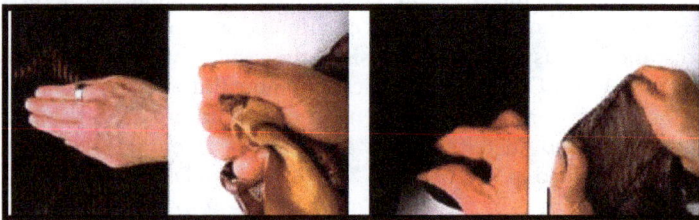

Fig. 6. Handle techniques (1. Touch stroke, 2.Rotating cupped, 3.Multiple finger, 4.Two handed rotation) (Moody et al., 2001)

	Handle Technique	Properties Evaluated
1	Touch-stroke	Surface quality (texture), temperature
2	Rotating cupped action	Stiffness, weight, temperature, comfort, overall texture, creasing
3	Multiple finger pinch: Rotating between the fingers action with one hand (thumb and 1 or 2 fingers)	Texture, stiffness, temperature, fabric structure, both sides of a fabric, friction, stretch (force-feedback)
4	Two handed rotation action	Stretch, sheerness

Table 2. Properties evaluated by different handle techniques (Moody et al., 2001)

The assessment conditions of subjective handle evaluation are critically important. Different skin hydrations of individuals affect notably the feel of a textile. A higher moisture level on the skin makes it more sensitive to the sense of touch (Mäkinen et al., 2005).

The evaluation is carried out in three different conditions: sight only, touch only and sight and touch together. During the assessment, in order to prevent the effect of colour and appearance on the assessments, wooden boxes with holes on the facing sides, through which the hands can easily go, can be used. Fabric samples are placed in these boxes prior to assessments. This helps the jury to assess the fabric without seeing it (Sülar &Okur, 2008b).

The first attempts of ranking and scaling the assessment methods in hand evaluation of textiles in an organized and quantitative manner were published as early as 1926 and have continued up to the present time and two basic procedures of subjective hand evaluation is proposed as follows: (Bakar, 2004; as cited in Howorth,1964):

a.   Direct method - is based on principle of sorting of individual textiles to defined subjective grade in ordinal scale (e.g., 0 - very poor, 1 - sufficient, 5 - very good, 6 - excellent).
b.   Comparative method - is based on sorting of textiles according to subjective criterion of evaluation (e.g., ordering from textiles with the most pleasant hand to textiles with the worst hand) (Bakar, 2004). If the number of samples is high, the second approach can be considered rather time consuming (Sülar & Okur, 2007).

It is preferable to use a paired comparison technique during assessment, the so-called bipolar pairs of sensory attributes, such as "thin/thick" or "soft/harsh" (Mahar et al., 1990; as cited in Brand, 1964). For the same reason, fabric hand attributes are measured on specific scales thus avoiding the intrinsic weakness of descriptive terminology. In case of using bipolar descriptors in the assessment of the handle, control fabrics are better to be used for the training of the panel members. Control fabrics are chosen according to the related objective properties. The test results for objective properties that are related to sensory attributes were sorted in ascending order for each parameter. The fabrics with minimum, maximum and medium values are determined and used for the initial tests for the selection of the control fabrics (Sülar & Okur, 2007).

It is crucially important to convert the subjective assessment results to the numerical values for finding a relationship between objective measurements to analyze statistical evaluation. Therefore, using these types of ranking scale is preferable (Table 3).

1	...	5	...	10
thinnest	...	medium	...	thickest
1	...	5	...	10
softest	...	medium	...	stiffest
1	...	5	...	10
smoothest	...	medium	...	rougest
1	...	5	...	10
proper	...	medium	...	most proper

Table 3. Handle components and the rating scale (Sülar & Okur, 2008a)

This subjective hand evaluation system requires years of experience and can obviously be influenced by the personal preferences of the assessor as mentioned before. A fabric may be felt light, soft, mellow, smooth, crisp, heavy, harsh, rough, furry, fuzzy or downy soft. So there is a need to replace the subjective assessment of fabrics by experts with an objective machine-based system which will give consistent and reproducible results (Hu, 2008).

4.2 Objective evaluation of fabric handle

For a long time, handle has been estimated by the organoleptic method. The producers and users of textile products try to formulate in words the impression of touching the flat textile product. In general, fabric hand is primarily assessed subjectively in a few minutes. Although this is a fast and convenient sort of quality control, the subjective nature of fabric handle leads to serious variations in quality assessment (Sülar & Okur, 2008a) and does not analyze the core of the problem connected with the influence of factors creating the particular sensations. This was why in the 1930s investigations were commenced into an objective measurement of the features which are decisive for handle. The common goal in objective measurement systems was to eliminate the human element in hand assessment and develop quantitative factors that could be measured in a laboratory (Kocik et al., 2005).

Peirce was a forerunner of such investigations with his works connected with determining the bending rigidity and compressibility of flat textile products (Kocik et al., 2005; as cited in Pierce, 1930). At the turn of the 1960s, researchers from the Swedish Textile Institute (TEFO) (Kocik et al., 2005; as cited in Eeg-Olofsson, 1957; as cited in Eeg-Olofsson, 1959; as cited in Olofsson, 1965; as cited in Olofsson & Ogucki; 1966; as cited in Lindberg et al., 1961; as cited in Lindberg et al., 1960) carried out intensive investigations into this matter. These research works led to determining the dependencies between the features of flat textile products subjected to bending, buckling, shearing, and compressing, and the susceptibility of these products to manufacturing clothing. Lindberg (Kocik et al., 2005; as cited in Lindberg et al., 1960) was the first researcher who applied the theory of buckling for estimating the behaviour of fabrics in the clothing manufacturing process. Kawabata and Niwa were followers of Peirce and the Swedish researchers, who since 1968 have conducted research into handle. These investigations have been crowned by the design and construction of a measuring system which serves for objective estimation of handle (Kocik et al., 2005; as cited in Kawabata et al., 1973; as cited in Kawabata et al., 1996).

Objective assessment attempts to find the relationships between fabric hand and some physical or mechanical properties of a fabric objectively. It quantitatively describes fabric hand by using translation result from some measured values of relevant attributes of a fabric. Techniques used for objective hand evaluations are based on special instruments for measuring handle related properties (Bakar, 2004).

Several attempts have been made to measure fabric handle properties objectively described simply as "Fabric Objective Measurement (FOM)", and also a number of items of equipment have been introduced for this purpose (Hasani & Planck, 2009; Bishop, 1996).

4.2.1 Objective measurement systems

The KES-F system (Kawabata's Hand Evaluation System for Fabrics) was developed in Japan by the Hand Evaluation and Standardization Committee (HESC, established in 1972) organized by Professor Kawabata. In this fabric objective measurement method, scientific principles are applied to the instrumental measurement and fabric low stress mechanical and surface properties such as fabric extension, shear, bending, compression, surface friction and roughness are measured. The fabric handle is calculated from measurements of these properties. Empirical equations for calculating primary hand values and total hand values were put forward by Kawabata and Niwa (Mäkinen et al., 2005; as cited in Kawabata, 1980; as cited in Shishoo, 2000).

The process of the subjective evaluation according to Kawabata can be given as follows (Bona, 1994):

Touch of fabric by hand	Detection of fabric basic mechanical properties such as bending, stiffness, etc.	Summarized expressions → about fabric characters by "primary hand"	Overall →judgement of fabric quality
Physiological sensing data processing in man's brain			

The first part of Kawabata`s work was to find the important aspects of handle and the contribution of each to the overall rating of the fabric. For each category such as stiffness, smoothness, etc. were identified and the title of primary hand values were give. The original Japanese terms of these primary hand definitions together with English meanings are given in Table 4. The primary hand values are combined to give an overall rating for the fabric categories such as man's summer suiting, man's winter suiting, lady's thin dress, and man's dress short and knitted fabrics for undershirts. The conversion of the primary hand values is done by using a translation equation for a particular fabric category determined empirically. This total hand value is rated on a five point scale, where five is the best rating (Kawabata, 1980).

The second stage of Kawabata`s work was to produce a set of instruments with which to measure the appropriate fabric properties and then to correlate these measurements with the subjective assessment of handle. The aim was that the system would then enable any operator to measure reproducibility the total hand value of a fabric (Saville, 1999).

The Kawabata Evaluation System for Fabric (KES-F) which has been widely used since the 1970's consists of four specialized instruments: FB1 for tensile and shearing, FB2 for bending, FB3 for compression and FB4 for surface friction and variation. A total of 16 parameters are measured at low levels of force (Table 5). The measurements are intended to simulate the fabric deformations found in use (Hu, 2008; Chen et al., 2001).

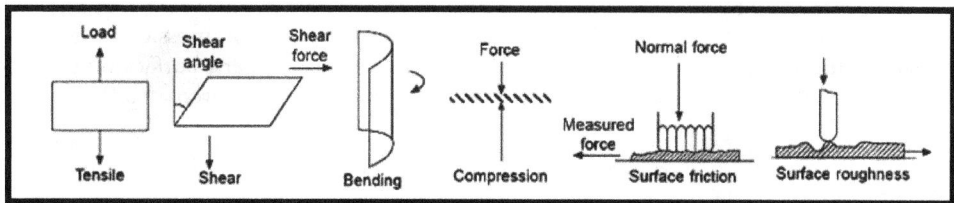

Fig. 7. Measuring principles of the KES system

Hand		Definition
Japanese	English	
Koshi	Stiffness	A stiff feeling from bending property. Springy property promotes this feeling. High-density fabrics made by springy and elastic yarn usually possess this feeling strongly.
Numeri	Smoothness	A mixed feeling come from smooth and soft feeling. The fabric woven from cashmere fiber gives this feeling strongly.
Fukurami	Fullness and softness	A bulky, rich and well-formed feeling. Springy property in compression and the thickness accompanied with warm feeling are closely related with this feeling (*fukurami* means 'swelling').
Shari	Crispness	A feeling of a crisp and rough surface of fabric. This feeling is brought by hard and strongly twisted yarn. This gives a cool feeling. This word means crisp, dry and sharp sound made by rubbing the fabric surface with itself).
Hari	Anti-drape stiffness	Anti-drape stiffness, no matter whether the fabric is springy or not. (This word means 'spread').
Kishimi	Scrooping feeling	Scrooping feeling. A kind of silk fabric possesses this feeling strongly.
Shinayakasa	Flexibility with soft feeling	Soft, flexible and smooth feeling.
Sofutosa	Soft touch	Soft feeling. A mixed feeling of bulky, flexible and smooth feeling.

Table 4. The definitions of primary hand (Kawabata, 1980)

The characteristic values are calculated from recorded curves obtained from each tester both in warp and weft direction. Tensile properties (force-strain curve) and shear properties (force-angle curve) are measured by the same apparatus. Bending properties (torque-angle curve) are measured by bending first reverse sides against each other and after that the face sides against each other. Pressure-thickness curves are obtained by compression tester. The measurements of surface friction (friction coefficient variation curve) and surface roughness (thickness variation curve) are made with the same apparatus using different detectors.

The tensile properties are measured by plotting the force extension curve between zero and a maximum force of 500 gf/cm, the recovery curve as the sample is allowed to return to its original length is also plotted to give the pair of curves shown in Figure 8a. From these curves the following values are calculated (Saville, 1999):

Tensile energy WT = the area under the load strain curve (load increasing)

Linearity $LT=WT$/area triangle OAB

Resilience RT=area under load decreasing curve /WT x 100

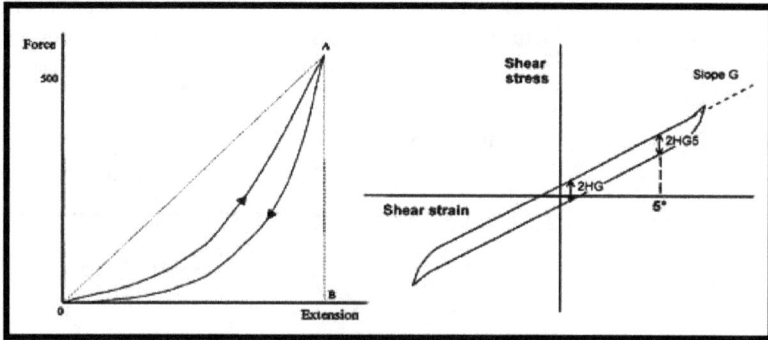

Fig. 8. (a) Load extension recovery curve (b) Hysteresis curve for shear (Saville, 1999)

Characteristic values measured in KES-F system			
		LT	Linearity of load-extension curve
	Tensile	WT	Tensile energy
KES- FB1		RT	Tensile resilience
		G	Shear rigidity
	Shearing	2HG	Hysteresis of shear force at 0.5° shear angle
		2HG5	Hysteresis of shear force at 5° shear angle
KES- FB2	Bending	B	Bending rigidity
		2HB	Hysteresis of bending moment
		LC	Linearity of pressure-thickness curve
KES- FB3	Compression	WC	Compressional energy
		RC	Compressional resilience
		MIU	Coefficient of friction
KES- FB4	Surface	MMD	Mean deviation of MIU, frictional roughness
		SMD	Geometrical roughness
Fabric	Weight	W	Weight per unit area
construction	Thickness	T	Thickness at 0.5 gf/cm²

Table 5. Characteristic values in KES-F system (Mäkinen et al., 2005; as cited in Kawabata, 1980)

In order to measure the shear properties, a sample in dimensions of 5cm x 20cm is sheared parallel to its long axis keeping a constant tension of 10 gf/cm on the clamp. The following quantities are then measured from the curve as shown in Figure 8b.

Shear stiffness G = slope of shear force-shear strain curve

Force hysteresis at shear angle of 0.5° 2HG = hysteresis width of curve at 0.5°

Force hysteresis at shear angle of 5° 2HG5 = hysteresis width of curve at 5°

In order to measure the bending properties of the fabric, the sample is bent between the curvatures -2.5 and 2.5 cm^{-1}, the radius of the bend is the reciprocal of the curvature as shown in Figure 9a. The bending moment required to give this curvature is continuously monitored to give the curve as shown in Figure 9b (Saville, 1999).

Compressibility is one of the most important properties in terms of fabric handle for the fabrics used in garment manufacture (Mukhopadyhay et al., 2002). The compression test for fabric is used to determine the fabric thickness at selected loads, and reflects the 'fullness' of a fabric (Hu, 2008).

The compression energy, compressibility, resilience and thickness of a specimen can be obtained by placing the sample between two plates and increasing the pressure while

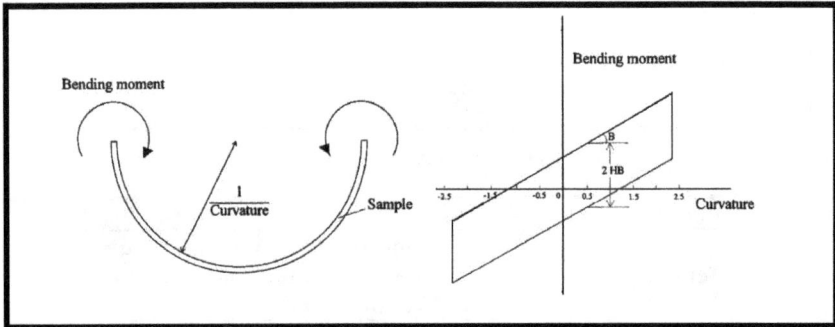

Fig. 9. a) Forces involving in fabric bending; b) Plot of bending moment against curvature (Saville, 1999)

continuously monitoring the sample thickness up to a maximum pressure of 50 gf/cm^2. A circular compressing board of 2 cm^2 attached with a sensor is used to apply the force on the fabric specimen (Figure 10) (Saville, 1999).

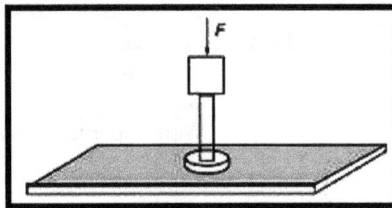

Fig. 10. Compression test on the KES-F system (Hu, 2008)

The surface friction is measured in a similar way by using a contactor which consists of ten pieces of the same wire as used in the surface roughness. A contact force of 50 gf is used in this case and the force required to pull the fabric past the contactor is measured. For the surface roughness, the contact force that the wire makes with the surface is 10gf (Chen et al., 2001).

Kawabata developed through extensive human subjective evaluations of a range of fabric types and the ranking of characteristics. The weighing factors are believed to be appropriate for the population within which the data were taken but there is some question as to the application of the same weighing factors in a different culture (Adanur, 2001). Critics still exist due to the high cost of the instrument. The system also requires experts for the interpretation of the resulting data. These deficiencies led to the development of another testing device called the FAST (Hu, 2004).

The Australian CSIRO designed and developed the FAST (Fabric Assurance by Simple Testing) set of instruments, as a simpler alternative to a KES system, which in terms of practicality and testing speed, go a long way towards meeting the requirements of garment makers, finishers and is designed to be relatively inexpensive, reliable, accurate, robust and simple to operate. Unlike the KES-F system, FAST only measures the resistance of fabric to deformation and not the recovery of fabric from deformation (Shishoo, 1995; Behery, 2005; Mazzuchetti et al., 2008; Potluri et al., 1995).

FAST gives similar information on the aesthetic characteristics of fabric as KES-F does, but in a simple manner, and is more suited to a mill environment. The FAST system includes FAST-1 for thickness, FAST-2 for bending, FAST-3 for extensibility and FAST-4 for dimensional stability. Through the objective measurements of fabric and a data set on a chart or 'fingerprint', manufacturers can identify fabric faults, predict the consequences of those faults and identify re-finishing routes or changes in production (Hu, 2008).

Characteristics measured in FAST system		Symbol	Unit	Device
Fabric weight		W	g/m^2	
Compression	Total thickness	ST	mm	FAST-1
	Surface thickness		mm	
Bending	Bending length	B		FAST-2
Tensile	Warp elongation		%	
	Weft elongation	E	%	FAST-3
	Crosswise elongation		%	
Dimensional stability	Relaxation shrinkage	RS	%	FAST-4
	Hygral expansion	HE	%	

Table 6. List of fabric properties measured using FAST (Saville, 1999)

FAST-1 is a compression meter enabling the measurement of fabric thickness and surface thickness at two predetermined loads (Hu, 2004). The fabric thickness is measured on a 10 cm^2 area at two different pressures, firstly at 2 gf/cm^2 and then at 100 gf/cm^2. This gives a measure of the thickness of the surface layer which is defined as the difference between these two values (Figure 11a). The fabric is considered to consist of an incompressible core and a compressible surface (Saville, 1999).

FAST-2 is a bending meter, which measures the bending length of the fabric. From this measurement, the bending rigidity of the fabric can be calculated. The instrument uses the cantilever bending principle described in BS: 3356. However, in FAST-2 the edge of the fabric is detected using a photocell. The bending rigidity, which is related to the perceived stiffness, is calculated from the bending length and mass/unit area. (Saville, 1999; Hu, 2004).

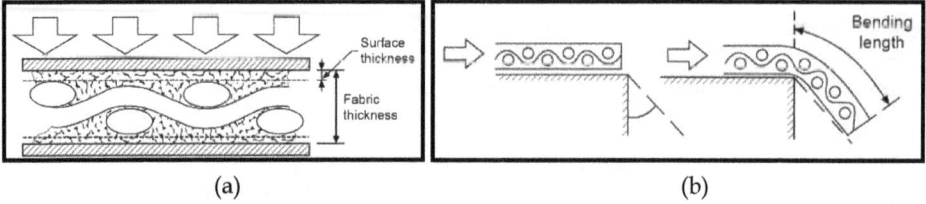

(a)   (b)

Fig. 11. (a) Measuring principle of the FAST-1 compression meter; (b) Measuring principle of the FAST-2 bending meter (Hu, 2004)

FAST-3 is an extension meter which operates on a simpler principle as shown in Figure 12a (Hu, 2004). The extension of the fabric is measured in the warp and weft directions at three fixed forces of 5, 20 and 100 gf/cm (sample size tested 100mm x 50mm). The extension is also measured on the bias in both directions but only at a force of 5gf/cm, this enables the shear rigidity to be calculated (Saville, 1999).

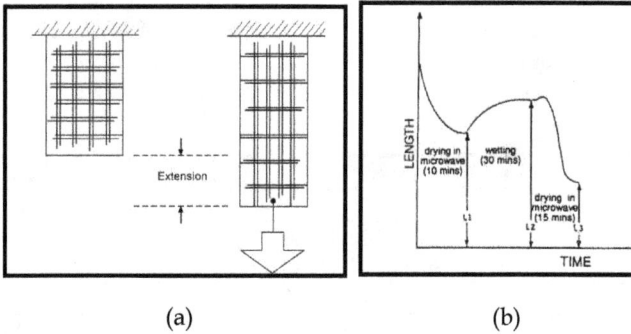

(a)   (b)

Fig. 12. (a) Measuring principle of the FAST-3 extension meter (Hu, 2004); (b) Dimensional stability curve (Bona, 1994)

The final component of FAST is a test method which measures the changes in the dimensions of fabrics that occur when the fabric is exposed to changing environmental conditions (Hu, 2004). A small amount of shrinkage (usually below 1%) is required for fabrics intended to be pleated. In order to measure dimensional stability the fabric is dried in an oven at 105°C and measured in both warp and weft directions to give the length L_1. It is then soaked in water and measured wet to give the wet relaxed length L_2. It is then re-dried in the oven and measured again to give the length L_3. The following values for dimensional stability are then calculated from these measurements for both warp and weft.

$$\text{Relaxation shrinkage} = \frac{L_1 - L_3}{L_1} x100(\%) \quad \text{Hygral expansion} = \frac{L_2 - L_3}{L_3} x100(\%)$$

Since the sensation is related to physical properties of the material, physical measurements constitute significant data in terms of objective evaluation. Disadvantages of the complex measuring systems such as high costs, difficulties in maintenance and reparation have resulted in conducting studies on improving simpler and individual instruments for each handle related objective fabric properties (Ozcelik et al., 2008).

4.2.2 Individual objective measurement testers

Shirley stiffness tester and circular bending rigidity tester for bending properties, cusick drape meter and sharp corner drape meter for drape properties, universal tensile testers for tensile and shear properties, thickness gauges for thickness and compression properties, universal surface tester and Frictorq for friction properties can be listed as commonly used simpler devices for measuring handle related properties of textile materials. Fabric extraction method and devices such as Griff-Tester (Kim & Slaten, 1999; Strazdienė & Gutauskas, 2005), robotic handling systems (Potluri et al., 1995) and various individual devices are some of the other objective measurement systems (Özçelik et al., 2008).

Cantilever stiffness tester supplies an easy way for measuring the fabric stiffness (Figure 13a). In the test, a horizontal strip of fabric is slid at a specified rate in a direction parallel to its long dimension, until its leading edge projects from the edge of a horizontal surface. The length of the overhang is measured when the tip of the specimen is depressed under its own mass to the point where the line joining the top to the edge of the platform makes a 41.5° angle with the horizontal. It is known as bending length (Figure 13b) and from this measured length, the flexural rigidity is calculated by using the formula given below (ASTM D 1388).

$G = 1.421 \times 10^{-5} \times W \times c^3$; where: G = flexural rigidity (μjoule/m), W = fabric mass per unit area (g/cm²) and c = bending length (mm).

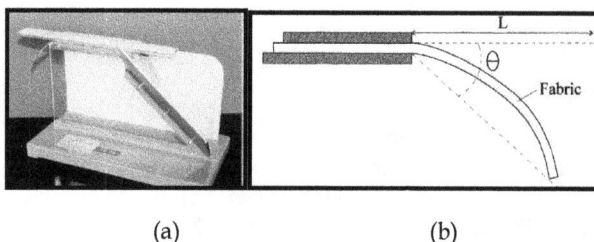

(a)   (b)

Fig. 13. (a) Cantilever stiffness tester, (b) Bending length (Saville, 1999)

Ring Loop: $l_0 = 0.3183\,L$ $\theta = 157°$ (d/l_0)
Bending length $C = L\,0.133\,f_2(\theta)$

Pear loop: $l_0 = 0.4243\,L$ $\theta = 504.5°$ (d/l_0)
Bending length $C = L\,0.133\,f_2(\theta)\,/\cos 0.87\,(\theta)$

Heart loop: $l_0 = 0.1337L$ $\theta = 32.85°$ (d/l_0)
Bending length $C = l_0\,f_2(\theta)$ $f_2(\theta) = (\cos\theta/\tan\theta)^{1/3}$

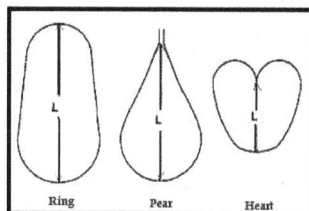

Fig. 14. Different shapes of hanging loops (Saville, 1999)

The cantilever method is not suitable for the fabrics that are too limp or show a marked tendency to curl or twist at a cut edge. The heart loop test can be used for these fabric types. A strip of fabric is formed into a heart-shaped loop. The length of the loop is measured when it is hanging vertically under its own mass (ASTM D 1388–08). The undistorted length of the loop l_o, from the grip to the lowest point is calculated (Saville, 1999; as cited in Peirce, 1930) for three different loop shapes: the ring, pear and heart shapes. If the actual length l of the loop hanging under its own weight is measured, the stiffness can be calculated from the difference between the calculated and measured lengths $d = l - l_0$.

Another instrument which has the same working principle with Kawabata KES-F system is TH 7 bending rigidity tester. The instrument has clamp, which firstly rotates 90° to the front and after that comes to the starting point and moves 60° to the backwards. The required forces to bend the sample in different angles are recorded (Ozcelik & Mertova, 2005).

In circular bending rigidity test, that gives fabric stiffness in all direction, a plunger forces a flat, folded swatch of fabric through an orifice in a platform (Figure 15). The maximum force required to push the fabric through the orifice is an indication of the fabric stiffness (resistance to bending). The circular bend procedure gives a force value related to fabric stiffness (ASTM D 4032 – 08).

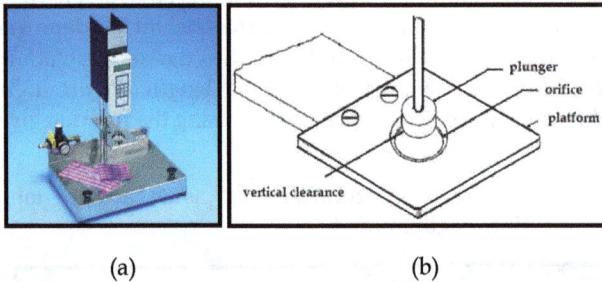

(a)   (b)

Fig. 15. (a) Circular bending rigidity tester (www.sdlatlas.com), (b) Platform and plunger of the tester (ASTM D 4032 – 08)

Drape is the term used to describe the way a fabric hangs under its own weight (Saville, 1999). Basically, fabric drape is not an independent fabric property. It relates to fabric bending, shear, tensile, fabric thickness and fabric weight (Hu, 2004; as cited in Niwa & Seto, 1986; as cited in Collier, 1991; as cited in Hu & Chan, 1998).

In cusick drape meter, the specimen deforms with multi-directional curvature and consequently the results are dependent to a certain amount upon the shear and bending stiffness properties of the fabric. In the test, a circular specimen is held concentrically between two smaller horizontal discs and is allowed to drape into folds under own weight (Saville, 1999). A light is shone from underneath the specimen as shown in Figure 16a and a fabric drape profile can be captured in a two dimensional image by using a digital camera (Figure 16b). The drape profile can be observed from the computer screen and drape coefficient can be calculated by using image analysis software. The stiffer fabric means that the area of its shadow is larger compared to the unsupported area of the fabric so the higher the drape coefficient is. It is considered that the drape coefficient by itself is not sufficient for the drape characteristic of a fabric (Stylios & Powell, 2003; as cited in Stylios & Zhu, 1997) and therefore a feature vector, consisting of the average minima and average maxima fold lengths and the evenness of the folds is defined (Stylios & Powell, 2003; as cited in Ballard & Brown, 1982).

Measurement of drape angle by means of a special tool (table) is carried out by moving this sample towards the sharp corner of the table, in such way that the axis of the 90°

angle coincides with the warp or weft direction. The fabric motion stops, when the peak of the corner reaches to the center of the sample. Then the fabric folds and forms a direct edge, whose inclination φ against the horizontal plane measured. The sin φ value measured by means of simple ruler (Figure 16c), then characterizes the level of drape (Hes, 2009). The fabric becomes harder as the drape angle gets smaller (Ozcelik et al., 2008; Hes, 2004).

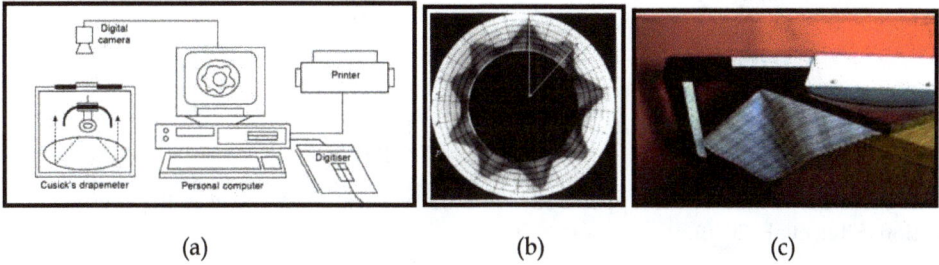

(a)   (b)   (c)

Fig. 16. The set up for the measurement of fabric drape profile: (a) Image analysis system, (b) Captured image on the drapemeter (Hu, 2004), (c) Sharp corner tester (Hes, 2004)

Friction coefficient is not an inherent characteristic of a material or surface, but results from the contact between two surfaces (Lima et al., 2005; as cited in Bueno et al., 1998). Two main ways are generally used to measure fabric friction. In one of these methods, as shown in Figure 17a, a block of mass (m) is pulled over a flat rigid surface, which is covered with the fabric being tested. The line connected to the block is led around a frictionless pulley and connected to an appropriate load cell in a tensile testing machine. This can measure the force (F) required both to start the block moving and also to keep it moving, thus providing static and dynamic coefficients of friction from the relation: Coefficient of friction $\mu = F / (m.g)$ (Figure 17a).

The second method used for measuring fabric friction is the inclined plane as shown in Figure 17b. The apparatus is arranged so that the angle of the plane can be continuously adjusted until the block begins to slide. At this point, the frictional force (F) is equal to component of the mass of the block parallel to the inclined plane (Saville, 1999).

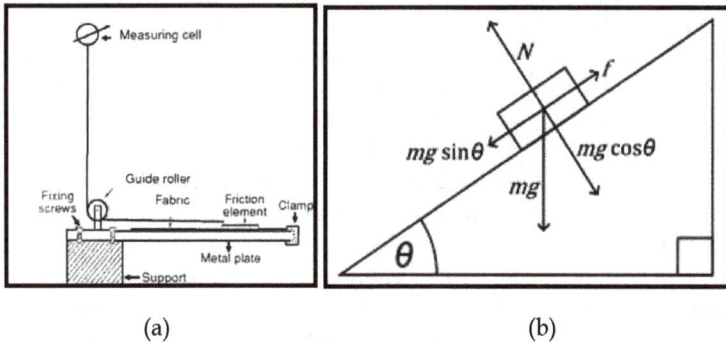

(a)   (b)

Fig. 17. Basic fabric friction measurement methods (Bona, 1994)

Frictorq is based on a new method to measure the coefficient of friction of the fabrics, using a rotary principle and, therefore, measuring torque. The upper body is a specially designed contact element, restricted to 3 small pads with an approximately square shape (covered by a number of calibrated steel needles), and placed over the fabric sample. This upper body is forced to rotate around a vertical axis at a constant angular velocity. Friction coefficient is again proportional to the torque measured with a precision torque sensor (Silva et al., 2010).

$$T = 3.F_a.r \, ,$$

$$F_a = \mu.N \, ,$$

$$N = P/3 \text{ and } \mu = T/ (P.r)$$

where, r is the radius of the upper body, P is the vertical load and μ is the coefficient of friction (Silva et al., 2010).

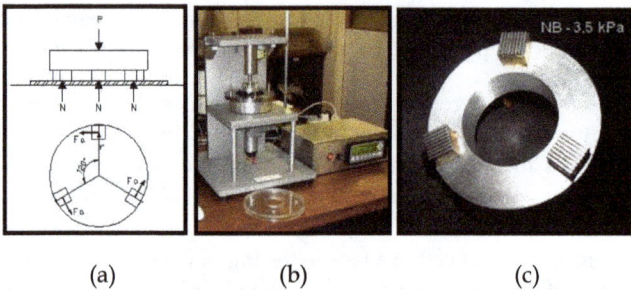

(a)   (b)   (c)

Fig. 18. (a) Loads in the measurement unit, (b) Frictorq instrument, (c) The upper body with 3 small pads (Silva et al., 2010)

Haptics, which derive from the Greek word haptesthai, means to touch and refers to simulate the feel of touch in the computer interface area (Govindaraj et al., 2003a). The other touch feedback systems do not have the sensitivity required for accurate simulation of fabric hand. PhilaU Haptic Device was developed to meet these requirements. During the development stage, the device called PHANToM® that uses a pen like probe to scan a virtual surface and generate the feel of surface, was used. By holding a pen with a stylus and moving the pen over a constructed surface in the virtual space, a feed back response was felt on the hand. The limitation of the device was that the contact with the virtual surface was over a line. However, it was possible to gain considerable information about a surface by moving a pencil-point across the surface, therefore it did not provide a tactile feeling (Govindaraj et al., 2003b, as cited in Katz, 1925). In order to overcome this limitation, the PhilaU Haptic Device (Figure 19) was designed as a combination force feed back and a tactile display. The device consists of a feeler pad at the end of an articulated arm joints, which is equipped with magnetic brakes, apply a force feed back to the hand holding the feeler pad assembly. The magnetic brakes get their input voltage proportional to surface friction of the fabric, while the tactile pins follow the contour. Together, the device provides a virtual fabric touch and feels (Govindaraj et al., 2003b).

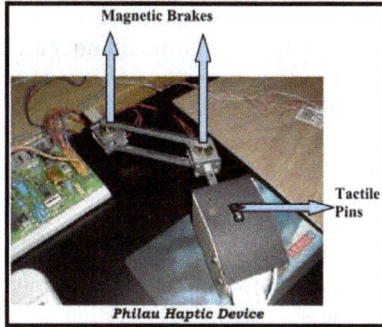

Fig. 19. PhilaU Haptic Device (Govindaraj et al, 2003a)

A robotic system developed by Potluri et al., designed for conducting all the fabric tests on a single sample, without operator intervention can be computed fabric properties such as tensile energy, shear stiffness, bending stiffness and compression energy. Uniform pressure is applied on the fabric sample by a manipulating device, attached magnetically to the robot arm, to avoid possible shear distortion or shear buckling (Potluri et al., 1995).

Several researches have been conducted for measuring the handle related mechanical properties of the fabrics by using universal tensile testers. The comprehensive handle evaluation system for fabrics and yarns (CHES-FY) is a kind of apparatus that is capable of measuring mass, bending, friction and tensile behavior just through one pulling-out test, and is able to characterize the handle of fabrics (Figure 20). The shape of a hung fabric was captured by a digital camera, and its weight was calculated. Then, a three-point bending in principle was utilized to model and analyze the bending properties of the fabric, and the corresponding formula was obtained for calculating the bending rigidity of the fabric (Du & Yu, 2007).

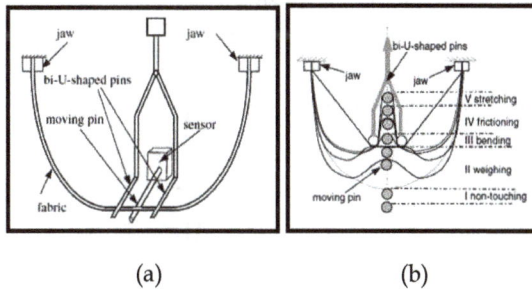

(a)   (b)

Fig. 20. (a) A schematic structure and (b) separated extraction steps of the CHES-FY (Du & Yu, 2007)

An alternative simple approach has been investigated by many researchers in order to predict fabric handle from the properties of fabric extraction through a ring or orifice (Wang et al., 2011; as cited in Alley & McHatton, 1978; Kim & Slaten, 1999). Extraction method is based on holding the sample at its centre and then pulling it through a ring of appropriate diameter by using a tensile tester (Yazdi & Shahbazi, 2006; as cited in Grover, 1993). For a properly designed nozzle, if the fabric extraction process is carefully examined, it will be found that during the process the sample is deformed under a very complex low stress state including tensile, shearing and bending as well as frictional actions, similar to the stress state, when handling the fabric (Figure 21a) (Pan, 2006). The behaviour of the fabric during testing is recorded on the load-elongation chart of the tensile testing machine (Yazdi & Shahbazi, 2006). Consequently, all the information related to fabric hand is reflected by the resulting load–displacement extraction curve.

A universal test unit (KTU-Griff-Tester) (Figure 21b) is recently developed as textile hand evaluation method based on pulling a disc-shaped specimen through a rounded hole operating together with either the standard tensile testing machine or an individual drive (Strazdienė & Gutauskas, 2005; as cited in Grover et al., 1993). It allows registration of the specimen pulling force-deflection curve and capturing of the shape variation images of the specimen (Strazdienė & Gutauskas, 2005).

Previously conducted researchers have used only one feature of the curve, e.g. the peak or the slope at a point (Pan, 2007; as cited in Alley, 1976), and discarded the rest of the information (Pan, 2006; as cited in Pan &Yen, 1984, 1992). The PhabrOmeter™ Fabric Evaluation System based on the research by Pan and his co-workers (Pan, 2006; Pan et al., 1993) was introduced. When compared to the KESF and FAST systems, the PhabrOmeter system uses a single instrument capable of testing the low-stress mechanical and physical properties of the fabrics related to the fabric handle. The objective data, obtained from extraction curves are sagging of unloaded fabric across orifice, slope of incline, height of curve peak, position deflection at peak height, post-peak height, width of peak, slope of decline, deflection post-peak height, work area underneath the curve within the triangle obtained from the PhabrOmeter tester (Figure 22). By using these objective parameters, a series of multiple linear regression models are developed and successfully validated to predict eight handle characteristics considered important for the handle of next-to-skin fabrics such as overall handle and primary handle characteristics, such as rough–smooth, hard–soft, loose–tight, hairy–clean, warm–cool and greasy-dry (Wang et al., 2011).

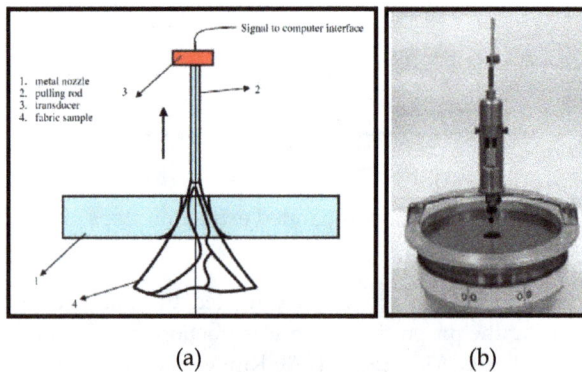

(a)   (b)

Fig. 21. (a) The fabric extraction technique (Pan, 2006) (b) KTU-Griff-Tester clamping device (Strazdienė & Gutauskas, 2005; as cited in Strazdiene et al., 2002)

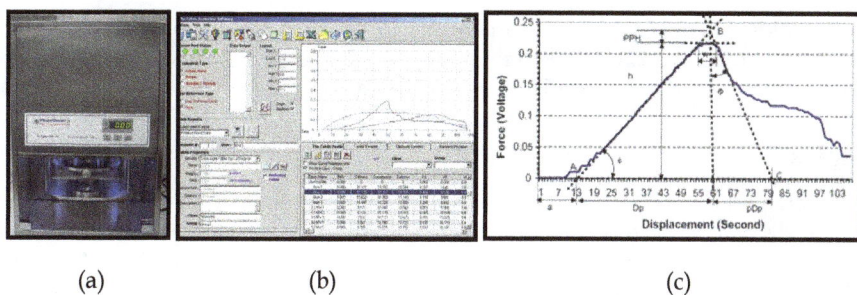

(a)   (b)   (c)

Fig. 22. (a) Hardware of PhabrOmeter model 3, (b) The user interface, (c) Extraction curve (Wang et al., 2011)

5. The relationship between the subjective evaluation and objective measurement of fabric handle

The subjective evaluation of fabrics leads to a set of linguistic terms strongly related to consumer preference but difficult to be quantized. It depends on many elements from raw materials to finishing processes. However, this evaluation restricts the scientific understanding of fabric performance for those who wish to design high-quality fabrics by engineering means. In the industry, the subjective evaluation is one of the main causes of conflict between producers and consumers on quality of products. Therefore, it is necessary to develop a normalized criterion representing the subjective evaluation or to replace it by an objective evaluation method. From any existing method of objective fabric evaluation, a set of precise quantitative data describing the fabric hand can be obtained but their relationship with the subjective evaluation is not completely discovered. Research has been done for modeling this relationship (Zeng & Koehl, 2003; as cited in Kawabata, 1996; as cited in Hu, 1993). However, progress in this field is rather slow because of the existence of uncertainties and imprecision in subjective linguistic expressions and the lack of mathematical models that constitute a nonlinear complex system for explaining the relationship between subjective and objective data, that, where no mathematical models are available (Zeng & Koehl, 2003).

Numerous methods such as Steven`s law, rank correlation, linear regression model, multiple-factor analysis, weighted euclidean distance, component analysis, decision and information theory, canonical correlation methods and as intelligent techniques fuzzy logic–based methods, neural network statistical models and mathematical models have been introduced for the generation of a quantitative criterion characterizing the quality of textile products and modeling relationships between the subjective fabric hand evaluation and objective numerical data. Since all these methods require tedious computations and are thus inappropriate for providing quick responses to consumers, in recent works fuzzy comprehension evaluation, neural network aggregation of data, classification methods are widely used. Advantages of these techniques can be stated as computing with numerical data and words, computing with uncertainty and imprecision, taking into account nonlinear correlation, computing with few numbers of data (Bishop, 1996; Hui et al., 2004; Bakar, 2004).

The modeling and the simulation of textile fabrics represent an important field of scientific research. Several disciplines involve in this field, such as mathematics, mechanics, physics, and informatics. This activity of research aims to produce simulations of textile fabrics behavior with more realism while remaining faithful to the physical and mechanical properties of this type of materials (Hedfi et al., 2011).

6. Conclusion

In this chapter, fabric sensorial comfort which has been studied by many researchers since the early 1900s, was dealt with in detail. As the studies are analyzed, it can be stated that, the previous studies generally focus on subjective evaluation whereas, in the last decades new objective evaluation methods and techniques simulating human tactile feeling and prediction methods are developed. It seems, as being an interesting research area, sensorial comfort will continue to attract the attention of researchers in the future.

7. References

Adanur, S. (2001). *Handbook of Weaving*, CRC Press, Taylor & Francis Group, Chapter 13: Fabric Structure, Properties and Testing, By Sulzer Textil Limited Switzerland

ASTM D 4032-08- Standard Test Method for Stiffness of Fabric by the Circular Bend Procedure

ASTM D 1388-08- Standard Test Method for Stiffness of Fabrics

Aliouche, D., Viallier, P. (2000). Mechanical and Tactile Compression of Fabrics: Influence on Handle, *Textile Research Journal*, Vol. 70 (11), pp. 939-944.

Bakar, B.A. (2004). *Subjective and Objective Evaluation of Fabric Handle Characteristics*, The University of Leeds, Department of Textile Industries, Master of Science

Barker, R.L. (2002). From Fabric Hand to Thermal Comfort: The Evolving Role of Objective Measurements in Explaining Human Comfort Response to Textiles, *International Journal of Clothing Science and Technology*, Vol. 14 Iss: 3/4, pp. 181 – 200

Behera, B.K., Chowdhry, S., Sobti, M. (1998). Studies on Handle of Microdenier Polyester Filament Dress Materials, *International Journal of Clothing Science and Technology*, Vol. 10 No. 2, pp. 104-113.

Behera, B.K. (2007). Comfort and Handle Behaviour of Linen-Blended Fabrics, *AUTEX Research Journal*, Vol. 7, No 1, March

Behery, H.M. (2005). *Effect of Mechanical and Physical Properties on Fabric Hand*, Woodhead Publishing Limited in Association with the Textile Institute, Published in North America by CRC Press LLC

Bensaid, S., Osselin, J-F., Schacher, L. & Adolphe, D. (2006). The Effect of Pattern Construction on the Tactile Feeling Evaluated Through Sensory Analysis, *The Textile Institute*, Vol.97, No.2, pp. 137-145.

Bishop, D. P. (1996). *Fabric: Sensory and Mechanical Properties*, The Textile Institute, Textile Progress, UK, Volume 6, No:3, ISBN 1 870812751

Bona, M. (1994) *Textile Quality*, ISBN: 1870812603, The Textile Institute

Chen, Y., Zhao, T., Turner, B. (2001). A New Computerized Data Acquisition and Analysis System for KES-FB Instruments, *Textile Research Journal*, Vol. 71 (9), pp. 767-770

Das, A., Ishtiaque, S.M. (2004). Technology and Managemet, Comfort Characteristics of Fabrics Containing Twist-less and Hollow Fibrous Assemblies in Weft, *Journal of Textile and Apparel*, Vol:3

Dillon, P., Moody, W, Bartlett, R., Scully, P., Morgan, R. & James, C. (2001). Sensing The Fabric Haptic Human-Computer Interaction, *Lecture Notes in Computer Science*, Vol. 2058, pp. 205-218

Dhinakaran M., Sundaresan S. & Dasaradan B.S. (2007). Comfort Properties of Apparels, *The Indian Textile Journal*, Vol.32, pp. 2-10

Domskienė J., Strazdienė E., (2005). Investigation of Fabric Shear Behaviour, *Fibres & Textiles in Eastern Europe*, April/June, Vol. 13, No. 2 (50), pp. 26-30

Du, Z., Yu, W. (2007). A Comprehensive Handle Evaluation System for Fabrics: I. Measurement and Characterization of Mass and Bending Properties, *Measurement Science and Technology*, Vol. 18, pp. 3547-3554

Frydrych, I., Dziworska, G., Bilska, J. (2002). Comparative Analysis of the Thermal Insulation Properties of Fabrics Made of Natural and Man-Made Cellulose Fibres, *Fibres & Textiles in Eastern Europe*, October/December, pp. 40-44

Grinevičiūtė, D., Daukantıenė, V., Gutauskas, M. (2005). Textile Hand: Comparison of Two Evaluation Methods, *Materıals Scıence (Medžıagotyra)*, Vol. 11, No. 1. 2005, pp. 57-63 ISSN 1392-1320

Govindaraj, M., Pastore, C., Upadhyay, A. K., Metaxas, D., Huang, G., Raheja, A. (2003a). S00-PH08 Haptic Simulation of Fabric Hand; *National Textile Research Annual Report*: November

Govindaraj, M., Garg, A., Raheja, A., Huang, G., Metaxas D. (2003b). Haptic Simulation of Fabric Hand, *Proceeding of Eurohaptic Conference*, pp. 253-260

Hasani, H., Planck, H. (2009). Analysis of the Physical Fundamentals of an Objective Integral Measuring System for the Determination of the Handle of Knitted Fabrics, *Fibres & Textiles in Eastern Europe*, Vol. 17, No. 6 (77) pp. 70-75

Hedfi, H., Ghith, A., Salah, H.B.H. (2011). Study of Dynamıc Drape Behaviour of Fabric Using FEM Part I: Model Formulation and Numerical Investigations, *International Journal of Engineering Science and Technology (IJEST)*, Vol. 3 No. 8 August 2011, pp. 6554-6563

Hes, L. (2004). Marketing Aspects of Clothing Comfort Evaluation, X. *International İzmir Textile Symposium Proceedings*

Hes, L. (2009). Fundamentals of Design of Fabrics and Garments with Demanded Thermo-physiological Comfort, *International Round Table, Clothing Comfort – Condition of Life Quality,* , Corp Tex 1, 25th -26th of September, Romania

Hu, J. (2004). *Structure and Mechanics of Woven Fabrics*, CRC Press Boca Raton Boston New York Washington, DC, Woodhead Publishing Limited Cambridge England

Hu, J. (2008*). Fabric Testing*, Woodhead Publishing Series in Textiles: Number 76

Hui, C. L., Lau, T. W., NG, S. F., Chan, K.C.C. (2004). Neural Network Prediction of Human Psychological Perceptions of Fabric Hand, *Textile Research Journal*, Vol. 74(5), pp. 375-383

Kawabata, S. (1980). *The Standardization and Analysis of Hand Evaluation*, Second Edition, The Hand Evaluation and Standardization Committee The Textile Machinery Society

Kim, J.O., Slaten B.L. (1999). Objective Evaluation of Fabric Hand Part I: Relationships of Fabric Hand by the Extraction Method and Related Physical and Surface Properties, *Textile Research Journal*, Vol.69 (1), pp. 59-67

Kocik, M., Żurek, W., Krucinska, I., Geršak, J., Jakubczyk, J. (2005) Evaluating the Bending Rigidity of Flat Textiles with the Use of an Instron Tensile Tester, *Fibres & Textiles in Eastern Europe*, April / June, Vol. 13, No. 2 (50), pp. 31-34

Le, C., Ly, N. Phillips, D., Boos, A.D., Characterising The Stability of the Surface Finish on Wool Fabrics, CSIRO Textile and Fibre Technology, Date of access:01.11.2011, web: http://www.scribd.com/doc/23116585/csiro

Lima, M., Hes, L., Vasconcelos, R., Martins, J. (2005) Frictorq, Accessing Fabric Friction with a Novel Fabric Surface Tester, *AUTEX Research Journal*, Vol. 5, No 4, December

Mahar, T. J., Wheelwright, P., Dhingra, R. C. & Postle, R. (1990). Measuring and Interpreting Fabric Low Stress Mechanical and Surface Properties, Part V: Fabric Hand Attributes and Quality Descriptors, *Textile Research Journal*, 60, pp. 7-17.

Majumdar, A., Saha, S.S. (2008). A Novel Fabric Thickness and Compression Tester Using Magnetic Inductance Principle, *Journal of the Textile Institute*, Volume 99, Issue 4.

Mamalis, P., Andreopoulos, A. Spyrellis, N. (2001). The Effect of a Durable Flame-Retardant Finishing on The Mechanical Properties of Cotton Knitted Fabrics, *International Journal of Clothing Science and Technology*, Vol. 13 No. 2, pp. 132-144.

Mazzuchetti, G., Demichelis, R., Songia, M. B.& Rombaldoni, F. (2008). Objective Measurement of Tactile Sensitivity Related to a Feeling of Softness and Warmth, *Fibres & Textiles in Eastern Europe*, Vol. 16, No. 4 (69) pp. 67-71.

Mäkinen, M., Meinander, H., Luible, C., Magnenat-Thalmann, N. (2005). Influence of Physical Parameters on Fabric Hand, *Proceedings of the HAPTEX'05 Workshopon Haptic and Tactile Perception of Deformable Objects*, Hanover, December

Militký, J., Bajzík, V. (1997). Influence of Washing/Ironing Cycles on Selected Properties of Cotton Type Weaves, *International Journal of Clothing Science and Technology*, Vol. 9 No. 3, pp. 193-199

Mitsuo, M. (2006). Fabric Handle and Its Basis Mechanical Properties, *Journal of Textile Engineering*, Vol.52, No: 1, pp. 1-8

Moody, W., Morgan, R., Dillon, P., Baber, C., & Wing, A. (2001). Factors Underlying Fabric Perception. In: *1st Eurohaptics Conference Proceedings*. Birmingham

Mukhopadyhay, A., Dash, A.K., Kothari, V.K. (2002). Thickness and Compressional Characteristics of Air-Jet Textured Yarn Woven Fabrics, *International Journal of Clothing Science and Technology*, Vol. 14 No. 2, pp. 88-99

Na, Y., Kim, C. (2001). Quantifying the Handle and Sensibility of Woven Silk Fabrics, *Textile Research Journal*, 71 (8) pp. 739-742

Namligöz, E.S, Özçelik, G.& Lima, M. (2008). The Effect of Softeners and Stiffeners on the Frictional and Handle Properties of Woven Fabrics, *The 3rd International Conference of Applied Research in Textile* (CIRAT-3) *Proceedings*, Sousse, Tunisia

Ozcelik, G., Fridrichová, L., Bajzík, V. (2005). The Comparison of Two Different Bending Rigidity Testers, *4th Central European Conference Proceedings*, September 7-10th, Liberec, Czech Republic

Ozcelik, G., Mertova, I., (2005), The Comparative Analysis of Bending Rigidity Methods, *12nd International Conference STRUTEX Proceedings*, Technical University of Liberec, Czech Republic

Ozcelik, G., Supuren, G., Gulumser, T. & Tarakcioglu, I. (2007). A Study on Subjective and Objective Evaluation of the Handle Properties of Shirt Fabrics, *AUTEX World Textile Conference Proceedings*, Finland, Tampere

Ozcelik, G., Supuren, G., Gulumser, T. & Tarakcioglu, I. (2008). A Study on Subjective and Objective Evaluation of the Handle Properties of Shirt Fabrics, *Fibres & Textiles in Eastern Europe*, Vol. 16, No. 3 (68)

Özdil, N, Özçelik, G. & Süpüren, G. (2008). The Effect of Boucle Yarn Characteristics on the Fabric Handle, *International Conference of Applied Research in Textile CIRAT-3 Preceedings*, Sousse, Tunisia

Özgüney A.T., Taşkin C., Özçelik G., Gürkan P.& Özerdem A. (2009). Handle Properties of the Woven Fabrics Made of Compact Yarns, *Tekstil ve Konfeksiyon*, Vol.2, pp.108-113

Pan, N., Yen, K.C., Zhao, S.T, Yang, S.R. (1988). A New Approach to the Objective Evaluation of Fabric Handle from Mechanical Properties, Part I; Objective Measure for Total Handle, *Textile Research Journal*, Vol 58, pp. 438-444.

Pan, N. (2007). Quantification and Evaluation of Human Tactile Sense Towards Fabrics, *Int. Journal of Design & Nature*. Vol. 1, No. 1,pp. 48–60

Park, S.W., Hwang, Y.G. (2002). Comparison of Total Hand of Single Knitted Fabrics Made from lincLITE® and Conventional Wool Yarns, *Textile Research Journal*, 72(10), pp. 924-930

Pavlinić, D.Z., Geršak, J. (2003). Investigations of the Relation between Fabric Mechanical Properties and Behaviour, *International Journal of Clothing Science and Technology*, Vol. 15 No. 3/4, pp. 231-240

Potluri, P., Porat, I.& Atkinson, J. (1995). Towards Automated Testing of Fabrics, *International Journal of Clothing Science and Technology*, Vol. 7 No. 2/3, pp. 11-23.

Saville, B. P. (1999). *Physical Testing of Textiles*, Woodhead Publishing Ltd, Cambridge England, ISBN 0849305683

Schwartz, P. (2008). *Structure and Mechanics of Textile Fiber Assemblies*, Woodhead Textiles Series No. 80, ISBN-13: 978 1 84569 135 6

Shanmugasundaram, O.L. (2008). Objective Measurement Techniques for Fabrics, *Asian Textile Journal*, 17(8): 63-67

Shishoo, R.L. (1995). Importance of Mechanical and Physical Properties of Fabrics In The Clothing Manufacturing Process, *International Journal of Clothing Science and Technology*, Vol. 7 No. 2/3, pp. 35-42.

Silva, L.F., Seabra, E., Lima, M., Vasconcelos, R., Alves, J., Guise, C., Martins, D. (2010). A Successful Partnership for the Development of a Laboratory Friction Testing Apparatus: A Project Review, *International Conference on Engineering Education Proceedings*, July 18–22, Gliwice, Poland

Sun, M.N., Cheng, K.P.S. (2000). The Quality of Fabric Knitted From Cotton Sirospun Yarn, *International Journal of Clothing Science and Technology*, Vol. 12 No. 5, pp. 351-359.

Supuren, G., Ozdil, N., Leskovsek, M., & Demsar, A. (2010). Surface Properties of Wool and Various Luxury Fibers, The Fiber Society Spring 2010, *International Conference on Fibrous Materials Proceedings*, Bursa-Turkey

Strazdienė, E., Gutauskas, M. (2005). New Method for the Objective Evaluation of Textile Hand, *Fibres & Textiles in Eastern Europe*, Vol. 13, No. 2 (50), pp. 35-38

Stylios, G.K., Powell, N.J. (2003). Engineering the Drapability of Textile Fabrics, *International Journal of Clothing Science and Technology*, Vol. 15, No. 3/4, 2003 pp. 211-217

Sülar, V., Okur, A. (2007). Sensory Evaluation Methods For Tactile Properties of Fabrics, *Journal of Sensory Studies*, 22, pp. 1–16

Sülar, V., Okur, A. (2008a). Objective Evaluation of Fabric Handle by Simple Measurement Methods, *Textile Research Journal*, Vol. 78: pp. 856-868

Sülar, V., Okur, A. (2008b). Handle Evaluation of Men's Suitings Produced in Turkey, *Fibres & Textiles in Eastern Europe*, April/ June, Vol. 16, No. 2 (67), pp. 61-68

Thordike, G.H., Varley, L. (1961). Measurement of the Coefficient of Friction between Samples of the Same Cloth, *Journal of Textile Insitute*, Vol.52, pp. 255-271

Yazdi, A.A., Shahbazi , Z. (2006). Evaluation of the Bending Properties of Viscose/Polyester Woven Fabrics, *FIBRES & TEXTILES in Eastern Europe*, Vol. 14, No. 2 (56), pp. 50-54

Yazdi, A.A., Özçelik, G. (2007). The Effect of the Yarn Twist Directions on the Shearing Behavior of the Woven Fabrics, *3rd International Technical Textile Congress*, 1-2 December

Yick, L. K., Cheng, K. P. S., & How, Y. L. (1995). Subjective and Objective Evaluation of Men`s Shirting, *International Journal of Clothing Science and Technology*, Vol. 7, No.4, pp. 17-29.

Wang, H., Mahar, T.J., Hall, R. (2011). Prediction of the Handle Characteristics of Lightweight Next-To-Skin Knitted Fabrics Using a Fabric Extraction Technique, *The Journal of The Textile Institute*, web:
http://www.tandfonline.com/doi/abs/10.1080/00405000.2011.602230#preview

Zeng, X.,Koehl, L. (2003). Representation of the Subjective Evaluation of the Fabric Hand Using Fuzzy Techniques, *International Journal of Intelligent Systems*, Vol. 18, pp. 355–366

http://www.sdlatlas.com

9

Novel Theoretical Approach to the Filtration of Nano Particles Through Non-Woven Fabrics

Nikola Jakšić[1] and Danilo Jakšić[2]
[1]*Turboinštitut*
[2]*University of Ljubljana*
Slovenia

1. Introduction

The motivation for this work comes from studying the composition of a medical face mask made of non woven fabrics and the mechanisms of filtrations inside the mask. Present medical mask filters passing air with the maximum efficiency of 99.97%, which might be seen as excellent. Main question is: "What presents this 0.03%?" If an aggressive virus is found in this tiny share, the effectiveness of the mask would be questionable.

The filtration mechanism in a non woven fabric is a complex process. It is obvious that air passing through a non woven fabric has to go through voids in the fabric. In order to understand the filtration mechanism, it is essential to be able to define fabrics porosity parameters, the nature of the air flow through voids in the fabric and the behaviour of the nano-particle caught inside the flow.

The classical theory of filtration is based on a single fibre in the fluid flow; Sharma (2000),Brown (1993), and Hutten (2007). It neglects neighbouring fibres and thus all porosity parameters. It was reasonable simplification due the fact that no easy method that would be reliable at the same time did exist until J-method for assessing the flat textiles porosity was introduced in Jakšić & Jakšić (2007) and Jakšić & Jakšić (2010). J-method enables us to determine all relevant porosity parameters, which makes this approach to filtration possible.

Medical face masks, which are made of textile fibres, are light, easy to use, relatively low-cost, and very efficient in combating air-borne infections. The mask design should provide tight fit to the skin of a face in order to ensure that air flows only through the mask. The design of the mask should also ensure filtration, not only dust particles, but also microbes and viruses. The masks are meant for a single use to avoid saturation of a mask. They are also suitable for use in a dust environment until saturation of a mask.

The mask composition must be proper regarding its use. The filtering layer must be protected with additional layers at its side faced towards a subject's face and outside side. Voids (pores) in the inner layer, which is actually filter passing air, should not lead air from the outer side to the inner side of the layer directly, as a channel. The dust particles, microbes and viruses might penetrate a mask designed in that way by being trapped in the unobstructed air flow. Pores in the inner layer, that enables air flow through mask, should be small enough and winded like a channel frequently changing its direction, in order to ensure maximum filtering efficiency.

The criteria can only be met with a non-woven fabric made of microfibers. The diameter of microfibers is normally between 1.5μm and 2.5μm. For comparison, the diameter of normal chemical fibres is normally around 20μm.

The medical face mask analysed here is made of three non-woven fabrics layers of different quality. The outer and inner fabric are intended for the partial filtration only and primarily for the protection of the filtration fabric in the middle. Fibres in all three fabrics are placed in the random pattern enhancing air-flow filtration capability.

The arguments focused on the nature of the air flow through a fabric and the behaviour of the nano-particle caught inside the flow were made based on analytical and numerical analysis - computational fluid dynamics (CFD).

The novel approach to the filtration of nano-particles through non-woven fabrics is based on a fabric porosity and consequently on the nature of the fluid flow through the fabric. It is thus essential to quickly recapitulate the method for estimating the porosity of flat textiles (J-method); Jakšić & Jakšić (2007) and Jakšić & Jakšić (2010). The design and the porosity parameters of the mask under consideration are presented next followed by description of the mechanisms of filtration in the mask.

1.1 J-method for determining porosity of flat textiles

The method is based on selectively squeezing the fluid in the pores out of the wet fabrics by air pressure and on the presumption that a pore is approximated with a cylinder. The selectivity is assured by the fact that the fluid is squeezed out of the pores with a certain hydraulic diameter providing that the precise value of the air pressure is applied. The air pressure is inversely proportional to the hydraulic pore diameter. Latter is important, while the process of squeezing out the fluid contained in the pores of the wet fabrics is under examination. There is always a small amount of the fluid that remains at the edges of pores if such edges exist.

We worked under two presumptions:

- The regime of the air flow through the dry and the wet sample is the same at same pressure difference regardless of the size of the open area of the wet sample.

- The number of the hydraulic pores is not the same as number of pores between threads of the warp and weft if the ratio of the rectangular sides, which represents real pore's cross-section, is at least 3:1.

The pressure difference p_i between the opposite surfaces of the flat textile, equation (1), results in squeezing the fluid out of the pores, which diameter is equal or larger than d_i. The fluid is characterized by the surface stress α.

$$d_i \geq \frac{4\alpha}{p_i} \tag{1}$$

The fluid is first squeezed out from pores, which have the largest hydraulic diameter. The flow of air will establish itself through these pores that are now empty. The volume flow rate of air through the flat textile can be described by equation (2)

$$V_i = A\,p_i = P\,a\,p_i^b = P\,v_i \tag{2}$$

where V_i stands for the air volume flow rate through the sample at the air pressure pi, A for a regression coefficient when fitting equation (2) to the measured dry data, P for the open surface, v_i for the linear air flow velocity, a for the coefficient and b for the exponent. The parameters a and P are unknown and they have to be estimated as well. The solution of the problem is enabled by equation (3) by putting the velocity v_i in the relationship with the air pressure p_i. The value for the exponent b is bounded between 0.5 and 1.0. The last part of equation (3) holds in the ideal circumstances, when all of the energy dissipation mechanisms are neglected.

$$v_i = a_0 \, p_i^b = 1.28 \, p_i^{0.5} \tag{3}$$

The selective squeezing out the fluid from pores enables us to compute the number of pores at each interval defined by the incremental pressure growth. The number of pores in the ith interval as

$$n_i = \frac{p_i^2}{4\pi a \alpha^2} \left(\frac{V_i}{p_i^b} - \frac{V_{i-1}}{p_{i-1}^b} \right) \tag{4}$$

The presumption of the equal regime of the air flow through the wet sample's open area and the dry one at the same pressure is taken into account. Small values of the Reynolds number in the extreme causes (maximal hydraulic diameter of pore), support that presumption. The air flow is either laminar through the open pores in the wet sample and through all pores in the dry sample, or the type of the air flow is same. This is the criterion for using the exponent b, which is estimated when equation (2) is fitted to the measured dry data, in the process of determining the pore distribution from the measured wet data. The corresponding coefficient a_j are determined by equation (5)

$$a_j = 1.28 \frac{n_{cj}}{n_t} = \frac{1.28}{a^*} \quad ; \quad a^* = \frac{n_t}{n_{cj}} \tag{5}$$

where n_t stands for the true number of pores, n_{cj} for the computed number of pores and a_j for the corrected a in equation (4). The values of theoretical limits, for exponent b ($b_0 = 0.5$) and coefficient a ($a_0 = 1.28$), that are used in the second procedure are shown in the last part of equation (3).

Four different samples were used for the method's testing, which practically encompasses all the fabric types that the method is suitable for. The basic design parameters of the woven fabrics are presented in table 1. They are made of monofilament, multifilament and cotton yarn. The measured average pore's hydraulic diameters of the textiles are in the interval of 18 up to 200 micrometers. The wide assortment of textiles is thus covered.

The results of the textile's porosity tests are presented in table 2. The first procedure is used for all four samples. The second procedure was used for porosity parameters estimation of samples (a) and (b) due to large value of the parameter a_0.

The nomenclature in table 2 - b stands for the exponent in equation (2), h [μm] for the width of the interval of the pore distribution, m for the number of the distribution intervals, n_t for the true number of pores between the threads of the warp and the weft per cm^2, n for the computed number of hydraulic pores between the threads of the warp and the weft per cm^2, when the true number of pores (or number of hydraulic pores) is unknown (second procedure), d for the average hydraulic diameter of pores, d_t for the optically measured average hydraulic pore diameter - for samples (b), (c) and (d); the pores are ill-defined in

Sample	Description	Measurement interval [μm]	Number of [pores per cm^2]	Warp / weft [threads per cm]
(a)	Cotton woven fabric	160 – 20	452	22 / 21
(b)	Thick monofilament fabric	80 – 10	2200	55 / 40
(c)	Multifilament woven fabric	270 – 140	960	32 / 30
(d)	Very thick monofilament woven fabric	24 – 12	32400	180 / 180

Table 1. Samples used in the testing of J-method

Porosity test procedure	Parameter	Sample (a)	(b)	(c)	(d)
Porosity parameters when the number of pores is known (the first procedure)	b	0.5794	0.6355	0.8329	0.7174
	h [μm]	14	10	13	2
	m	10	7	10	7
	n_t	452*	2200	960	32400
	d [μm]	45.04	31.37	200.45	18.84
	d_t [μm]	53**	30	199	18.78
	P [%]	0.98	2.07	31.32	9.06
	P_t [%]	1.00**	3.43	29.84	8.98
	a_0	9.4074	2.8236	0.28	0.8791
Porosity parameters when the number of pores is unknown (the second procedure)	d [μm]	45.00	32.77		
	P [%]	6.96	6.17		
	n	3314	6213		
	a_0	1.28	1.28		

Table 2. Parameters of porosity estimated with J-method for all four samples. * - the number corresponds to the product of the warp and weft. ** - corresponds to the 452 measured pores - between the threads of warp and weft only one typical pore was measured in each void between the threads of warp and weft.

Statistic parameters	w [μm]	l [μm]	d_t [μm]	$P_{real} = w\,l$ [μm^2]	P_{hydr} [μm^2]	$\frac{l}{d_t}$
Mean	20.13	66.06	30.00	1364.67	786.29	2.50
Standard deviation	7.77	13.75	10.19	664.13	461.38	1.19
Minimum	4.76	23.81	8.82	287.12	61.07	1.04
Maximum	34.92	87.3	46.91	2519.68	1727.43	6.86
Total				68233.43	39314.56	

Table 3. Results of the scanning-electron microscope pore's shape and open area measurement on 50 pores of the sample (b).

sample (a), P [%] for the average open hydraulic flow area, P_t [%] for the average open flow hydraulic area computed on the bases of the optical experiment, a_0 for the coefficient a, equation (2), at presumption that exponent b has minimal value ($b = 0.5$).

When dealing with the sample (b), the average ratio l/w in table 3 is $66.06/20.13 = 3.23$. Hence, the criterion of having more than one hydraulic pore in a pore between threads of the warp and weft is thus met. The maximum value of the ratio between the value of the longer rectangular side l and the hydraulic diameter of the same pore is 6.84 and the mean value is 2.5. Hence, the true number of the hydraulic pores of the specimen (b) is $2200 \cdot 2.5 = 5500$, see table 2. The estimated number of hydraulic pores is 6213, table 2, which is only 13% more than the true number of hydraulic pores in specimen (b).

The porosity of the non-woven flat fabrics is extremely difficult to characterize due to their irregular structure. The structure makes them better, more effective filtration media in comparison to the woven fabrics. Hence, the challenge is to estimate their porosity parameters. The experimental results presented in table 2 and 3, especially sample (b), proved that the porosity of the non-woven flat textiles can be estimated by J-method.

2. Medical mask

A medical mask fabric is composed of three layers. The outer layer and the layer suite on the face of a subject are composed of fibres which diameter is $18\mu m$. The inner layer is composed from microfibers which diameter is $2\mu m$, see figures 2 and 3. In this layer the fibres are arranged in 37 sub-layers.

2.1 Initial state of a medical mask

When discussing the mechanisms of filtration the initial state of the mask before usage must be defined:

1. The mask is new and in the original package.
2. The mask is placed over a subject's face in a way that it covers respiratory organs and air can pass only through the mask.
3. The mask assures 100
4. The laminar flow of air through mask is observed.
5. A mask user is moderately active; volume flow: 40-60 l/min, breathing rate: 12-20 breaths per minute.
6. The mask thickness is 285 μm and specific mass is 57.5g/m2.
7. The fibres for the inner mask layer are randomly orientated and spread over the whole area, see figures 2 and 3.
8. There are no fibres in the direction perpendicular to the mask face, which means that the non-woven fabric was not needled.

2.2 Virus

Viruses are smaller than dust particles and microbes and thus the most difficult to filter out. They come in different shapes: spherical, see figure 1, helical, icosahedra, etc. and sizes, which varies between 10nm and 300nm. The average density of a virus is approximated to v = 1200kg/m^3. We will focus on the spherical or near spherical viruses.

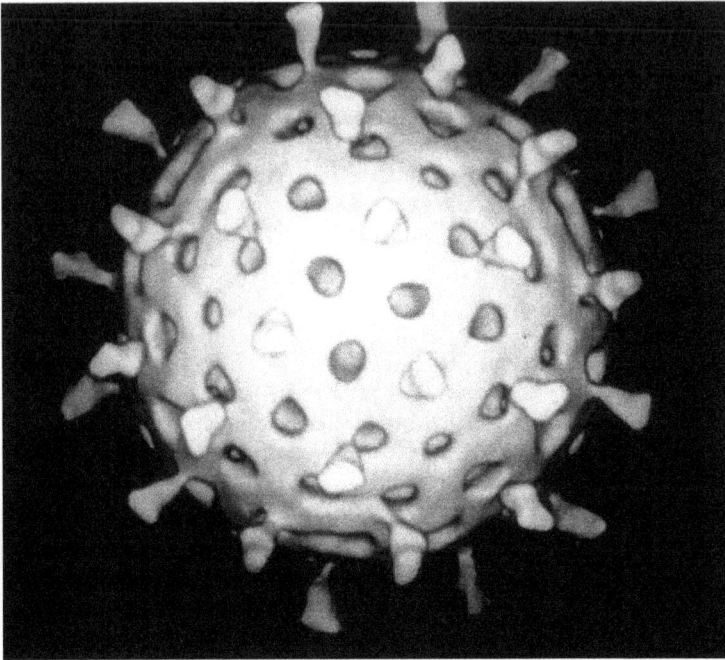

Fig. 1. Computer-aided reconstruction of a rotavirus based on several electron micrographs; Wikipedia (2011)

2.3 Porosity of the medical masks

We have applied J-method to characterise the porosity of a medical mask. The walls of pores are defined by fibres. In contrast to a woven fabric, where pores are straight from one surface to the opposite one and where the length of pores is equal to thickness of the fabric, the pores in the non-woven fabric changes its direction and are thus much longer than the fabric's thickness. It is this property that makes them an excellent filtration media and at the same time, very difficult to characterise. Even though the viruses are much smaller than the hydraulic diameter of pores, the configuration of pores allows for high filtration efficiency.

Non-woven layer	Mass $[g/m^2]$	Thickness $[\mu m]$	Number of fibre layers	Thickness of fibres $[\mu m]$	Active surface* $[\%]$	Total surface $[cm^2]$
Outer	17.6	92	5	18	87	160
Inner	20.4	74	37	2	100	160
Outer on the subject face	19.1	120	7	18	75	160

Table 4. Comparison of chosen physical parameters of the fibres from which the mask is made. * - not blocked with fibres pressed together

All three layers are made of polypropylene fibres. The area of the mask that allows the air flow is around 160cm², table 4. The active surface of the outer layers is smaller than the active surface of the inner layer. Both outer layers are strengthened for their ability to retain shape by pattern of dots where the fibres are partially pressed together. In this way, the parts of the other layers are transformed into a foil. Nevertheless, it is the open surface P that allows the airflow, table 5. The inner layer sub-layers are not pressed nor melted together and hence the air can flow through sub-layers and pores that are actually blocked by the outer layers areas transformed into the foil. Hence, 100% of active surface of the inner layer reported in table 4 holds.

Parameters of porosity	Outer non-woven layer	Inner non-woven layer	Outer on the subject face	Mask (all three layers)
The biggest pore [μm]	305	38	211	30
d_{max} [μm]	275	28	195	26
d_{min} [μm]	15	8	15	9
d_p [μm]	83.2	12.46	76.3	13.91
b	0.6183	0.7313	0.6143	0.7521
A	0.249	0.0889	0.2925	0.0921
P [%]	27.71	8.43	25.32	8.42
Width of the classes [μm]	13	2	18	2
Number of the classes	13	10	10	10
Number of pores/cm^2	3745	68414	4506	45294

Table 5. Parameters of porosity for all three non-woven layers of mask; the surface of samples: 1cm^2; liquid in the pores: n-butanol

Fig. 2. The inner non-woven layer of the medical mask zoomed 200 times

The mask porosity parameters are presented in table 5. Number of the pores on 1cm^2 is 45294. The maximal diameter of pores is 30μm. The open area (free for air flow) is (8.42%. The coefficient a (regression equation 2 - flow air through dry sample), is 0.0921. The exponent b

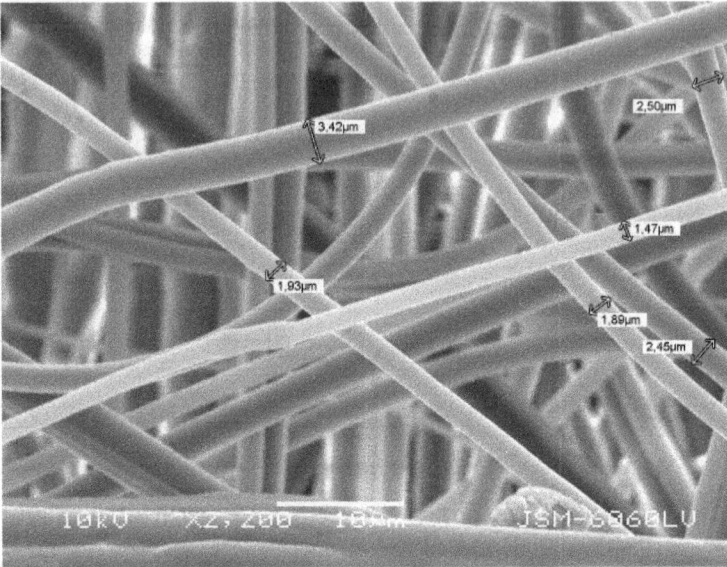

Fig. 3. The inner non-woven layer of the medical mask zoomed 2200 times

(regression equation 2 - flow air through dry sample) is 0.7521. Mean hydraulic diameter of pores is 13.91 μm.

The nomenclature of table 5 is as following: d_{max} stands for the average pore diameter of the first interval (the largest pores), d_{min} stands for the average pore diameter of the last interval (the smallest pores), d_p stands for the average pore diameter of the sample and P stands for the average open hydraulic flow area. The pore distribution is presented in table 6.

Limits of the classes [μm]	Hydraulic diameter of pores [μm]	Pressure [Pa]	Volume flow [m^3/s]·10^{-6}	Number of pores	Portion of pores [%]
25 – 27	26	3815	5.556	230	0.51
23 – 25	24	4133	13.100	329	0.73
21 – 23	22	4509	16.704	139	0.31
19 – 21	20	4959	66.931	2814	6.21
17 – 19	18	5510	120.306	3135	6.92
15 – 17	16	6199	233.431	6979	15.41
13 – 15	14	7089	323.569	6880	15.15
11 – 13	12	8266	516.463	16637	36.73
9 – 11	10	9919	652.269	8171	18.04

Table 6. Parameters of porosity for mask (for all three non-woven layers) - number of pores according to the pore size intervals

2.4 Influence of the mask layered structure on the filtration

There are some differences in the functioning of the mask layers during its usage and during the porosity parameter measurement that had to be noted. As already noted, both outer layers

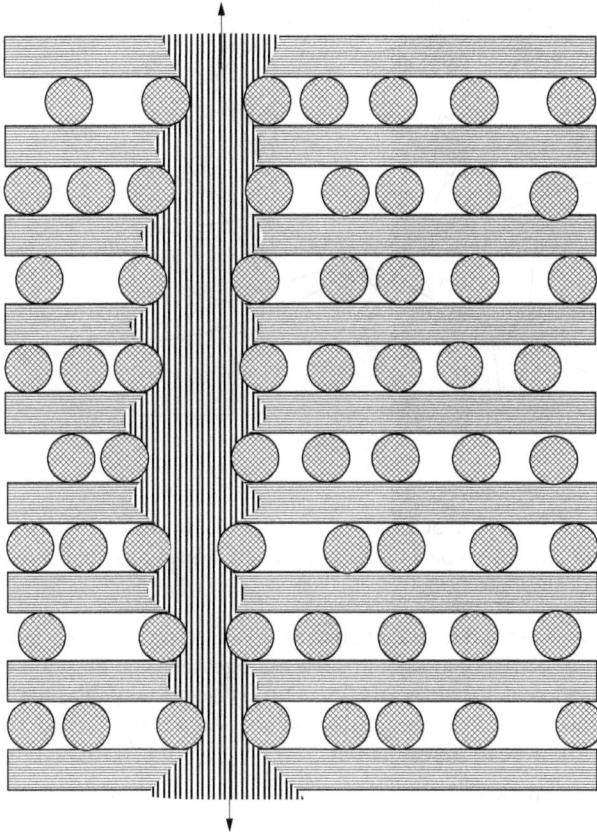

Fig. 4. Model of the stream of air through the non-woven mask due to the respiration

has smaller active surface due to their additional function to provide mechanical support for the inner layer.

The sample, which is undergoing the porosity test, is taken from the mask and placed in the apparatus head in exactly the same way. The area of the head is 1cm². The layers in the sample are pressed together during the test to avoid side leakage. This boundary effect affects relatively small area and may be neglected. True difference comes with the pressure at which the test is carried out. The test pressure at test compresses all three layers of a mask together. This effect obstructs free air passage beneath the areas of the outer layers that are

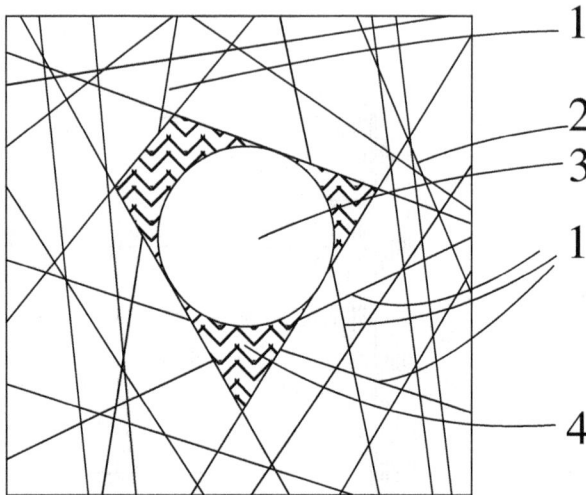

Fig. 5. Pore and a hydraulic pore of non-woven fabric. 1 - fibres that are narrowing the pore, 2 - fibre, 3 - hydraulic pore, 4 - a part of the pore where liquid has not been squeezed out during experiment of the J-method.

pressed together, which is not the case when a mask is used on a subjects face. In latter case all three mask layers are relatively loosely positioned together. The pressure caused by human breathing at supposed velocity of the air flow (1m/s) is so low (20Pa) that it does not affect the position of a mask layer in respect to the others. As a consequence, the experimentally estimated open area may be underestimated.

The fluid flow rate, open area and fluid velocity are linked by equation (2) and the pressure (pressure difference between both sides of a mask) and air flow velocity are linked by equation (3). It is clear from these relations that if the same fluid flow rate is achieved in reality and at experiment, larger pressure inducing larger velocities would be needed at experiment, due to smaller open area, which is the consequence of mask layers being pressed together during the experiment.

The open surface was estimated to be 8.42% of the total active area. It has to be stressed that the estimate refers on the hydraulic area, defined by hydraulically active section of pores, see figure 5, in interval down to pore hydraulic diameter of 8μm. The limitation is due to the maximal pressure that can be achieved during experimentation and the capacity of the volume flow. The pore distribution presented in table 6 indicates that there may be a lot of pores smaller than 8μm. These pores play an important role in filtration process due to relative proximity of the pore wall to the microbe or virus on one hand and they also let air through the mask. Due to latter we can assume that the open area is somewhat larger than estimated.

3. Classical filtration theory

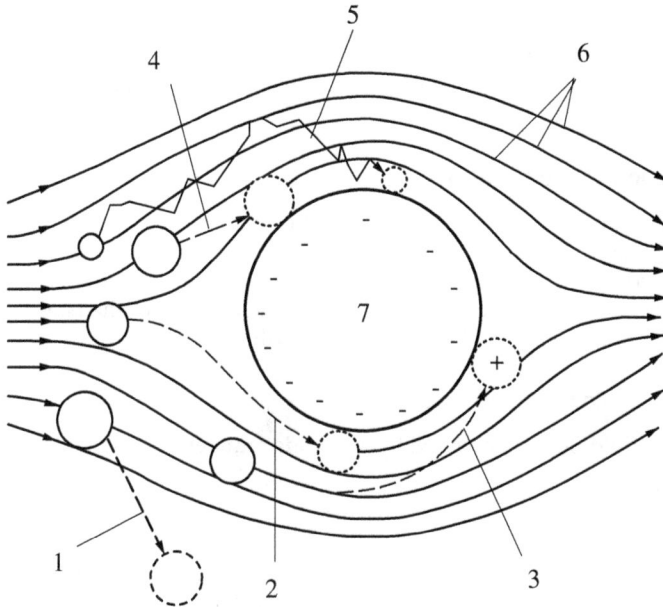

Fig. 6. Classical theory of the particles filtration on one fibre in the case of the laminar air flow: 1 - filtration of particles because of gravitation; 2 - filtration of the smaller particles on the surface of the fibres when they change their position from central part of the air flow to the virtual surface of the flow; 3 - electro filtration of the particles; 4 - filtration particles because of their steadiness; 5 - filtration very small particles by diffusion; 6 - air flow; 7 - fibre.

The classical filtration theory setting is presented in figure 6 – Sharma (2000), Brown (1993) and Hutten (2007), where a single fibre is put in the laminar fluid flow. The different sized particles are depicted with their supposed paths in order to depict different mechanisms of filtration. They are described in the following list:

1. The filtration due to gravitational field is applicable to largest and heaviest particles. This filtration mechanism can be observed in the outer fabric of the mask and does not apply to microbes and viruses.

2. The caption of a particle on a surface of the fibre due to its the collision with the fibre is generally applicable to the particles of all sizes. Unfortunately, the smaller the particle, the larger is the probability for the fluid flow to carry the particle pass the fibre due to the laminar nature of the air flow. The inertia of the particle in this case is negligible in comparison to fluid induced forces.

3. The electro filtration is a valid concept. However it is not applicable to the mask under consideration.

4. The inertia of the particle may cause the particle to leave the streamline of the flow around the fibre and collide with it. In contrast with the point 2 the inertia of the particle is not negligible.

5. The presumption here is that the small particles (smaller than 1.5μm) are moving by diffusion like paths as in the still air (Brownian motion) with one predominant direction in the direction of the fluid flow. We believe that this presumption is unjustified if the fabric porosity is taken into account.

4. Novel approach to the filtration

The classic theory of filtration is based on the flow around a single fibre. We argue that air is channelled through pores in the inner layer of the mask and hence, the modelling of filtration as a fluid flowing through a tube is justified. The justification is supported by the measurement of porosity parameters of the mask, which presumes the laminar flow through a sample Jakšić & Jakšić (2007) and Jakšić & Jakšić (2010).

4.1 Nature of the air flow through a mask

The nature of the fluid flow is described by Reynolds number, equation (6).

$$Re = \frac{\rho v d}{\mu} \tag{6}$$

where ρ stands for the fluid density, v is the mean velocity of the object relative to the fluid, or vice versa, d for the characteristic linear dimension (pore hydraulic diameter or particle diameter or fibre diameter) and μ for the dynamic viscosity of the fluid.

The value of the air density is $\rho = 1.2$kg/m^3 and its dynamic viscosity $\mu = 18\mu$Pa·s. The characteristic linear dimension is defined according to the object of interest. If the flow through a pore is studied, the dimension is its hydraulic diameter. On the other hand, if the flow around stationary particle is in question, the dimension is its diameter.

The velocity of the fluid (air) through a mask can be deduced from mask's porosity parameters and human physiology. The air velocity at inhaling or exhaling depends on breathing intensity. Latter depends on an activity and the intensity of the activity of the mask user. The exhaled air is normally denser than inhaled one due to its increased humidity. Let's suppose that a subject inhales from 12 litres of air per minute at normal pressure and during light activity and 60 l/min during moderate exercise. Further on, suppose that the subject inhales 12 times per minute. The size of the active surface of the mask inner layer is 160cm^2, and the open area of the layer is 8.42% of the active surface, table 4. The open area is thus 13.5cm^2. The air flow velocity through mask inner layer is thus approximately 0.17m/s during light activity and 0.86m/s during moderate exercise Zuurbier et al. (2009). As a conservative approach the value of 1m/s is taken in order to compute Re number.

The maximal pore hydraulic diameter is 30m, table 4. The Reynolds number for fluid flow through the largest pore is $Re = 2 << 2300$, which ensures laminar air flow through the inner mask layer.

The mask is designed to filter all viruses, which come in different sizes: from 10nm to 300nm. When inhaling (or exhaling) the air starts to flow, but a virus is not following instantaneously. The Reynolds number at the moment when a virus is stationary and the air is already moving is computed with maximal air velocity, even though that the virus is picked up by the air flow, to stay at conservative approach. In this case the $Re = 0.02 << 0.1$, the limit for the laminar flow around the object in the fluid flow is much stricter than the one for the flow in a tube.

We can conclude that the filtration in the inner mask layer is done in the laminar air flow.

4.2 Determination of the physical domain of the problem

The process of filtration that is carrying out in the medical masks is at the micro and nano scale. The size of the pores of the mask inner layer is at the micrometer scale and the size of viruses is at the nanometer scale. The Kundsen number (Kn) is used to determine whether the classical mechanics of continuum is still valid approach. Kundsen number is defined as

$$Kn = \frac{k_B T}{\sqrt{2}\,\pi\sigma^2 pL} \tag{7}$$

where k_B stands for the Boltzmann constant, T for the thermodynamic temperature, σ for the particle hard shell diameter and p for the total pressure and finally L for the representative physical length scale. The value of Kn for maximal pore and virus size is $Kn \approx 10^{-9}$, which is much less than 1. Hence, the mechanics of continuum is applicable.

4.3 Numerical modelling of the virus behaviour in the straight pore

In addition to the mechanisms of filtration described by the classical filtration theory the filtration caused by laminar flow through pores is described. A pore in the mask inner layer is approximated by a tube or a channel. The inner layer is formed in complicated way - it is composed of 37 sub-layers. Each of these sub-layers is composed of fibres randomly placed in random direction, making a pore meandering through all sub-layers a complex path for air to take, see figure 4. We can suppose that:

1. A pore changes its direction 35 times, when leaving on sub-layer and enters the other.
2. The change of the direction is random.
3. At the direction change the pore may split. Nevertheless, during the experiment for determining porosity parameters, we have actually taken into account these newly formed pores.
4. The length of a pore is at least twice the width of the inner layer, which means that the length of the pore is at least 570 μm, together with both outer layers.
5. Estimated pore diameter in the experiment is the smallest diameter of the pore over a whole width of the mask. Hence, the pore is acting as a tube or a channel.
6. The particle can be filtered on the obstacle (classical filtering theory) or in a void after the obstacle or at the pore's wall.

Based on the presumptions about shape of the pore the simple numerical model was built in order to show different possibilities of filtration and to asses some postulates of the classical filtration theory. The shape of a virus is idealised with the sphere/circle.

A simple 2D computational fluid dynamics (CFD) model was used for the simulation purposes of the virus behaviour in the laminar fluid flow in straight tube as a first approximation. A virus, represented by a circle, is placed in a rectangular, which is representing 2D model of a tube (pore), figure 7. The top and bottom walls are defined as no-slip wall where fluid velocity is zero due to the surface roughness. The virus boundary is defined in roughly the same way. The radial velocity of the fluid at the virus boundary is zero (no fluid penetration is allowed) and the circumference fluid velocity at the virus boundary is set to an arbitrary value to simulate possible virus rotation. The initial velocity with the uniform profile is prescribed at the inlet (left tube boundary) and the environment pressure is prescribed at the outlet (right tube boundary), figure 7. The results of the laminar model obtained with our target solver (ANSYS-FLUENT) were checked with the results of the turbulent model of the same solver and with the laminar model results of the 3D model in ANSYS-CFX solver. The results of the different models/solvers were in the excellent agreement and we proceeded with the 2D laminar modelling in ANSYS-FLUENT solver.

The aim the numerical modelling is gaining a qualitative picture of the mechanical system that is governing the particle flow through a medical mask. On this base the mesh of the model was relatively course, figure 8. The boundary layer at the tube walls were not in focus of this analysis and so the free meshing was used at whole region. The virus surroundings are meshed with the structured mesh with the smallest elements at the virus surface.

The tube diameter is $30\mu m$, which correspond to the maximal pore hydraulic diameter. The length of the tube is $120\mu m$, which is enough for the complete laminar velocity profile formation before striking the virus in the middle of the tube, figure 9. The laminar velocity profile, equation (8), was established approximately one tube diameter from the inlet.

$$v(r) = \frac{3}{2}v_o \left(1 - \frac{r^2}{R^2}\right) \tag{8}$$

where v_o stands for the initial velocity at the inlet, r for the radius of the streamline and R for the radius of the tube Widden (1966). The maximal velocity is $v_m = 1.5v_o$.

4.3.1 Fluid induced motion of the particle in the axial direction

The fluid induced force was estimated for the stationary virus at the centre of the tube cross-section ($r = 0\mu m$) for different initial velocities as a function of the maximal flow velocity. The circumferential velocity of the fluid at the virus surface is set to zero value - no virus rotation allowed. The fluid induced force is actually the driving force in this case, since the fluid is moving and the virus is stationary. It was found out that the fluid induced force F_a scales linearly with the fluid velocity v_m.

$$F_a = C v_m \tag{9}$$

where C stands for the constant that depends on the virus size. $C_{300} = 4.8 \cdot 10^{-12} \text{Ns/m}$ for the maximal virus size of 300nm, figure 10, and $C_{10} = 32.5 \cdot 10^{-15} \text{Ns/m}$ for the minimal virus size of 10nm. It is obvious that the forces acting on a virus are extremely small even if the virus is kept stationary.

If the virus is not stationary, the velocity v_m in equation (9) represents the difference of the velocity of the fluid flow and the virus. The second Newton law is used to compute virus velocity in the axial direction.

$$C(v_m - v) = m\frac{dv}{dt} \tag{10}$$

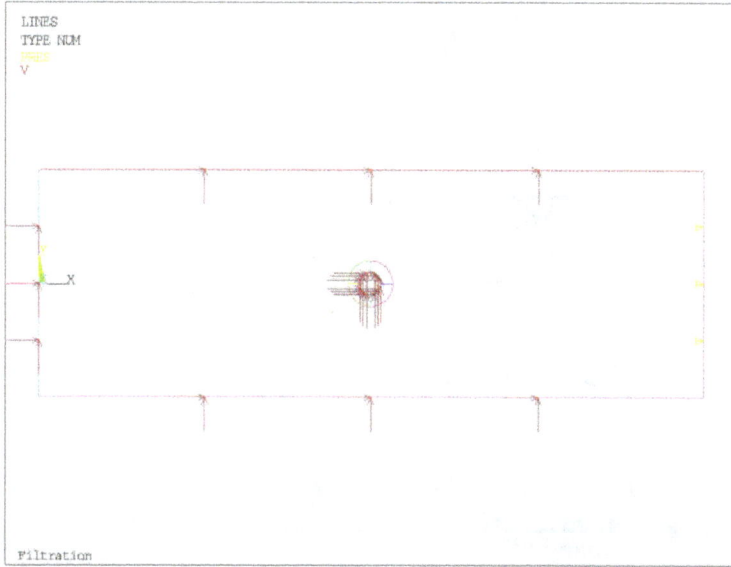

Fig. 7. 2D model of the virus in the pore. The size of the largest virus is magnified 10 times here for the presentation purpose only.

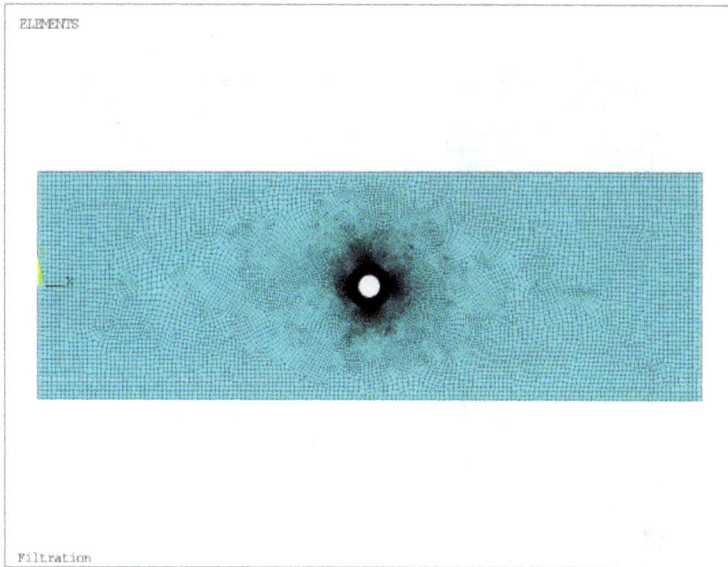

Fig. 8. CFD mesh. The size of the largest virus is magnified 10 times here for the presentation purpose only.

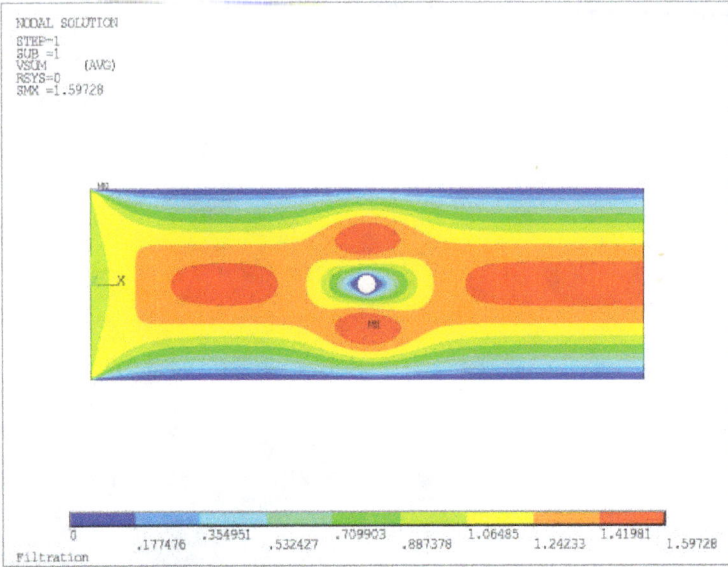

NODAL SOLUTION
STEP=1
SUB =1
VSUM   (AVG)
RSYS=0
SMX =1.59728

Filtration

| 0 | .177476 | .354951 | .532427 | .709903 | .887378 | 1.06485 | 1.24233 | 1.41981 | 1.59728 |

Fig. 9. Velocity contours. The size of the largest virus is magnified 10 times here for the presentation purpose only.

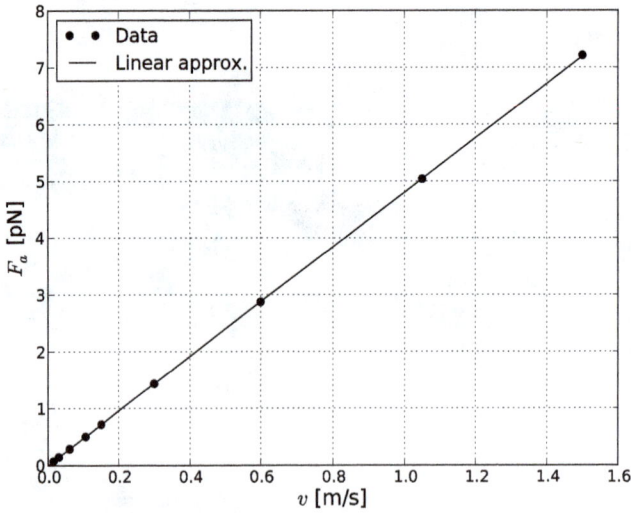

Fig. 10. The fluid induced force on a larger virus.

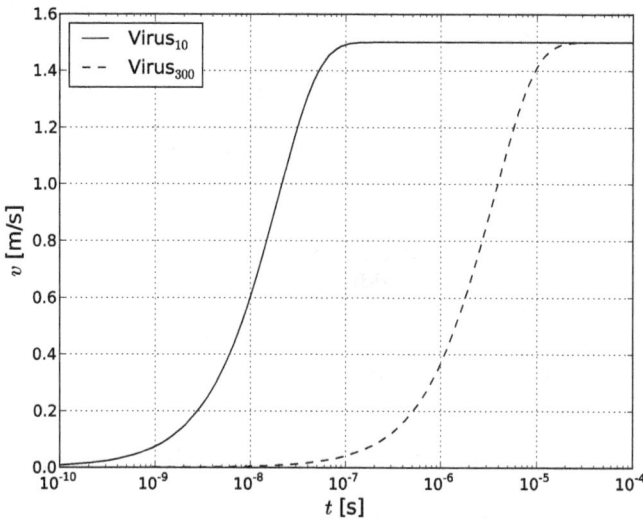

Fig. 11. The virus velocity as a function of time.

where v stands for the virus velocity and m for its mass. The mass of the virus of the maximal size can be estimated to $m_{300} = 17.0 \cdot 10^{-18}$ kg and the mass for the smallest can be estimated to $m_{10} = 0.63 \cdot 10^{-21}$ kg. Equation (10) can be resolved analytically and the results are presented in figure 11.

$$v(t) = v_m \left(1 - e^{-\frac{C}{m}t}\right) \tag{11}$$

Further on, the axial displacement of the virus can be computed by integrating the equation (11) once again.

$$x(t) = v_m \left(t + \frac{m}{C}\left(e^{-\frac{C}{m}t} - 1\right)\right) \tag{12}$$

The smaller virus reaches 99% of the fluid velocity in $8.9 \cdot 10^{-8}$s and 99.9% in $1.3 \cdot 10^{-7}$s. The larger virus's inertia is larger and so does the time. The 99% of the fluid flow is reached in $1.6 \cdot 10^{-5}$s and 99.9% in $2.4 \cdot 10^{-5}$s.

4.3.2 Fluid induced motion of the particle in the radial direction

It might appear that the predominant direction of motion is in the axial direction since the forces acting on a absolutely stationary virus in the perpendicular direction to the fluid flow are much smaller. The analysis providing us with the axial force on a stationary virus provides us also with the radial one. There is no radial force if the virus is placed in the tube centreline, due to the symmetry. If the stationary virus is moved towards the tube top boundary, the radial force is pointing towards the tube centreline. The same is true if the stationary virus is moved towards bottom tube boundary. The radial forces are at least one order of the magnitude smaller than the axial ones depending on the virus position. It appears that the largest radial force relative to the axial one can be found at roughly $\frac{1}{4}$ of the tube diameter from the top or bottom boundary.

So far the numerical results showed that the filtering in the straight tube is not possible because the stationary viruses are forced toward the fluid flow centreline. A look at moments acting on the virus reveals that the moment (torque) is present due to fluid viscosity. The torque is acting in the negative (CW - clock wise) direction if the virus is placed between the centreline and top boundary and in the positive (CCW - counter clock wise) direction if the virus is placed between the centreline and bottom tube boundary, figure 18. The virus is actually free to rotate and the redial force changes its direction towards tube boundaries, figure 14, due to the Magnus effect. The model shows this phenomenon, figure 14, and hence the filtering is possible even if the tube is perfectly straight.

The natural way of causes and consequences tells us that the moment drives the virus's rotation. The model is set in a way to address this phenomenon in the inverse approach. The circumferential velocity of the virus is prescribed as boundary condition for the fluid and the moment acting on the virus is estimated in the analysis. The results for the large virus placed at $\frac{1}{4}$ of the tube diameter from the top boundary, figure 12.

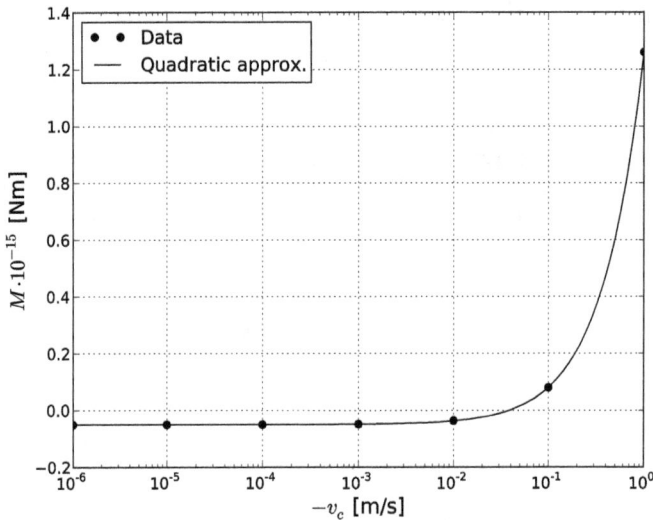

Fig. 12. The fluid induced moment acting on large virus as a function of the virus rotation.

The moment is at first negative and quite constant. It increases with the increasing circumferential velocity of the virus. It is obvious that it cannot reach the positive value at high circumferential velocities, due to the simple fact that the moment is driving the virus rotation and the change in the moment sign indicates, that the fluid flow, that generates the moment and was previously acceleration the rotation, is slowing down the virus rotation in such a case. The expectation is that the virus will rotate with the constant angular velocity, which value is defined by the zero moment. The moment is approximated with the polynomial of the second order, equation (13), as presented in figure 12.

$$M(\omega) = a\omega^2 + b\omega + c \qquad (13)$$

where M stands for the fluid induced moment acting on virus, ω for the angular velocity of the virus and a, b, c for the approximation parameters. The second Newton's low for rotation of the rigid body is used to generate the equation of motion for the virus rotation.

$$\frac{d\omega}{dt} = \frac{M(\omega)}{J} = A\omega^2 + B\omega + C \tag{14}$$

where J stands for the virus moment of inertia and A, B, C for the approximation parameters divided by J. The moment of inertia J of the largest virus is $1.53 \cdot 10^{-31} \text{kgm}^2$ and for the smallest is $6.3 \cdot 10^{-39} \text{kgm}^2$. Equation (14) can be resolved analytically.

$$w(t) = \frac{D}{2A} \tanh\left(\text{arctanh}\left(\frac{B}{D}\right) - t\frac{D}{2}\right) - \frac{B}{2A} \tag{15}$$

where t stands for time, A, B, C for the approximation parameters divided by J and D for the discriminate, $D = \sqrt{B^2 - 4AC}$. The virus terminal angular velocity is reached in the limit case $t \to \infty$ as

$$\omega_{max} = \frac{-D - B}{2A} \tag{16}$$

The terminal circumferential velocity is computed as $v_{c_{max}} = \omega_{max} R$, where R stands for the virus radius. The values of the terminal (maximal) circumferential velocity for the larger virus is $v_{c_{max}} = -0.0386 \text{m/s}$ and for the smaller one $v_{c_{max}} = -0.0146 \text{m/s}$ which is around 4% and 1.5% respectively of the air velocity at the inlet. What is really important is that viruses begins to rotate with the terminal angular velocity in a very short time, almost instantly, see figure 13.

The radial force was computed and approximated in the same way as the moment, see figure 12. Hence, by combining it with the angular velocity, the radial force can be plotted as function of time, figure 14. The force becomes constant with the virus reaching the terminal angular velocity.

Although figure 14 presents the resultant force, we can say that there are two competing mechanisms of delivering the radial driving force to the virus. It has been stated that the virus is driven towards the tube centre if it does not rotate due to the fluid flow velocity profile. The same velocity profile also drives the rotation of the virus which in turn adds radial force (Magnus effect, where the rotation is self induced - we named it as J-effect) in the direction of the tube/channel boundary. If the rotation is large enough, which is the case here, the radial force due to the rotation is larger than one due to the fluid profile. The force partition described here is a simplification of a complex fluid velocity and pressure distribution in the vicinity of a virus and serves only as an intuitive description of the physical behaviour due to the fact that an object in the tube/channel distorts the fluid flow profile and the way that fluid flows around a virus.

The analysis were repeated for other virus initial positions - position at $\frac{1}{8}$ of the tube diameter from the top boundary and for position at $\frac{3}{8}$ of the tube diameter from the top boundary. The terminal radial forces are presented with dots for both viruses in figures 15 and 16 together with their quadratic approximations, equation (17).

$$F_r(y) = ay^2 + by + c \tag{17}$$

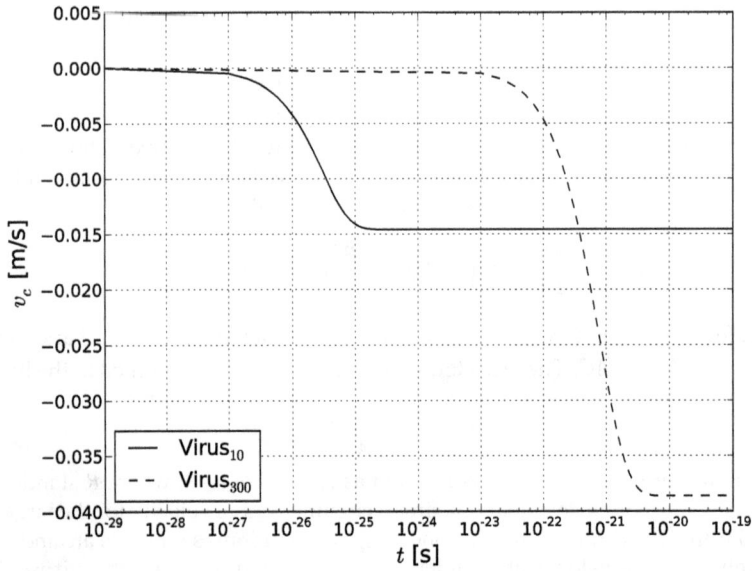

Fig. 13. The virus circumferential velocity as a function of time.

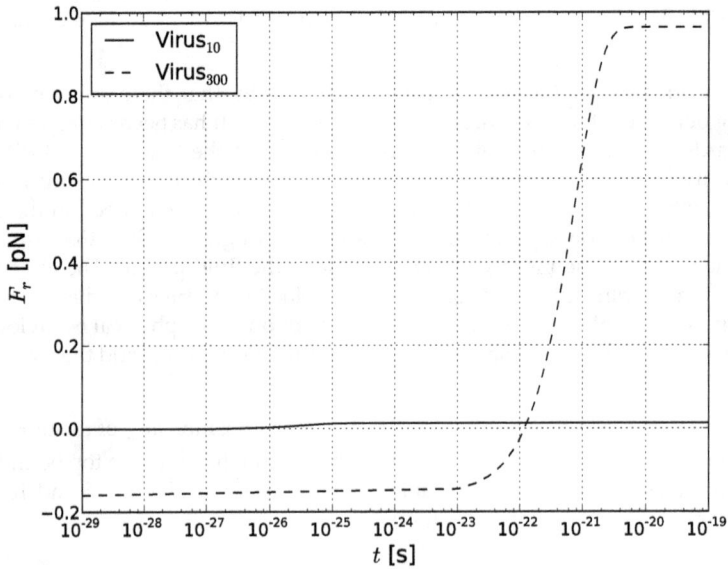

Fig. 14. The radial force as a function of time.

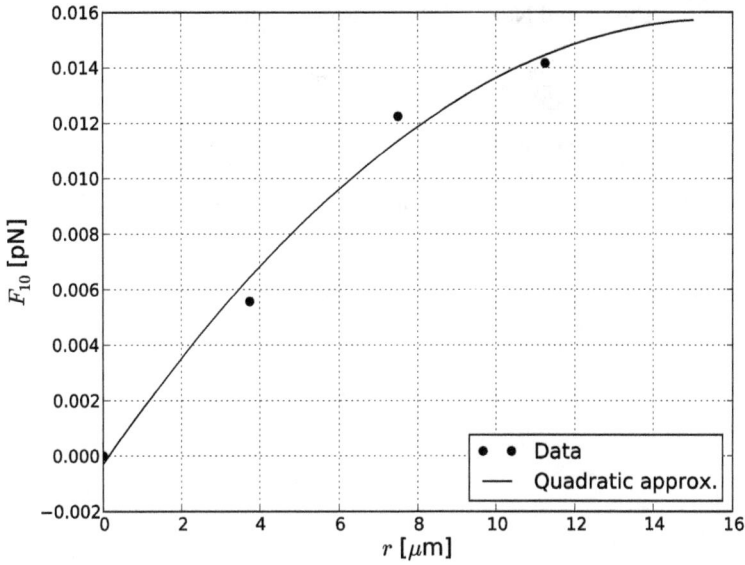

Fig. 15. The radial force at virus positions form the centreline up for the small virus.

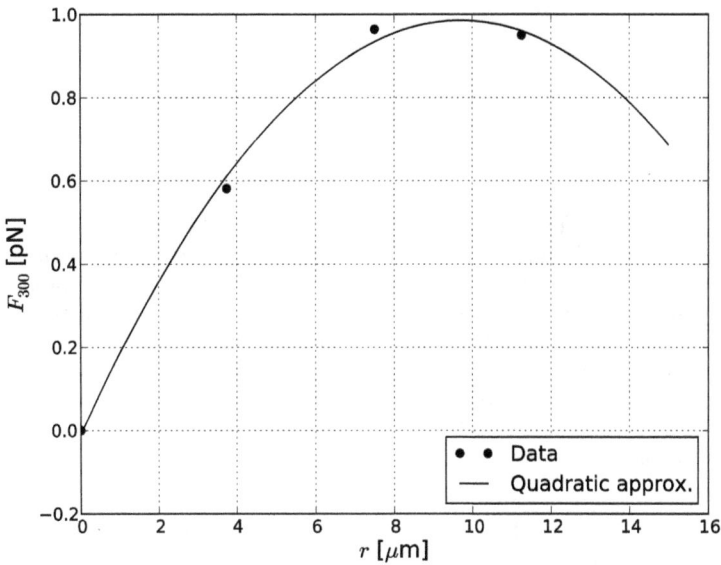

Fig. 16. The radial force at virus positions form the centreline up for the large virus.

where F_r stands for the radial force on virus, y for the virus displacement in the radial direction and a, b, c for the approximation parameters. This data is used to compute the virus kinematics in the radial direction. The second Newton's law was used again.

$$\frac{d^2y}{dt^2} = \frac{F_r(y)}{m} = Ay^2 + By + C \tag{18}$$

where m stands for the virus mass and A, B, C for the approximation parameters divided by m. The equation can be solved analytically. Nevertheless, the solution is awkward to use and the numerical integration of the differential equation (18) presented in figure 17 was used.

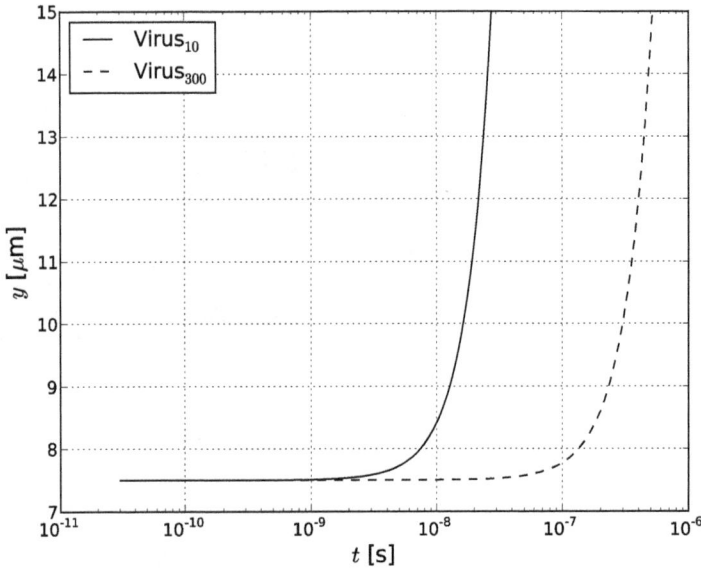

Fig. 17. The radial virus path.

It is interesting to compare figures 11 and 17. We can conclude for the larger virus that it is captured by the tube wall in $5.1 \cdot 10^{-7}$s and for the smaller virus that its capturing time is approximately $2.7 \cdot 10^{-8}$s. The axial velocity of the larger virus at the capture time is only 0.15m/s, which is about 10% of its maximal axial velocity and its axial displacement is approximately $4.0 \cdot 10^{-8}$m, which is three orders of the magnitude less than its displacement in the radial direction. The picture is somewhat different with the smaller virus. It reaches 0.85m/s, which is already about 57% of its maximal axial velocity. It travels approximately $1.4 \cdot 10^{-8}$m in the axial direction, which is again three orders of the magnitude less than its displacement in the radial direction.

The mechanism of filtration presented here, which is based on fluid induced rotation of a round object, is closely tied to the Magnus effect. It is valid also in the case of the curved tube, due to the laminar nature of the fluid flow. Hence it is an efficient way of filtration of viruses and microbes.

4.4 Criticism of the classical theory of filtration

The fluid flow that flows around a single fibre is the base upon the classical filtration theory is built, as presented in figure 6. The flow is divided by the fibre and the steam lines that are curved around the fibre, get together again behind the fibre.

The situation, described by the classical filtration theory, is actually occurring during the air flow through the mask. However, the amount of air experiencing it is much smaller than the amount of air that is flowing through pores; see figures 2, 3 and 4, particularly due to the reason that the porosity measurement takes into account split pores.

The shortcomings of the classical filtration theory can be summarized as follows:

1. The fibres have a role of the pore boundary at non-woven fabrics. There are caverns in between the fibres where the air does not flow, see figure 5. The virtual surface is formed between the fluid flow and the void when fluid is in motion. The pore caverns, where there is never a significant air flow, are not explained by the classical filtration theory.

2. The classical filtration theory foresees the Brownian motion with one predominant direction, see figure 6, for particles smaller than $0.15\mu m$. Such particles supposedly hit the fibre by being thrown out of the flow by the diffusion mechanism. They may be forced by a collision with a larger particle or supposedly by air molecules. The mechanism of the collision with the larger particle is feasible. On the contrary, the model of the mechanics of continuum does not support a mechanism of filtration based on the Brownian motion of the small particle in the fluid flow. The Brownian motion of the small particles plays an important role in the caverns, where there is predominantly still air.

3. The classical filtration theory considers the diameter of a fibre as the only characteristic linear measure when computing the Reynolds number. It has been shown that, if the maximal pore diameter is taken into account, the Reynolds number is the order of the magnitude larger. The Reynolds number for the flow around an object in the flow is much smaller, but the limit where turbulent flow starts is also much smaller than in the case of the flow in a tube.

4.5 Mechanisms of filtration of the medical masks

The main difference between the classical filtration theory and the theory presented here is that the air flowing through the medical mask is supposed to flow through channels. This claim is supported by the theoretical and experimental findings of the J-method for the flat textile porosity assessment. The laminar flow through a channel (or a tube) establishes parabolic fluid flow velocity profile. The velocity profile enables the self-induced rotation of the spherical or near-spherical particles. The angular velocity reaches its limit value almost instantly causing the radial force to point towards the channel boundary due to the Magnus effect. Thus, the effective filtering is enabled and the whole phenomenon is named as the J-effect.

Based on the developed theory, numerical simulations and observations of the filtration of the industry waste gases, where needled flat textile is used, the mechanisms of filtration can be summarized in following points.

1. The particle, which is positioned at the centerline of the channel (tube), stays there until it collides with another particle or until the channel changes the direction.

2. When the particle is captured by the fiber surface or is separated from the flow into a cavern, then its kinetic energy is too low for the particle to be sucked into the flow again.

3. The particle, which velocity is lagging from the flow one, does not exhibit Brownian motion.

4. Small particles, which are positioned in the caverns, exhibit Brownian motion and are filtered out by being captured at the fibers' surfaces at the caverns' boundaries in the process of diffusion.

5. The volume of the caverns and the channels become an unified space during the undisturbed period between the inhalation and exhalation. Thus the viruses can move freely between the caverns and channels.

6. The particles having spherical or near-spherical shape starts to rotate, if they are not placed at the channel centre.

7. The particle rotation is caused by the velocity-pressure distribution around the particle surface which is the consequence of the parabolic velocity profile (J-effect). The angular velocity of the particle reaches stationary value almost instantly and this triggers the radial force that is driving the particle towards the channel boundary (Magnus effect). The main difference with the normal comprehension of the Magnus effect is that the rotation of the virus is of the endogenous origin (self-induced) and not of the exogenous one (J-effect).

8. The particles that have not been filtered during the inhalation or have migrated from the caverns to the channels during the undisturbed period, will, during the exhalation, behave in the same way as they would during the inhalation. Only the direction of the fluid flow changes.

9. The air flow through a channel is at least partly accelerated and decelerated by breathing of a subject which enhance the fluid induced forces acting on a virus. Only stationary state has been taken into account at the analytical/numerical analyses.

10. The channel configuration is complex, changing its direction and diameter many times. The geometry of the channel is not known to us except for its minimal diameter, which is identified by the J-method.

11. The cylindrical (helical) viruses generally do not rotate. The probability for them to be positioned in the channel just right to enable J-effect is extremely small. Hence, this viruses are difficult to filter in a straight channel (tube). Their size is normally much larger than 10nm and their inertia plays an important role at filtering those viruses at the channel's bends.

12. Except for the criticism expressed in this text, the mechanisms of the classical filtration theory still applies and they are not discussed further.

This analysis was limited to a straight channel (tube) and perfectly spherical virus, see figures 7 – 9 and figure 18. We have also established, that the channel length is large enough to support the stationary fluid flow. There are more parameters that affects the complex process of filtration. Some of them are difficult to assess, some are left aside to show the basic mechanism of filtration of the spherical or near-spherical particles in a medical mask. The interesting parameters that influence the filtration process are:

1. the concentration of the particles,

2. the particle interactions,

3. the size distribution of the particles,

4. the different shape of the particle,

5. the different thicknesses of the sub-layers of the mask filtering layer,

6. the thickness of the fibres,

7. the fibres' cross-section,

8. the maximal hydraulic pore diameter and the pore distribution,

9. the size of the caverns and

10. the breathing pattern as a function of a subject condition.

4.6 Illustration of the mechanisms of filtration of the medical masks

The virus behaviour of filtration (J-effect) is depicted in figure 18. If a virus is placed at the channel centerline, then the distribution of the fluid velocity and pressure around the virus is symmetrical, no virus rotation and consequently no radial force is present. The virus B stays on the centerline. Such a virus can be filtered at the channels' bends or if it is thrown out by a collision with another particle.

The virus A (or C) does not rotate at first and the radial force drives it toward the centerline, figure 18 (I). When the virus begins to rotate and this happens almost instantly and the direction of the radial force changes toward the channel boundary, figure 18 (II). It has to be stressed here that the velocity profile changes locally at the virus location and the parabolic profile is locally lost.

The process of filtration during and between the inhalation and the exhalation is shown in figures 19 – 21, where denotations are: A - cross-section, B - view along the channel, 1,2,3,4 -

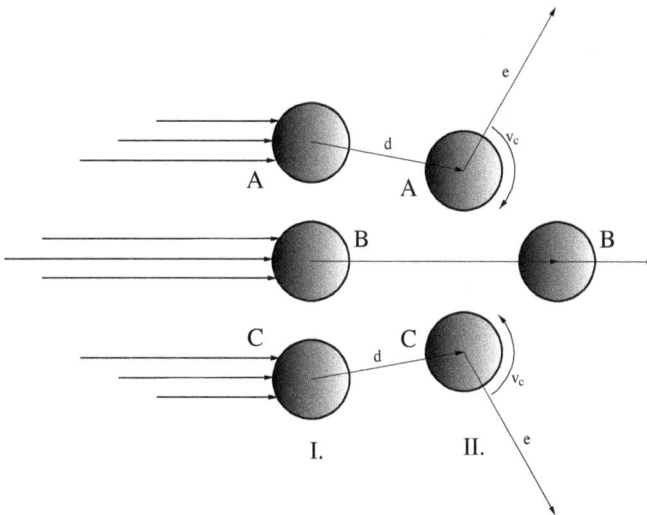

Fig. 18. Illustration of a virus in the fluid flow in a channel.

fibre, 5 - particle, that has been captured by a fibre surface, 6 - particle, that has been separated into a cavern, 7 - streamline, 8 - airflow velocity profile, 9 - mouth of the mask user.

The filtration process during inhalation is shown in figure 19. There are five particles shown at the entrance of the channel, one of them is placed at the channel centreline (position at the beginning). This virus travels the fastest in the axial direction due to the airflow velocity profile. The other viruses travels also into the axial directions due to the J-effect, it is clearly visible in the second position, which presents the second time snapshot. In the third snapshot, the viruses, most distant to the centreline, have already travelled to the vicinity of the fibre surface. They are captured by the surface in the next time snapshot. In the last snapshot the other viruses are filtered out to the caverns and only the virus on the centreline stays in the

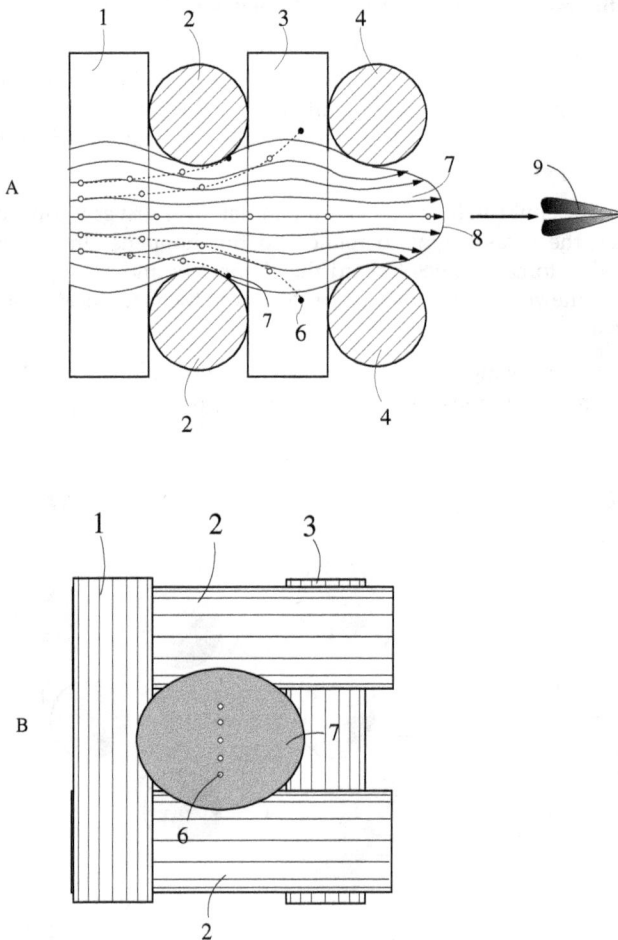

Fig. 19. Inhalation - illustration of a movement of small particles, which would exhibit Brownian motion in a still air.

airflow. This means, that there is no particles larger than 10nm left in the airflow that might be filtered out in the diffusion process. It is clear that the diffusion process does not apply here.

The frozen state between the inhalation and exhalation (undisturbed period) is shown in figure 20. There are no streamlines and the channel and cavern form a single space and thus enabling the diffusion exchange of the particles between the channel and cavern and filtering based on the Brownian motion.

In the figure 21 the phenomenology is the same as in figure 19, just the direction of the airflow is reversed.

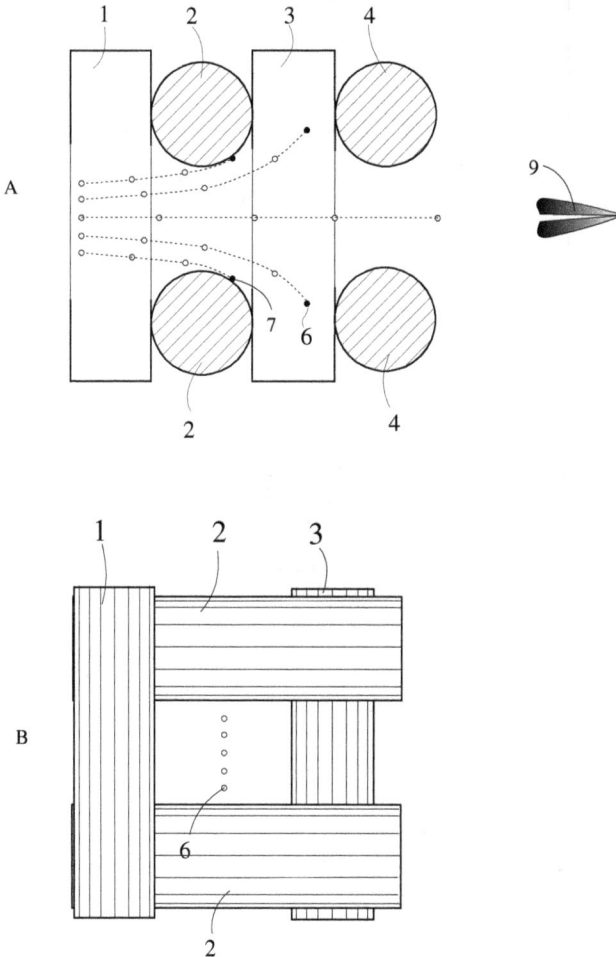

Fig. 20. Undisturbed period between the inhalation and exhalation.

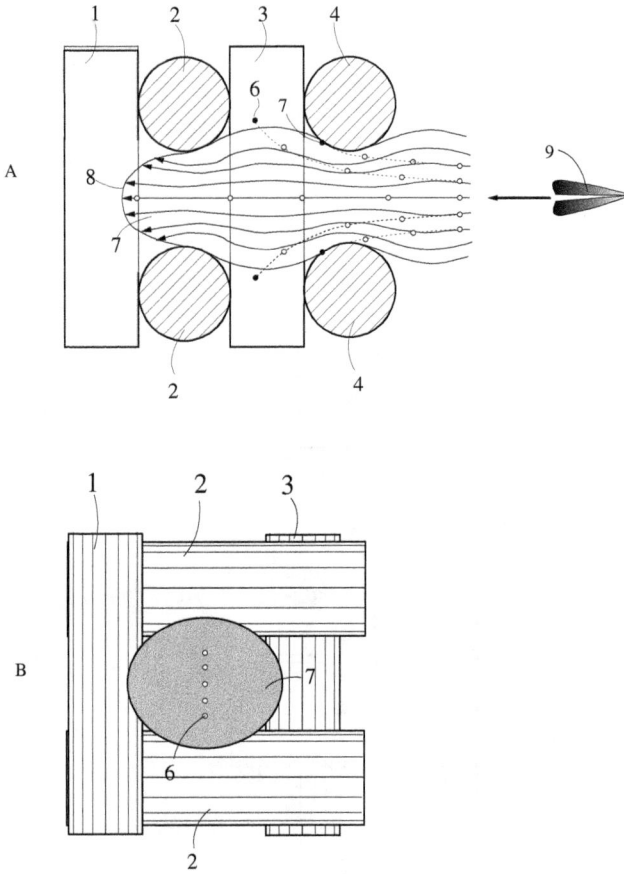

Fig. 21. Exhalation - illustration of a movement of small particles.

5. Conclusions

The novel approach to the filtering mechanisms of nano-particles through medical masks is presented in this chapter together with the criticism of the classical filtration theory.

The novel view on the filtration problem was enabled by development and usage of the J-method, which is the method for determining the porosity parameters in the flat textiles. The pores' hydraulic diameter distribution is one of the method's results, which considers that the channels are formed from one surface of a flat textile to the other one. The channel cross-section size is defined by the pore hydraulic diameter distribution. We showed that the air flow through channels is laminar and that the problem size is in domain of the continuous mechanics. The laminar air flow through channel forms distinctive velocity profile, which is responsible for driving the spherical virus rotation. The Magnus effect appears due to the self-induced rotation of the virus. The axial force is thus generated that is driving a virus toward the channel boundaries. This is named as J-effect.

The novel view on the filtration problem was enabled by development and usage of the J-method, which is the method for determining the porosity parameters in the flat textiles. The pores' hydraulic diameter distribution is one of the method's results, which considers that the channels are formed from one surface of a flat textile to the other one. The channel cross-section size is defined by the pore hydraulic diameter distribution. We showed that the air flow through channels is laminar and that the problem size is in domain of the continuous mechanics. The laminar air flow through channel forms distinctive velocity profile, which is responsible for driving the spherical virus rotation. The Magnus effect appears due to the self-induced rotation of the virus. The axial force is thus generated that is driving a virus toward the channel boundaries. This is named as J-effect.

The numerical investigation, using the computational fluid dynamics approach and the classical Newtonian mechanics, showed that the J-effect is a phenomenon that appears fast and is therefore an efficient mechanism of filtering spherical or near-spherical viruses even in the straight channel.

On the other hand, the Brownian motion of the particles trapped in the air flow as advocated by the classical filtration theory could be dismissed as unrealistic.

There are some questions that have not been addressed here but are also important for understanding of the complete filtration phenomenon. Some of them could be partially addressed by the classical filtration theory. The questions that arise with this analysis could be summarised as:

1. What is happening at the channel bend?
2. What is happening in the caverns?
3. How is the particle separated into a cavern?
4. What share of the particles is separated into cavern?
5. How the particle interaction is affecting the filtering process?

6. References

Brown, R. (1993). *Air Filtration: An Integrated Approach to the Theory and Applications of Fibrous Filters*, Pergamon Press.

Hutten, I. (2007). *Handbook of nonwoven filter media*, Elsevier.

Jakšić, D. & Jakšić, N. (2007). Assessment of porosity of flat textile fabrics, *Textile Research Journal* 77(2): 105 – 110.

Jakšić, D. & Jakšić, N. (2010). *Woven fabric engineering*, Sciyo, cop., chapter Porosity of the flat textiles (Chapter 14), pp. 255 – 272.
    URL: *http://www.intechopen.com/books/show/title/woven-fabric-engineering*

Sharma, A. (2000). *Penetration, pressure drop, and wicking characteristics of niosh certified p-100 and p-95 filters under heavy dop holding*, Master's thesis, College of Engineering and Mineral Resources at West Virginia University.

Widden, M. (1966). *Fluid Mechanics*, Macmillan Press.

Wikipedia (2011). http://en.wikipedia.org/wiki/rotavirus.
    URL: *http://en.wikipedia.org/wiki/Rotavirus*

Zuurbier, M., Hoek, H., Hazel, P. & Brunekreef, B. (2009). Minute ventilation of cyclists, car and bus passengers: an experimental study, *Environmental Health* 8(48): 0 – 0.
    URL: *http://www.ehjournal.net/content/8/1/48*

10

Microbial Degradation of Woven Fabrics and Protection Against Biodegradation

Beata Gutarowska[1] and Andrzej Michalski[2]
[1]*Technical University of Lodz,
Institute of Technology Fermentation and Microbiology, Lodz,
[2]Spółdzielnia Inwalidów ZGODA, Konstantynów Łódzki,
Poland*

1. Introduction

The textile industry is one of the most important and fastest developing industries in the world. An significant problem encountered by manufacturers is that of ensuring that the fabrics produced are of suitable quality and durability. Particular attention needs to be paid to the destructive action of microorganisms present in the environment.

In favourable conditions these can rapidly destroy material, rendering it entirely unusable and causing substantial economic losses.

In 1960, in the UK, annual losses due to biological degradation of cotton fibres were put at 110 000 tonnes of cotton, which at time was 1% of output (Howard & Mc Cord). According to estimates by Hueck-van der Plas (1971), the process of biodeterioration affected 2% of annual production of natural and artificial fibres (Zyska, 1977).

At the start of the 21st century annual world consumption of unwoven fabrics (for practical and technical uses) stood at 46 million tonnes, of which synthetics and cotton accounted for 49% and 42% respectively (with an upward trend in subsequent years), wool for 5%, and other fabrics 4% (linen, sisal, silk and others, with a downward trend) (Central Statistical Office Yearbooks – *Roczniki GUS*, Poland 2008). If 2% of the global value of fibre production is assumed, the problem of microbiological decomposition may affect 920 000 tonnes of fabric annually.

Not all losses can have a price attached to them: museum fabrics are particularly rapidly damaged by microorganisms, and the artistic and cultural value of these items cannot be recreated.

Microorganisms which attack textile products not only have a destructive effect, but also pose a significant danger to human health. Particularly dangerous are the pathogenic microorganisms present on fabrics which come into direct contact with the human body, such as on dressings and surgical masks; this may lead to skin infection, and even heart disorders and pneumonia.

It is a significant challenge for manufacturers to produce fabrics with antimicrobial properties – namely bioactive fabrics, containing biocides to provide protection against pathogenic microorganisms.

A separate issue is the protection of finished textile materials against biodegradation through proper storage, and possibly the use of an appropriate process of disinfection which can effectively eliminate microorganisms without affecting the material's strength properties.

2. Microbial degradation of fibers

Textiles are easily attacked by microorganisms, which means that they quickly become damaged. Microorganisms pose a threat to textile materials at all stages of their production – from the obtaining of raw material (for example on plantations), through to the transportation and storage of the raw material and of the finished product.

Microbial degradation of fabrics depends primarily on their chemical composition. Fabrics of natural origin are particularly susceptible to attack by microorganisms.

The decomposition of natural plant-based fibres caused by the presence of fungi was known as early as 1926–1928, and was described by Smith and Morris. Research into the mechanism of the decomposition of such fibres by microorganisms has continued for 80 years (Zyska, 1977). The main component of plant fibres is **cellulose**. The content of cellulose depends on the type of fibre – in cotton it reaches 94%, in linen fabric around 80%, and in others from 63% to 77% (jute, sisal, hemp). Cellulose is a polysaccharide composed of molecules of β-glucose linked by 1,4-β-glycoside bonds. The number of glucose molecules in a chain ranges from 7 to 10 thousand. Chains may be arranged in parallel, forming a crystalline structure, or tangled to form an amorphous structure. Cellulose is broken down by microorganisms through a process of enzymatic hydrolysis. This mechanism involves a multistage decomposition of cellulose to glucose, brought about successively by the enzymes 1,4-endo-β-D-glucan cellobiohydrolase (EC 3.2.1.91) (also called exoglucanase, cellobiohydrolase), endo-1-4-β-D-glucan glucanohydrolase (EC 3.2.1.4) (endoglucanase, β-glucanase) and glucohydrolase of β-D-glucosides (EC 3.2.1.21) (cellobiose, β-glucosidase) (Evans, 1996; Jeffries, 1987; Szostak-Kot, 2005).

The intensity of cellulose decomposition is indicated by the appearance of differently coloured stains on fabrics (carotenes, anthraquinones, excreted by the microorganisms), reduction in the degree of polymerization, breakage of the fibre structure and reduction in tearing strength. In extreme cases the cellulose may decompose completely.

Plant fibres also contain small quantities (up to 10%) of such compounds as hemicellulose and lignin, which give the fibres rigidity, and pectins, which act as a kind of glue. Many microorganisms are capable of producing enzymes which decompose hemicelluloses and pectins (xylanase, galactosidase, mannosidase, glucuronidase, pectinesterase, glycosidase and others) (Bujak & Targoński, 1990; Szostak-Kot, 2005). Lignin is the least rapidly decomposed component of plants, because of its structure – phenylpropane compounds are linked by ether and carbon bonds and are very resistant to enzymatic decomposition. In spite of this there are certain species of fungi and bacteria which are capable of decomposing lignin (*Chaetomium, Paeciliomyces, Fusarium, Nocardia, Streptomyces, Pseudomonas, Arthrobacter* and others) (Szostak-Kot, 2005; Targoński & Bujak 1991).

The rate of decomposition of natural plant-based fibres depends on their chemical composition. Among cellulose-based fibres, the slowest to decompose is jute (35% noncellulose substances, including 25% lignin) (Basu & Ghose, 1962; Szostak-Kot, 2005). The rate also depends on many other factors: apart from environmental factors and the type of microorganisms, there is also an effect from thickness, type of weave, degree of crystallinity (amorphous cellulose is more easily degraded) and degree of orientation (namely the angle made by the fibrils with the long axis of the fibre – highly oriented fibres are less susceptible to biodeterioration) (Pedersen et al., 1992; Salerno-Kochan & Szostak-Kotowa, 2001; Szostak-Kot, 2005; Tyndal, 1992).

Artificial cellulose fibres include regenerated fibres (rayon) and cellulose acetate. Rayon usually has a lower degree of crystallinity, polymerization and ordering than cotton. It is also highly hygroscopic (its capacity to absorb water in normal conditions is 9.8–13%), which is a reason for its common use in making woven and knitted fabrics and as an additive to natural and synthetic fibre products. Its rate of microbiological decomposition is comparable to that of cotton. Cellulose acetate is produced by the acetylation of cellulose with acetic anhydride, as a result of which the product has a maximal degree of acetylation, and the fibre becomes more resistant to microbiological decomposition than cellulose (Buchanan et al., 1993, Buschle-Diller et al., 1994; Salerno-Kochan & Szostak-Kotowa, 2001; Szostak-Kot, 2005).

Wool is characterized by high strength, thermal insulation properties and hygroscopicity (it can absorb 50% moisture without feeling wet). Chemically, wool is built from three types of keratins: low-sulphur, high-sulphur and high-tyrosine. Low-sulphur keratins primarily are linked with each other and to proteins of the matrix by numerous bonds – sulphide bridges, covalent bonds and hydrogen bonds, and in the presence of water also hydrophobic bonds. Due to the presence of these bonds and the network structure of wool, it is resistant to stretching and tearing and to environmental factors, including enzymatic degradation.

The biodeterioration of woollen fabrics involves microorganisms with mainly proteolytic and keratinolytic enzymes. So far 299 species of fungus with keratinolytic properties have been described, of which 107 are pathogenic to humans (Błyskal, 2009). Decomposition of a woollen fabric proceeds by way of deamination, sulphitolysis and proteolysis (Kunert, 1992, 2000). The first stage involves the splitting of disulphide bridges, which are the source of keratin's resistant strength. This is followed by the enzymatic decomposition of proteins by proteolytic enzymes (proteases) into oligopeptides, and these are then broken down by peptidases into amino acids, which are used in metabolic processes of oxidative deamination with the release of ammonia (Gochel et al., 1992; Kunert, 1989; Szostak-Kot, 2005). Characteristic symptoms of the microbiological decomposition of wool include the variously coloured stains on the fabric surface, a distinctive smell (in anaerobic conditions H_2S is produced), and loss of stretching strength.

During the technological process the woollen raw material is subjected to mechanical, chemical and photochemical action, which increases the susceptibility of the fibres to biodegradation. Many problems have been reported and described resulting from the development of microorganisms on woollen textiles, for example when carpets are in storage (Gochel et al., 1992; Hoare, 1968; Simpson, 1987). In favourable conditions of temperature (37°C) and humidity of the material (25–75%), the number of fungi may increase to as much as 109 CFU/1g of wool over 20 days (Zyska, 2001).

Natural **silk** is a fibre produced from the cocoon shell of the mulberry silkworm. Silk is characterized by high strength, elasticity, thermal insulation properties and hygroscopicity (in natural conditions silk contains approximately 11% moisture, and it can have a moisture content of 30% without feeling wet).

Raw silk consists of protein fibres – fibroins – stuck together with the protein sericin. The chains are linked by disulphide bridges, which give the fibre its strength; there are also hydrogen bonds within and between molecules. This polypeptide has a crystalline structure, and around 90% of it consists of four amino acids: alanine, glycine, serine and tyrosine. Textile manufacturers formerly used raw silk, which was resistant to the damaging action of light (chiefly ultraviolet) and was stronger, although the fabric yellowed with time. Fabrics are now made from degummed silk (with the sericin removed) – this material does not yellow under the action of light, and is more resistant to microbiological decomposition (Becker et al., 1995; Kaplan et al., 1994; Szostak-Kot, 2005). Microorganisms probably assimilate sericin more easily than fibroin. The decomposition of sericin involves mainly proteolytic enzymes of microorganisms (Forlani et al., 2000). In vitro tests have also confirmed the degradation of fibroin by protease (Horan et al., 2005).

Synthetic fibres are obtained by means of polymerization. The most commonly used types are polyamide, polyester, polyurethane and polyacrylonitrile fabrics. Synthetic fabrics are resistant to biodeterioration as a rule, and if the process occurs, it is a long-lasting one.

Synthetic fibres which have undergone a process of biodegradation become less resistant to stretching (by as much as 20–30%), undergo swelling (increase in diameter by up to 20%), and change colour due to microbially produced dyes and acidic products which react with the dyes present in the fabrics (Zyska, 2001)..

Mechanisms of biodegradation involve physical damage to fibres and chemical decomposition due to numerous metabolites produced by microorganisms (ammonia, nitrates, hydrogen sulphide, organic acids) or by an enzymatic route (activity of lipases, esterases, proteases, ureases) (Lucas et al., 2008).

Polyamide fibres contain amide groups in the main chain of their macromolecules. Greatest interest is shown in aliphatic polyamides, and among them, polyamide 6 (Steelon, Perlon) and polyamide 6.6 (Nylon). Polyamides are resistant to microbiological decomposition, although research is carried out using various strains of microorganisms which contribute to that process. It has been found that some bacterial and fungal oxidases and hydrolases (for example manganase peroxidase from white rot Basidiomycetes) decompose aliphatic polyamides, leading to their depolymerisation (Friedrich et al., 2007; Lucas et al., 2008).

In the textile industry there are two types of **polyurethane** fibres used: high-crystalline types with a linear structure, and highly elastic segmental fibres of the Spandex type. High-crystalline fibres have a similar structure to polyamides, and display high rigidity. The highly elastic type of fibres contain a minimum of 85% polyurethane polymer with a segmental structure. This fibre has a very large extension at rupture, colour permanence, and resistance to radiation and ageing. Microbial degradation of polyurethanes occurs by way of chemical hydrolysis, as a result of the extracellular action of esterase enzymes (Akutsu et al., 1998; Allen et al., 1990; Ruiz et al., 1999).

Polyester fibres are produced from large-molecule compounds with repeating ester bonds in the main chain. The type most commonly used in the textile industry is poly(ethyl terephthalate) (PET), while among aliphatic polyesters polylactic acid (PLA) is beginning to take on great importance. Polyester fibres containing terephthalate are resistant to microbiological decomposition, although in research into the effect of soil microflora such fabric displayed changes in fibre structure, which may indicate the possibility of biodegradation over a long period (Salerno-Kochan & Szostak-Kotowa, 1997). The processes of decomposition of PLA may involve enzymes such as proteinase K (Li & Vert, 1995).

Polyacrylonitrile fibres are produced from polyacrylonitrile, or else are copolymers of acrylonitrile with other monomers containing groups capable of reacting with reactive dyes. These fibres have high resistance to atmospheric effects, a pleasant feel, good strength and resistance to chemical and biological agents. Polyacrylonitrile, as well as dipolymers and terpolymers of acrylonitrile, are resistant to microbiological decomposition, although at high air relative humidity (90%) mould attack on the surface of polyacrylonitrile has been described (Zyska, 2001).

Biodeterioration of fabrics is mainly caused by filamentous fungi, and to a lesser extent by bacteria. Microorganisms capable of degrading natural and artificial fibres are listed in Table 1. (based on a survey of the literature).

3. Conditions favourable for biodegradation of fibres and fabrics

The rate of microbiological decomposition of fabrics is affected by environmental factors such as air relative humidity, temperature, light, and the properties of the fabrics, chiefly their chemical composition, fibre structure, density and thickness of weave, and the type of substances used in the finishing of the unwoven fabric (Szostak-Kot, 2005).

High humidity in a fabric is the most important factor affecting the development of microorganisms. The absorption of water by a fabric depends, among other things, on its hygroscopicity and porosity. A level of fabric relative humidity above 65% increases fibre swelling and favours the development of microorganisms, particularly moulds, on the fabric. The development of bacteria requires a high fabric relative humidity, above 95% (Szostak-Kot, 2009). At the **temperature** used for fabric storage (20–35°C) many microorganisms develop on the fabrics, and the range within which microorganisms develop is significantly greater (4–50°C, excluding extremophilic microorganisms).

All fibres are sensitive to photo-oxidation caused by **light radiation** (particularly ultraviolet and infrared). Ultraviolet radiation in cellulose fibres, such as cotton, causes breakage of the cellulose chain and leads to its decomposition. Wool and silk are also susceptible to photochemical degradation, particularly in the presence of oxygen – for example the photodegradation of fibroin in silk occurs as a result of the breakage of hydrogen bonds and oxidation of tyrosine. Biodegradation of silk may be favoured by prior photodegradation under the action of ultraviolet (Sionkowska & Planecka, 2011). The action of infrared radiation on textile material causes overheating of the surface and leads to many physicochemical changes. Light, increased temperature and atmospheric impurities additionally speed up the process of ageing, and in such conditions fabrics may also be more sensitive to attack by microorganisms (Szostak-Kot, 2009).

Physical features of fabrics, such as fabric thickness and density of weave, may enable the spread of microorganisms and processes of fabric destruction (thinner fabrics with a looser weave are subject to more rapid decomposition). The microbiological decomposition of a fabric is also affected by the substances added to the fabric, such as dyes, glues and treatments. These may provide an additional source of food for microorganisms, or else may have a negative effect on their development (Szostak-Kot, 2005). Many substances currently used in the textile industry are characterized in terms of susceptibility to microbiological decomposition or effect on microorganisms.

4. Protection of fibres against microbial degradation

Control of environmental conditions during storage, transportation and use is an effective method for protecting fibres against biodeterioration. This involves the maintenance of constant environmental conditions which are unfavourable to microorganisms, with the use of ventilation and air-conditioning devices. The temperature in storage rooms should be maintained at 18–20°C, and the air relative humidity should not exceed 60% (Szostak-Kot, 2005).

To combat the development of microorganisms, chemical compounds known as **biocides** are used. These are added at various stages of fibre production. Biocides make it possible to eliminate microorganisms effectively, but if used improperly may cause damage to the health of the user.

Modern biocides are expected to satisfy several basic criteria (Figure 1): high effectiveness at low concentrations (of the order of ppm) against a wide spectrum of microorganisms, absence of increased immunity of microorganisms, lack of effect on the properties of the fabric, absence of toxic or allergenic action or irritant action to the skin and mucous membranes in humans and animals, high biodegradability following application, good water solvency, low volatility, absence of smell, high stability (durability), absence of corrosive action on technical materials, and favourable price.

Nonvowens	Microorganisms isolated from nonwovens and/or able to biodegradation of nonwoven	Author , year
Cotton	**Fungi:** *Aspergillus* sp. (*A.versicolor, A.flavus, A.fumigatus, A.niger, A.terreus, A.nidulans, A.ustus, A.fischerii, A.flaschentraegeri);Penicillium* sp. (*P.notatum, P.citrinum, P.funiculosum, P.cyclopium, P.janthinellum*); *Cladosporium* sp. (*C.macrocarpium, C.herbarum*); *Cheatomimum* sp. (*Ch.globusum, Ch.cochlioides*); *Alternaria* sp. (*A.tennuis, A.geophila*); *Trichoderma* sp. (*T.viride, T.reesei*); *Fusarium nivale; Myrothecium* sp.; *Memnoniella* sp.; *Stachybotrys* sp.; *Verticillum* sp.; **Bacteria:** *Cytophaga* sp.; *Cellulomonas* sp.; *Bacillus* sp.; *Clostridium* sp.; *Sporocytophaga* sp.; *Microbispora bispora*	Abdel-Kareem et al., 1997; Bartley et al., 1984; Evans, 1996; Flannigan et al., 2001;Kowalik, 1980; Kubicek et al.,1988

Nonvowens	Microorganisms isolated from nonwovens and/or able to biodegradation of nonwoven	Author , year
Flax	**Fungi:** *Aspergillus* sp. (*A.flavus, A.fumigatus, A.niger, A.terreus, A.nidulans, A.ustus, A.fischeri, A.auratus, A.carbonarius, A.proliferans, A.spinulosus*); *Penicillium* sp. *P.funiculosum,P.rajstrickii, P.biforme, P.soopi*) *Trichoderma viride; Alternaria alternata; Cheatomium cochlioides; Fusarium nivale*	Abdel-Kareem et al., 1997
Wool	**Fungi:** *Aspergillus* sp. (*A. cervinus, A. fischeri, A.flavus, A. fumigatus, A.nidulans, A.niger, A.rapier, A.sparsus, A.spinulosus, A.ventii*); *Chrysosporium* sp.; *Penicillium* sp. (*P.canescens, P.cyclopium, P.granulatum, P.lanoso, P.paxilli, P.soopi*); *Microsporum* sp.; *Trichopchyton* sp.; *Fusarium* sp.; *Rhizopus* sp.; *Cheatomium* sp.; *Alternaria* sp.;*Ulocladium* sp.; *Stachybotrys chartarum; Scopulariopsis brevicaulis; Acremonium* sp.; **Bacteria:** *Bacillus* sp. (*B.mesentericus, B. subtilis, B.cereus, B..mycoides*); *Pseudomonas* sp.; *Streptomyces* sp. (*S.fradiae*)	Abdel-Kareem et al., 1997; Abdel-Gawada, 1997; Agarwal & Puvathingal, 1969; Błyskal, 2009; Kowalik, 1980; Lewis, 1981; McCarthy & Greaves, 1988; Nigam & Kushwaha, 1992; Safranek & Goos, 1982
Silk	**Fungi:** *Aspergillus* sp. (*A.flavus, A.niger, A.rapei*); *Penicillium* sp. (*P.canescens,P.paxilli*); *Chaetomium* sp.; *Cladosporium* sp.; *Rhizopus* sp.; **Bacteria:** *Bacillus megaterium; Pseudomomas* sp. (*P.aureofaciens, P.chlororaphis, P.paucimobilis P.cepacia*); *Serratia* sp.; *Streptomyces* sp.; *Variovorax paradoxus*	Abdel-Kareem et al., 1997; Forlani et al., 2000; Ishiguro & Miyashita, 1996; Nigam et al., 1972; Sato, 1976; Seves et al., 1998
Polyamide	**Fungi:** *Aspergillus* sp. (*A.niger*); *Penicillium* sp. (*P.janthinellum*); *Blennoria* sp.; *Monascus* sp.; *Tritirachium oryzae; Absidia* sp.; *Trichosporon* sp.; *Rhodotorula* sp.; white rot *Basidiomycetes* **Bacteria:** *Pseudomonas* sp. (*P.aeruginosa*); *Protaminobacter* sp.; *Achromobacte* sp.; *Brevibacterium* sp.; *Flavobacterium* sp.; *Alcaligenes* sp.; *Bacillus* sp. (*B.pallidus*); *Corynebacterium* sp.	Bailey et al., 1976; ; Cain, 1992; Denizel et al.,1974; Ennis et al., 1978; Nigam et al., 1972; Prijambada et al., 1995; Szostak-Kotowa, 2004; 2005
Polyurethane	**Fungi:** *Aspergillus terreus; Penicillium* sp.; *Cladosporium* sp.; *Paecilomyces* sp.; *Alternaria* sp.; *Trichoderma* sp.; *Stachybotrys* sp.; *Chaetomium globusom; Curvularia senegalensis; Fusarium solani; Aureobasidium pullulans; Glicoladium roseum; Stemphylium* sp. **Bacteria:** *Pseudomonas* sp.; *Acinetobacter calcoaceticus; Arthrobacter globiformis*	Halim El-Sayed et al., 1996; Howard, 2002; Szostak-Kotowa, 2004; Wales & Sagar, 1988
Polyacrylo-nitrile	**Fungi:** *Aspergillus* sp.; *Penicillium* sp.; *Stachybotrys* sp. **Bacteria:** *Arthrobacter* sp.	Szostak-Kotowa, 2004; Yamada et al., 1979; Zyska, 2001

Table 1. Fibre-degrading microorganisms

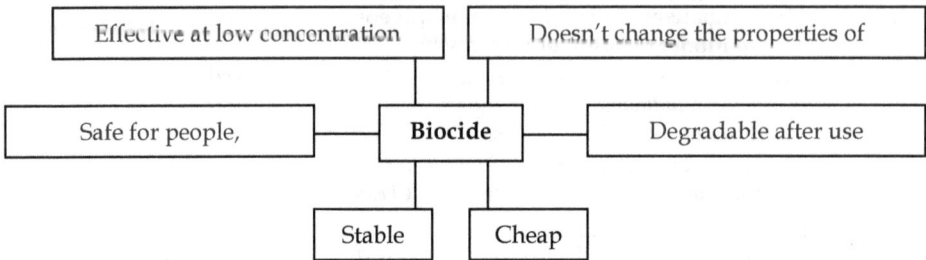

Fig 1. Features of good biocide for nonwovens

There are few active compounds that meet all of these requirements, and therefore work is still being done to find substances for use in fabrics with the desired properties.

5. Biocides approved by the EU for use in the textile industry

Biocides which can be used in the textile industry belong to biocidal compounds category II and to group 9 on the list of biocidal compounds (under Directive 98/8/EC), and include 134 active substances. The active substances in category II which can be used in the textile industry belong to eight groups of chemical compounds: inorganic compounds, compounds of nitrogen, phenol and their derivatives, compounds of halogens and their derivatives, oxidizing compounds, alcohols, aldehydes, organic acids and their derivatives (Table 2).

Mechanisms of action on microorganisms depend on the type of compound, and often take multiple routes. Biocides cause disturbance of the functioning of the cytoplasmic membrane and cell wall, inactivation of proteins, slowing of DNA synthesis, and many other types of damage to the cells of microorganisms (Brycki, 2003).

Examples of commercial preparations containing the listed chemical compounds, which are currently used frequently in the textile industry, are listed in Table 3. On the international market many firms also offer ready-made fibres with antimicrobial properties, containing biologically active substances – examples of these are given in Table 4.

However, the mass use of chemical preparations and ready-made fibres containing biologically active substances lead to an increase in microorganisms' resistance to biocides. There are also increased requirements in terms of high effectiveness against a wide spectrum of microorganisms. For this reason new solutions are constantly being sought – new compounds or mixtures of compounds, and methods for stabilizing them and applying them to fabrics. The development of new technologies related to the production of fibres with antimicrobial action has proceeded particularly rapidly in the last decade. An overview of selected research into production and antimicrobial activity of natural and artificial fibres containing biocides is presented in Table 5. Extensive data concerning polymeric materials can be found in the survey paper by Munoz -Bonilla and Fernandez- Garcia (2011).

Among the biocides used in both natural and artificial fibres, there is a high level of interest in quaternary ammonium salts and phosphonium salts. The role of biological agent is played by the fibre additive chitosan, as well as antibiotics tetracycline, cephalosporin, vinyloimidazol, ciprofloxacin, and antifungal clotrimazol, ketokonazol. There has recently also been great interest in inorganic nanoparticles as agents with antimicrobial properties.

The obtaining of nanoparticles of metals or nanostructured fibres has, thanks to the increase in surface area, led to the achievement of new characteristics desirable in the textile industry, and significantly greater effectiveness in destroying microorganisms. High activity have: TiO_2 nanoparticles, metallic and non metallic TiO_2 nanocomposites, titania nanotubes (TNTs), silver nanoparticles, silver-based nanostructured materiale, gold nanoparticles, zinc oxide nanoparticles and nano-rods, copper nanoparticles, metallic and inorganic dendrimers nanocomposite, nanocapsules cyclodextrins containing nanoparticles. New methods of obtaining such fibres with the addition of nanoparticles are constantly being developed, and the stabilization of nanoparticles on the surface of fibres is also of great importance (Dastjerdii et al., 2009; Dastjerdi & Montazer, 2010, Silver, 2003).

One of the conditions which a biocide is required to satisfy is its safety, and therefore much attention is currently being paid to substances of natural origin which are not toxic or allergic and are easily biodegradable. Some natural dyes and substances extracted from plant seeds and fruit contain active substances which slow the development of microorganisms and can be used to produce biologically active fabrics (Table 6).

The active substances	Antimicrobial activity	Mechanisms of antimicrobial activity
Inorganic compounds metals such as silver, zinc, copper, metal oxides such as titanium dioxide, metal salts	gram positive, gram negative bacteria, fungi, viruses	Inhibition of DNA replication, denaturation of proteins, abnormal functioning of the cytoplasmic membrane, outflow of the low molecular masses intracellular components from cell, disruption of transport of electrons and protons
Nitrogen compounds aliphatic amines such as N-(3-aminopropyl)-N-dodecylopropano-1 ,3-diamine; bis (3-aminopeopylo) octylamine; quaternary alkyl ammonium salts such as chloride, didecyl dimethyl ammonium chloride, alkylobenzylodimetyloamomoniu m chloride; guanidine, alkyl of aza compounds, oksaaza, tiaaza aromatic compounds	gram positive, gram negative bacteria, fungi, viruses	Damage and dysfunction of cytoplasmic membrane and cell wall, outflow of the low molecular masses intracellular components from cell, guanidine - inhibition of DNA replication, protein denaturation
Phenol and its compounds mono cyclic compounds such as chlorocresol, chloroxylenol, bis-phenol compounds, triclosan, dichlorophen, biphenyl-2-ol	gram positive, bacteria (including *Mycobacterium tuberculosis*) , gram negative bacteria, viruses	Inhibition of DNA replication, denaturation of proteins, damage and dysfunction of cytoplasmic membrane and cell wall, outflow of the low molecular masses intracellular components from cell
Halogens and their compounds inorganic: chlorine, iodine, sodium	gram positive, gram negative	Denaturation of proteins, damage and dysfunction of cytoplasmic

The active substances	Antimicrobial activity	Mechanisms of antimicrobial activity
and calcium hypochloride, sodium chlorate, chlorine dioxide; organic: chloroarylamides, halohydantoin, chloroisocyanuric acid	bacteria	membrane and cell wall, outflow of the low molecular masses intracellular components from cell
Oxidizing compounds peracetic acid, , peroxyoctanoic acid, hydrogen peroxide, 2-butanone peroxide	gram positive, gram negative bacteria, viruses	Transformation of sulfhydryl groups to di sulfide bridges in protein - protein deactivation
Alcohols propan-2-ol, 2-phenoxyethanol, benzyloxymethanol, 2,4-dichlorobenzyl alcohol-	gram positive, gram negative bacteria	Inhibition of DNA replication, denaturation of proteins
Aldehydes formaldehyde, dialdehydes glyoxal, glutaraldehyde, orthophthalic aldehyde	gram positive (including *Mycobacterium tuberculosis*, and bacterial spores), gram negative bacteria, viruses,	Inhibition of DNA replication, denaturation of proteins (by joining the amino groups), dysfunction of cytoplasmic membrane and cell wall, outflow of the low molecular masses intracellular components from cell
Organic acids and their compounds aliphatic: carboxylic acids: formic, glycolic, lactic, nonanoic; aromatic acids: benzoic, salicylic karbaminic	gram positive, gram negative bacteria	Inhibition of DNA replication, denaturation of proteins, damage and dysfunction of cytoplasmic membrane and cell wall, outflow of the low molecular masses intracellular components from cell

Table 2. Biocides used in protection of nonwovens and their antimicrobial activity (based on: Brycki, 2003)

Chemical preparations (trade name)	The active substances
Afrotin ZNK 10 & ZNL	Zinc pyridinate
Actifresh RT-87-11	Alkaline mixture of halogenated organic compounds
Armesan A	Phenyloxy-chloro-phenol
Biocide PB 940	2,2 '-dihydroxy-5 ,5-dichloro-diphenyl-mono-sulfide
Cuniculate 2419-75;Mystox 8	8- copper quinolinate
Densil P	Dithio-2, 2'-biobenzomethylamide
Esterol 100 CD; Mystox LPL	Pentachlorophenol laurate
Fungitex ROP Dichlorophen	Bis (chlorohydroxyphenyl) methane
GivGard DXN	6-acetoxy-2 ,4-dimethyl-1 ,3-dioxane
Kathon LM	2-octyl-4-isotiazolino-3-one
Metanit 55-61	Carbendazim + diuron
Myacide	2-bromo-2-nitropropane-1 ,3-diol
Myacide SP	2,4-dichloro-benzyl alcohol

Chemical preparations (trade name)	The active substances
Nuodex zinc naphtenate	Zinc naphthenate
Nuodex copper naphtenate	Copper naphtenate
Preventol GD	2,2 '-dihydroxy-5, 5'-dichloro-diphenylmethane
Preventol O extra	2-hydroxy-biphenyl
Preventol R80 & R50	Quaternary ammonium salts
Sanitized BSC	Tiobendazol
Sanitized DET 8530	Quaternary ammonium salts
Tolcide C30	2'-(thiocyanomethylthio) benzothiazole

Table 3. Chemical preparations with antimicrobial activity used in the textile industry (based on: McCarthy, 1995; Evans, 1996)

Bioactive nonwovens/ producer	Antimicrobial properties of nonwovens
Navaron/ Taogoesei Chemical Industry Co. (Japan)	natural origin fibers such as cotton with antibacterial and antifungal activity
Amicor AB, Amocor AF, Amicor Plus / Acords (England)	polyacrylic fibers with antibacterial and antifungal activity
Fibra K/ Asach Chemical Ind.Co (Japan)	viscose silk with colloidal sulfur with antibacterial activity
Gymlene/F.Drake Fibres (England), Microban/ Filament Fiber (USA)	polypropylene with antibacterial activity
Rhodia/ Rhodia Technical Fibres (Germany), Livefresh/ Kanebo (Japan)	polyamide with antibacterial activity
Huvis Corp. (Korea), Bacterkiller/ Kanebo (Japan), Kuraray/ Kuraray (Japan) , Trevira Bioactive / Trevira (Germany)	polyester with antibacterial activity
Rhovyl' AS/ Rhovyl (France)	PVC with antibacterial activity

Table 4. Nonwovens with antimicrobial activity (available on the global market)

Nonvowens	Antimicrobial agents	Antimicrobial activity	Author, year
Cotton	phosphonium salts with alkyl chains; chloroacetyl chloride, naphtylacetic acid, alginate-quaternary ammonium complex	active against *Staphylococcus aureus, Escherichia coli*	Jantas & Górna, 2006;Kanazawa et al., 1994; Kim et al., 2010
Polyurethane, Polyglicidyl methacrylate	quaternary ammonium salts with aliphatic triisocyanate, phosphonium salts	active against *S.aureus, E.coli, Pseudomonas aeruginosa, Bacillus subtilis, B.cereus, Shigella* sp., *Salmonella typhi,*	Kenawy et al., 2002; Kenawy & Mahmoud, 2003; Nurdin et al., 1993

Nonvowens	Antimicrobial agents	Antimicrobial activity	Author, year
		Trichophyton rubrum, Candida albicans, Aspergillus flavus, Fusarium oxysporium	
Polypropylene, Polypropylene-cotton	glycidal methacrylate, β-cyclodextrin, quaternary ammonium – chitosan complex, chitosan,	polypropylene with glycidal methacrylate, β-cyclodextrin, quaternary ammonium –chitosan complex active against *Lactobacillus plantarum, S.aureus, E.coli;* polypropylene with chitosan active against *S.aureus, E.coli, Proteus vulgaris,* not effective against: *Klebsiella pneumoniae, P.aeruginos;.* polypropylene-cotton with chitosan active against *Fusarium oxysporum, Verticillium alboatrum, Alternaria alternata, Clavibacter michiganensis, Pseudomonas solantacearum*	Abdou et al., 2005; Kim et al., 2010
Cotton	chitosan	active against *S.aureus*	Lim & Hudson, 2004
Cotton	silver, nanosilver	active against *Candida albicans, C.tropicalis, S.aureus, E.coli, K.pneumoniae, Streptococcus faecalis*	Gorensek & Recelj, 2007; Hipler et al., 2006; Sachinvala et al., 2007
Wool	silver and nanotitanium dioxide photo-induced	active against *S.aureus, E.coli,*	Montazer et al., 2011
Nylon, silk	nanosilver	active against *S.aureus*	Dubas et al.,2006
Poliacrylonitryle, poli(N-vinyl-pyrrolidone), PVC, cellulose acetate, Poliester, Polycaprolactone Polyurethane, Polipropylene	nanosilver, lidocaine, gold, zinc oxide nanotitanium dioxide	active against *S.aureus, E.coli, P.eruginosa*	Jain & Pradeep, 2005; Lala et al., 2007; Radetic et al., 2008, Yu et al., 2003
Phosphate glass fiber	copper (CuO)	active against *Staphylococcus epidermidis*	Abdou-Neel et al., 2005

Nonvowens	Antimicrobial agents	Antimicrobial activity	Author, year
Polypropylene , polypropylene with cotton	4-vinyl pyridine, radiation-induced	active against *E.coli*, depended on the structure and content of pyridinium groups, not bactericidal, but bacteriostatic	Tan et al., 2000
Cotton	N-halamine	active against *S.aureus*, *E.coli*,	Ren et al., 2009
Poly(L,L-lactide) on viscose	triclosan	active against *S.aureus*, *E.coli*	Goetzendorf-Grabowska et al., 2004
Polypropylene, polyacrylonitryle	tetracycline hydrochloride, vinyloimidazol, ciprofloxacin	active against *S.aureus*, *E.coli*, *K.pneumoniae*	Gupta et al., 2007, 2008
Poly(ethylene terephtalate)	cephalosporin	active against *S.aureus*, *E.coli*, *P.aeruginosa*	Bucheńska et al., 2003
Polyamide, polypropylene, polyester	clotrimazol, ketokonazol	active against *C.albicans*, *Penicillium funiculosum*, *P.mycetomagenum*, *Aspergillus niger*, *A.repens* *T.mentagrophytes*	Struszczyk et al., 2003

Table 5. Nonwovens with antimicrobial agents (based on the scientific researches)

Nonvowens	Antimicrobial agents	Antimicrobial activity	Author, year
Silk	Dyes from plants *Morinda citrifolia; Terminalia catappa, Artrocarpus heterophyllus, Tectona grandis* (contain of: flavonoids, quinonoids, indigoids, tannins)	active against *E.coli*, *K.pneumoniae*, *C.albicans*, *A.niger*	Prusty et al., 2010
Wool	Dyes Catechu from *Acacia catechu* (main component catechin)	active against *E.coli*, *S.aureus*, *C.albicans*, *C.tropicalis*	Khan et al., 2011
Wool	Dyes from *Acacia catechu, Kerria lacca, Quercus infectoria, Rubia cordifolia, Rumex maritimus*	active against *E.coli*, *B.subtilis*, *K.pneumoniae*, *Proteus vulgaris*, *P.aeruginosa*	Singh et al., 2005
Wool	Dye curcumin from *Curcuma longa*	active against *E.coli*, *S.aureus*,	Han & Yang, 2005
Cotton	Dyes from *Acacia catechu, Kerria lacca, Mallotus philippinensis, Punica granatum, Quercus infectoria, Terminalia chebula, Rheum emodi*	active against *E.coli*, *K.pneumoniae*, *Proteus vulgaris*,	Gupta et al., 2004

Nonvowens	Antimicrobial agents	Antimicrobial activity	Author, year
Chitosan and viscose	Flavonoids (flavanols, flavonol, flavone, flavanone, isoflavanone)	active against *B.subtilis*, *P.aeruginosa*	Sousa et al., 2009
Wool, cotton	Dye: Citrus grandis Osbeck extract	active against *S.aureus*, *K.pneumoniae*,	Yi et al., 2010
Cotton	Neem seed extract from *Azadirachta indica* (contain of:: azadirachtin, nimbin,, nimbidin, salannin, nimbidol, gedunin)	active against *B.subtilis*, *P.vulgaris*,	Joshi et al., 2007

Table 6. Nonwovens with natural origin antimicrobial agents (based on the scientific researches)

6. Factors affecting on the activity of biocides in the fibers

The activity of biocides depends on many factors, of which the most important include time of contact with the microorganisms, concentration of active substance, type of microorganism, presence of organic and inorganic impurities, temperature, humidity and pH.

The most important factors for affecting biocidal activity are **time** of contact between the active substance and the microorganism cells, and the biocide **concentration** (Brycki, 2003).

The product of the concentration and time of action for specified groups of active substances is a constant value, expressed in terms of Watson's equation:

$$c^{\eta} \times t = \text{const.} \tag{1}$$

where c denotes concentration, t denotes time, and η is a concentration coefficient determined empirically for a given substance.

Example values of the concentration coefficient (η) are 10 for alcohols, 6 for phenols, and 1 for quaternary ammonium salts.

The use of this relationship is important from a practical standpoint – it tells us that given an appropriate concentration of biocide, a biocidal effect will be achieved in a precisely specified time. With a preparation based on alcohol, for example, if it were diluted to half of the concentration, the length of time required to obtain the same effect would increase by 1024 times. In the case of phenol it would increase by 64 times, and for quaternary ammonium salts it would merely double. Because of their properties, alcohols work effectively for a short time, and hence their use is limited to short-lasting disinfection (Brycki, 2003).

With regard to the uses of bioactive fabrics, either a short time of action on microorganisms is required (for example, in the case of protective masks the time should not exceed 8 hours), or the time may be extended to 24–48 hours (filtration and technical materials, etc.).

The effective action of a biocide also depends on the **type of microorganisms**, chiefly the structure of the cell wall and the presence of genetic resistance mechanisms. For this reason, research into the antimicrobial activity of fabrics should include evaluation with respect to different species of microorganisms (Table 7).

Microorganisms	N_t After 6 h of incubation	
	Mean	SD
Escherichia coli	0.000	0.000
Pseudomonas aeruginosa	0.001	0.000
Klebsiella pneumoniae	0.001	0.000
Staphylococcus aureus	0.000	0.000
Micrococcus flavus	0.000	0.000
Bacillus subtilis	**0.550**	**0.348**
Candida albicans	0.000	0.000
Aspergillus niger	**0.036**	**0.006**
Penicillium chrysogenum	**0.004**	**0.001**

SD – standard deviation
*needle-punched nonwoven; polypropylene-silver (in the form of master batches) + acrylic fiber-biocide

N_t: $N_t = \dfrac{N}{N_0}$ where N_0 – the number of microorganisms on the sample of the textile material for time t
= 0, N – the number of microorganisms on the sample of the textile material for time t_n

Table 7. Microorganisms Survival Index (N_t) for various microorganisms after 6 hours incubation with bioactive nonwoven* (based on Majchrzycka et al., 2010)

Microorganisms sensitive to the action of biocides are bacteria – Gram-positive cocci and Gram-negative bacilli. The most resistant organisms, with high survival rates, include spore-forming bacteria and moulds (Majchrzycka et al., 2010). This is because the activity of biocides added to fabrics is dependent on the physiological state of the microorganisms: the most sensitive are cells in a phase of vegetative growth, while resistance is shown by endospore of bacteria and the spores of moulds (Gutarowska et al.,2010).

Organic contaminants present on the fabric may reduce the biological effect. Proteins are substances that protect microorganisms, sugars and fats may be a source of food and lead to the development of microorganisms, and moreover those compounds may react with the biocides, reducing their effectiveness. Research into antimicrobial activity in the presence of artificial sweat (inorganic compounds) did not reveal any significant effect on the bioactivity of fabrics (Majchrzycka et al., 2010).

Increased **temperature** generally strengthens the antimicrobial activity of chemical agents, due to the increased reactivity of the active substances as well as synergy between the destructive effects of the substance and temperature (Brycki, 2003).

Increased **humidity** strengthens the antimicrobial activity of fabrics containing biologically active substances. The presence of water makes it possible for the biocide to penetrate into the cells of microorganisms in the form of ions, and for these to act effectively. Hence fibres with **hydrophilic** properties containing biocides will be more effective than hydrophobic fibres containing the same active substances (Gutarowska et al., 2010). Comparative studies on the antimicrobial action of hydrophobic PAN fabrics containing quaternary ammonium salts and the same fabrics containing biocide on an inorganic medium – perlite – with hydrophilic properties showed a significant improvement in the biocidal effectiveness of hydrophilic fabrics with added perlite (Table 8). Bioactivity improved with increasing concentration of perlite, which changed the properties of the fabric to hydrophilic, and with increasing humidity of the fabric (Table 9).

Amount of bioperlite in the nonwoven (%)	Amount of alkylammonium microbiocides in the nonwoven (%)	Number of bacteria (CFU/sample)					
		Escherichia coli			Staphylococcus aureus		
		Incubation time		Reduction %	Incubation time		Reduction %
		0 h	6 h		0 h	6 h	
Control without bioperlite	0	Mean: 6.07×10^6 SD: 3.52×10^6	Mean: 1.59×10^6 SD: 1.53×10^5	73.80	Mean: 3.33×10^6 SD: 2.91×10^6	Mean: 1.13×10^6 SD: 1.02×10^6	66.0%
Nonwoven bioperlite 5 %	0.23	Mean: 3.72×10^6 SD: 2.35×10^6	Mean: 2.76×10^5 SD: 1.49×10^5	95.45	Mean: 5.01×10^6 SD: 3.35×10^6	Mean: 2.34×10^5 SD: 2.05×10^5	92.97
Nonwoven bioperlite 10 %	0.46	Mean: 7.23×10^5 SD: 1.13×10^5	Mean: 0 SD: 0	100	Mean: 5.29×10^6 SD: 4.58×10^6	Mean: 0 SD: 0	100
Nonwoven bioperlite 15 %	0.69	Mean: 8.67×10^5 SD: 1.05×10^5	Mean: 0 SD: 0	100	Mean: 1.84×10^7 SD: 3.12×10^6	Mean: 0 SD: 0	100
Nonwoven bioperlite 20 %	0.93	Mean: 7.05×10^5 SD: 1.20×10^5	Mean: 0 SD: 0	100	Mean: 2.43×10^6 SD: 1.56×10^6	Mean: 0 SD: 0	100

SD – standard deviation

Table 8. The influence of bioperlite concentration (with alkylammonium microbiocides) in the nonwoven on antimicrobial activity against *E.coli* and *S.aureus* (based on Gutarowska et al., 2010)

Nonwoven mass humidity level (%)	Number of microorganisms (CFU/sample)		Reduction (%)
	Incubation time		
	0 h	6 h	
Nonwoven (5%) control*	Mean: 2.04×10^4 SD: 1.80×10^4	Mean: 4.15×10^3 SD: 5.60×10^3	79.66
Nonwoven (9.5%)	Mean: 3.52×10^4 SD: 3.08×10^4	Mean: 6.96×10^3 SD: 6.03×10^3	80.23
Nonwoven (43%)	Mean: 2.04×10^4 SD: 1.78×10^4	Mean: 3.54×10^3 SD: 3.50×10^3	82.64
Nonwoven (213%)	Mean: 2.04×10^4 SD: 1.80×10^4	Mean: 3.44×10^3 SD: 3.68×10^3	83.14
Nonwoven (1274%)	Mean: 2.04×10^4 SD: 1.80×10^4	Mean: 1.87×10^2 SD: 1.71×10^2	99.08

*without the addition of water; SD – standard deviation

Table 9. The influence of the humidity level of a nonwoven with 8% bioperlite on antimicrobial activity against *E.coli* (Gutarowska et al., 2010)

The impact of **pH** on biocidal activity depends on the chemical nature of the compound; it may have both positive and negative effects. In the case of phenol compounds an increase in pH causes a reduction in antimicrobial activity, although such a change causes an increase in the activity of quaternary ammonium salts (Brycki, 2003).

Of significant importance for the biological activity of fabrics is the way in which the biocide is introduced into the fabric. The **carriers** for active substances are highly significant. Sample tests with the use of several mineral carriers for silver have shown significant differences in the antimicrobial activity of the resulting fabrics (Gutarowska & Michalski, 2009). In these studies, it was observed the best biocidal effects against test microorganisms (*E.coli, S.aureus, C.albicans, A.niger*) characterized the nonwovens containing silver on TiO2 and BaSO4 carriers (BT nonwovens) and nonwovens with silver on TiO2 and ZnO carriers (TL nonwovens) (Fig.2).

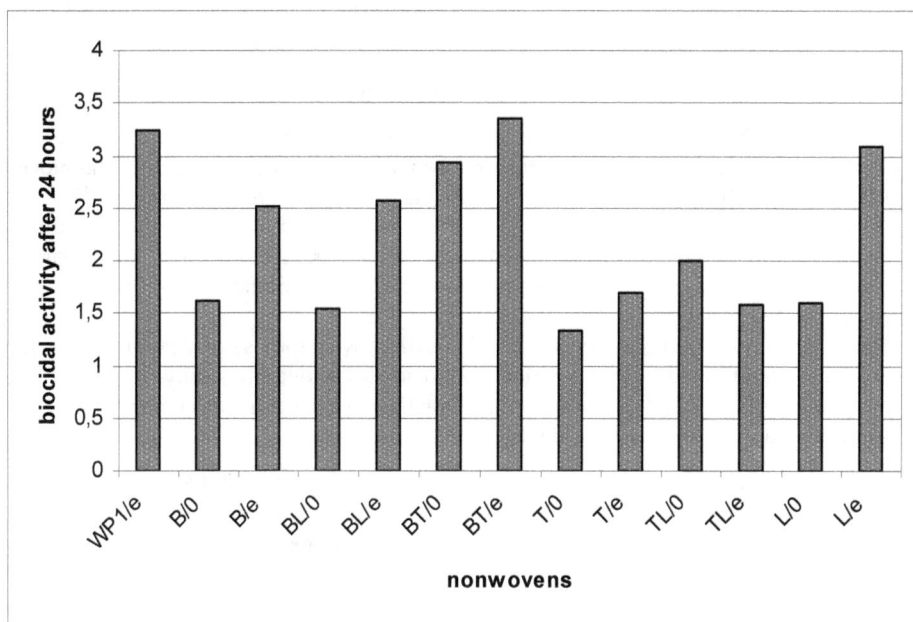

Fig. 2. Biocidal activity of nonwoven with biocides (Ag), to bacteria E.coli after 24 hours incubaction with nonwoven (Gutarowska & Michalski, 2009)

Legend:

Sample code	Added concentrate containing 30% Ag/AgCl with a carrier/ Presence of static charge (no: -; yes:+)
WP1/0	Control without concentrate, (-)
WP1/e	Control without concentrate, (+)
B/0	BaSO$_4$ (-)
B/e	BaSO$_4$ (+)
T/0	TiO$_2$ (-)

T/e TiO_2 (+)
L/0   ZnO (-)
L/e   ZnO (+)
BT/0 $BaSO_4$ + TiO_2 (-)
BT/e $BaSO_4$ + TiO_2 (+)
BL/0 $BaSO_4$ + ZnO (-)
BL/e $BaSO_4$ + ZnO (+)
TL/0 TiO_2 + ZnO (-)
TL/e TiO_2 + ZnO (+)

Generally it was observed that appropriate selection of two carriers improves the effectiveness in comparison with nonwovens in which a single carrier was used. Good effect was reflected both by high biocidal activity and by reduced time off effective contact of microorganisms with the nonwoven. High activity was obtained for the majority of nonwovens with electrostatic charge against bacteria (BL/e, BT/e, T/e, L/e) and for all nonwovens with charge against fungi.

Active substances can be added to fabrics in different ways:

1. Physical modification – introduction of an active compound into the spinning solution or molten fibre-forming polymer and closure within the fibre (occlusion). The biocidal substance then diffuses to its surface, where it acts on the microorganisms.
2. Chemical modification – chemical reactions on the finished textile product, bonding of the biocide through the formation of chemical bonds, e.g. introduction of metal particles to zeolites added during fibre formation, addition of antibiotics to modified fibres by way of grafted copolymerization.
3. Finishing – application of a poorly soluble coating, with the use of a polymeric or low-molecular-weight medium with which the biocide is bonded physically or chemically.
4. Microencapsulation – the introduction into textiles of microcapsules containing volatile substances, dyes with antimicrobial action (Nelson, 2002).

In the case of the first method the biocides must be chosen to have suitable properties so that the technological process (high temperature) does not cause inactivation of the compound: many chemical substances display volatility at high temperatures. This method gives a long-lasting biological effect, as the biocides are permanently fastened to the fibre matrix. Chemical modification of a polymer by acetylation/phosphorylation makes it possible to obtain fibres with permanent antimicrobial properties. However due to the high costs of the production technology, and frequent change in the strength parameters of fabrics, these methods are rarely used. The most popular method is the application of a biocidal finishing layer. The use of a finish on the surface of the finished product favours high antimicrobial activity, although such a product does not retain its properties for a long time, losing them during successive washing cycles (Szostak-Kot, 2004).

The choice of method of producing a fabric should depend on its intended use. Textiles meant for repeated use (socks, bed linen, aprons, underwear, towels) should be highly wash-resistant; in these case the biocides must be permanently joined to the fibre matrix, in contrast to disposable items (aprons, masks, filters, bandages, dressings, gauzes and hospital foot coverings).

7. Methods for evaluating anti-microbiological activity of nonwovens

The need to produce bioactive textiles containing biocides has led to the development of methods for evaluating antimicrobial activity. The final result of such a test is highly dependent on the testing method and the choice of test microorganism. Methods of evaluating antimicrobial properties can be divided into quantitative and qualitative methods (Dymel et al., 2008; Gutarowska et al., 2009).

Evaluation of the antimicrobial activity of textile products by qualitative methods is based on observation of the growth of microorganisms under and around a sample placed on an agar medium with a culture of the microorganisms. The effect of antimicrobial activity is indicated by the variously sized area in which the growth of the microorganisms is suppressed (Photographs 1-3).

Qualitative methods make it possible to evaluate the biocidal action of textiles both in the form of flat products, namely unwoven, woven and knitted fabrics, and in the form of fibres, threads, etc. The hydrophilic or hydrophobic nature of the textiles also has no effect on the final result. The only criterion for a textile product to be tested by qualitative methods is the diffusion of the active substance into the medium. Products must demonstrate at least minimal diffusion of the active component.

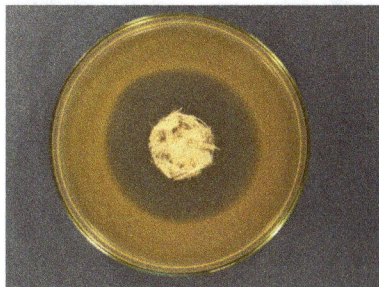

Photo 1. Growth inhibition zone around the *S. aureus;* polypropylene fibers containing 2% Ingaguard- method according to SN 195 920

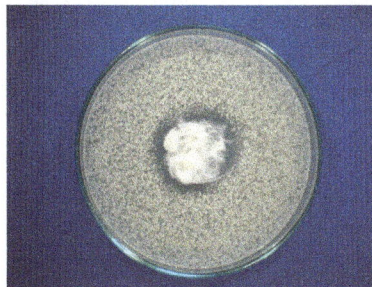

Photo 2. Growth inhibition zone around the *C. albicans* polypropylene fibers containing 2% Ingaguard- method according to SN 195921

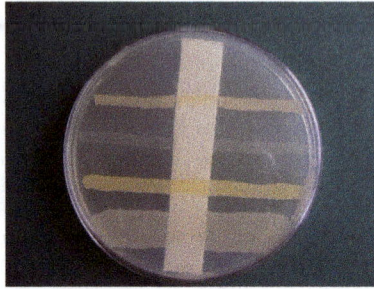

Photo 3. Growth of bacteria respectively from the top: *S. aureus, E. coli; M.flavus, B. licheniformis* under polymer with nano-silver - method according to AATCC 147

Table 10 lists the most commonly used qualitative methods, including Swiss (SN), American (AATCC), Japanese (JI) and European (EN ISO) methods.

Method / standards	Standard number
Antifungal activity, assessment of textile materials: Mildew and rot resistance of textile materials	AATCC 30
Antibacterial activity of fabrics, detection of: Agar plate method	AATCC 90
Antimicrobial activity assessment of textile materials: Parallel streak method	AATCC 147
Antimicrobial activity assessment of carpets	AATCC 174
Standard Test Method for Using Seeded-Agar for the Screening Assessment of Antimicrobial Activity In Carpets	ASTM E2471-05
Standard Test Method for the Assessment of Antimicrobial Activity In Carpets; Seeded-Agar Overlay Screen	ASTM WK4757
Resistance of Textiles to Microbiological Attack. Textiles – Determination of the antibacterial activity – Agar plate diffusion test	CEN/TC 248/WG13
Testing for antibacterial activity and efficacy on textile products	JIS L 1902
Textile fabrics: Determination of the antibacterial activity: Agar diffusion plate test	SN 195920
Textile fabrics: Determination of the Antimycotic Activity: Agar Diffusion Plate Test	SN 195921

Table 10. Qualitative methods for assessing antimicrobial activity of bioactive nonwovens

Qualitative testing methods are similar to each other. They involve pouring out a layer of agar inoculated with a bacteria culture or fungal spores of specified density, or the application of microorganisms on an agar plate via linear inoculation. The tested material and a control sample of specified size are then placed on the inoculated medium. Following incubation, the action of the biocide is evaluated by measuring the area of suppression of growth, compared with a control sample not containing active antibacterial agent.

Quantitative methods are based on the general principle of inoculating the tested sample of material with a suspension of microorganisms of specified density, and then incubating them with the fabric. After some time, based on the number of microorganisms which survived contact with the fabric, the activity of the biocide in the sample is determined relative to a control sample not containing biocide. Quantitative methods are superior to qualitative ones, as the numerical results obtained for the biological activity of unwoven fabrics and textiles can be compared, to select the most effective solution for eliminating microorganisms. Table 11 lists the quantitative methods used for determining the antimicrobial activity of bioactive unwoven fibres and textile products.

Method / standards	Standard number
Assessement of antibacterial finishes on textile materials	AATCC 100
Testing for antibacterial activity and efficacy on textile products	JIS L 1902
Testing hygienically-treated textile products for effectiveness against bacteria. Textile products hygienic finish council	Shake Flask Method
Properties of textiles-Textiles and polymeric surfaces having antibacterial properties. Characterization and measurement of antibacterial activity	XP G39-010
Textile fabrics: Determination of the antibacterial activity: Germ count method	SN 195924
Testing for antibacterial activity	ISO/TC 38/-/WG23
Antimicrobial products – Test for antimicrobial activity and efficacy	JIS Z 2801:2000

Table 11. Methods for quantitative assessment of antimicrobial activity of bioactive nonwovens and textiles

The choice of method cannot be a random one; it is chiefly dependent on such criteria as the type of fabric, its properties and intended use, and the time of action on microorganisms. Based on these criteria and on analysis of the quantitative methods for evaluating antimicrobial properties of bioactive fabrics, a decision chart has been drawn up to enable the selection of an appropriate testing method (Figure 3).

The test results Method AATCC 100 and Shake Flask Method are stated relative to the surface area or mass of the sample, in terms of reduction in the quantity of microorganisms:

$$\% \text{ reduction} = (N_0-N)/N_0 \times 100\% \tag{2}$$

where:

N_0 is the number of microorganisms per sample at time t_0 with the bioactive fabric, and N is the number of microorganisms per sample after a time t_n of exposure with the bioactive fabric.

A positive evaluation is given to fabrics on which the reduction in microorganisms is greater than 85%.

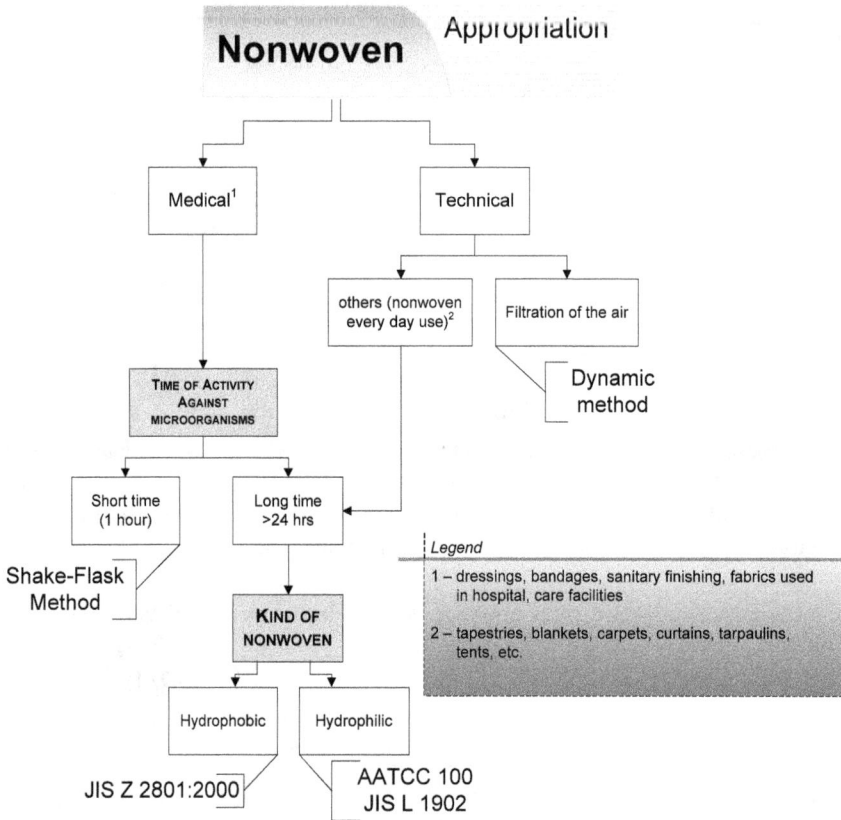

Fig. 3. A decision tree for choose the method of quantitative evaluation of antimicrobial activity of bioactive nonwoven (based on Gutarowska et al. 2009)

The result may be given in the form of bactericidal activity and bacteriostatic activity (Method JIS L 1902, JIS Z 2801:2000).

Biostatic activity is calculated from the formula:

$$\text{biostatic activity (S)} = \log N_k/N \qquad (3)$$

where:

N_k is the number of microorganisms per sample after a time t_n of exposure with the control fabric, and N is the number of microorganisms per sample after a time t_n of exposure with the bioactive fabric.

Biocidal activity is calculated analogously:

$$\text{biocidal activity (L)} = \log N_0/N \qquad (4)$$

where:

Microorganism	Pathogenicity	Characteristic
Escherichia coli ATCC 11229	digestive disorders, urinary tract infections	reference strain, gram negative rods, a significant resistance to biocides
Pseudomonas aeruginosa	pathogen, various types of infection, inflammation of the skin and nosocomial infections	reference strain, gram negative rods, a significant resistance to biocides
Klebsiella pneumoniae	pathogen, pneumonia, transmitted by air, nosocomial infections	reference strain, gram negative rods
Staphylococcus aureus ATCC 6538	pathogen, dermatitis, pneumonia, venous blood clots, ulcers, myocarditis, transmitted by air, common carriers in the nasal cavity and throat, nosocomial infections	reference strain, gram positive coccus
Staphylococcus epidermidis	saprophyte, harmless to health, sometimes skin infections	gram positive coccus, exists on the skin
Micrococcus flavus	saprophyte, harmless to health	gram positive coccus, often isolated from the air, high resistance to UV and disinfectants
Bacillus subtilis	saprophyte, harmless to health, sometimes causes digestive disorders	gram positive bacilli produces spores, often found in the environment (air, soil),
Candida albicans	a potential pathogen, systemic infections, skin, nail mucous membranes infections, hypoallergenic	reference strain, yeast, widespread in the environment (mucous membranes, air, skin)
Rhodotorula rubra	saprophyte, harmless to health, sometimes skin infections	yeast, widespread in the environment (air, food)
Aspergillus niger	saprophyte, harmless to health, sometimes respiratory, cornea and skin infections	mould, reference strain for testing of technical material resistance, present in the air
Penicillium chrysogenum	saprophyte, harmless to health, sometimes upper respiratory tract infections, ear and nail infections, allergies	mould, often isolated from air
Alternaria alternata	saprophyte, harmless to health, hypoallergenic	mould, often isolated from air
Trichophyton mentagrophytes	pathogen, infections of hair, skin and nails	mould, reference strain
Scopulariopsis brevicaulis	pathogen, nail, skin and mucous membranes infections	mould
Epidermophyton floccosum	pathogen, infections of hair, skin and nails	mould

Table 12. Characteristics of test microorganisms for determination of the antimicrobial activity of bioactive nonwovens (based on: Gutarowska et al., 2009)

N_0 is the number of microorganisms per sample at time t_0 with the bioactive fabric, and N is the number of microorganisms per sample after a time t_n of exposure with the bioactive fabric.

A sample is taken to have bactericidal properties if the value of the coefficient of bactericidal activity (L) is greater than zero, and to have bacteriostatic properties if the value of the coefficient of bacteriostatic activity (S) is greater than 2 (Yu, 2003), which denotes a 100 fold reduction in the number of microorganisms.

The evaluation of activity is made with respect to selected potentially pathogenic (from Pure Culture Collections ATCC, NCTC) or saprophytic microorganisms occurring naturally in the human environment. Table 12 lists test microorganisms used for evaluation of the bioactivity of textiles and for their description.

The fundamental criterion for the selection of microorganisms for testing of antimicrobial activity is the intended use of the fabric. In the case of therapeutic fabrics, coming into contact with the human skin, or intended for use in hospitals and care centres, the microorganisms chosen for testing are those which are pathogenic and which are particularly resistant to chemical disinfection and antibiotic treatment, leading to hospital infections, for examples: *Pseudomonas aeruginosa, Klebsiella pneumoniae, Staphylococcus aureus, Escherichia coli, Bacillus licheniformis, Corynebacterium xersosis, Trichophyton mentagrophytes, Candida albicans.* Technical fabrics for uses such as air filtration, and for everyday uses (upholstery, blankets, carpets, net curtains, tarpaulin, etc.) usually come into contact with saprophytic microorganisms, not hazardous to human health, which are constantly present in the air in the form of bioaerosols. Such fabric is tested against the fungi: *Aspergillus niger, Penicillium chrysogenum, Alternaria alternata, Cladosporium cladosporioides* and bacteria: *Micrococcus flavus, Bacillus subtilis.*

8. Conclusions

Biodeterioration of textile materials, mainly natural origin is a serious global economic problem. It requires long-term protection of these materials against destructive activity of microorganisms. At the same time the high standards of hygiene in some areas, primarily medicine, at the work places and others, requires the use of textile materials with antimicrobial properties. In recent years the number of studies on the new biocides and technology of textiles production with antimicrobial activity has increased. The requirements for modern fabrics with antimicrobial properties include high efficiency. In this area the effective methods for proper localization of chemical preparations have been developed, eg microencapsulation or by increase of the surface of preparation, eg by using the active agent in the form of nanoparticles. Most research has been focused on the searching for the new agent - biocides with high efficiency, which are not only effective but also safe, which don't cause the skin irritation, respiratory allergy. Future application will be concentrated on the natural origin substances. The attention also should be done on the biodegradability and environmental protection.

9. References

Abdel-Gawada K.M. (1997) Mycological and some physiological studies of keratinophilic an other moulds associated with sheep wool. *Microbiological Research* 152, pp. 181-188.

Abdel-Kareem O.M.A., Szostak-Kot J., Barabasz W., Paśmionka I. & Galus A. (1997) Fungal biodeterioration of ancient Egyptian textiles. Part I. Surveying study for the most dominant fungi on ancient Egyptian Textiles In: *Microorganisms in environment, occurrence, activity and significance.* pp. 279-290. Agricultural University in Kraków Publ., Kraków, Poland

Abdou - Neel E.A., Ahmed I., Pratten J., Nazhat S.N. & Knowles J.C. (2005) Characterisation of antibacterial copper releasing degradable phosphate glass fibres. *Biomaterials* 26, pp. 2247-2254.

Agarwal P.N. & Puvathingal J.M. (1969) Microbiological deterioration of woolen materials. *Textile Research Journal* 39 (1) pp.38.

Akutsu Y., Nakajima-Kambe T., Nomura N. & Nakahara T. (1998) Purification and properties of a poliester polyurethane-degrading enzyme from Comamonas acidovorans TB-35. *Applied Environmental Microbiology* 64, pp. 62-67.

Allen A., Hilliard N. & Howard G.T. (1990) Purification an characterization of soluble polyurethane degrading enzyme from Comamonas acidovorans. *International Biodeterioration and. Biodegradation* 43, pp. 37-41.

Bailey W.J., Okamoto Y., Kuo W.C., Naria T. (1976) Biodegradable polyamides. *Proceedings of 3rd International Biodegradation Symposium* J.M. Sharpley, A.M.Kaplan (Ed), pp. 765-773, Applied Science Publ.Ltd. London

Bartley T., Waldrom C. & Eveleigh D. (1984) A cellobiohydrolase from temophilic actinomycete Micrbiospora bispora. *Applied Biochemical Biotechnology* 9, pp. 337.

Basu S.N. & Ghose R. (1962) Microbiological study on degradation of jute fiber by microorganisms. *Textile Research Journal* 32(11), pp. 932.

Becker M.A., Williams P. & Tuross N.C. (1995) The USA first ladies gowns: a biochemical study of silk preservation. *Journal of the American Institute for Conservation* 34, pp. 141-152.

Błyskal B. (2009) Fungi utilizing keratinous substrates. *International Biodeterioration and. Biodegradation* 63 pp. 631-653.

Brycki B. (2003) Chemiczne inhibiotory biodeterioracji (Chemical inhibitors of biodeterioration) *Proceedings of IV Symposium of Microbial Corrosion of Technical Materials,* pp. 272-292, Łódź 2003, Poland. (in polish)

Buchanan C.M., Gardner R.M. & Komarek R.T. (1993) Aerobic biodegradation of cellulose acetate. *Journal Applied Polymers Science* 47, pp. 1709-1719.

Bucheńska J., Słomkowski S., Tazbir J. & Sobolewska E. (2003) Antibacterial poly(ethylene terephthalate) yarn containing cephalosporin type antibiotic. *Fibers and Textiles in Eastern Europe* 11(1), pp. 41-47.

Bujak S. & Targoński Z. (1990) Mikrobiologiczna degradacja hemiceluloz. *Postępy Mikrobiologii* 29 (1-2), pp. 77-90

Buschle-Diller G., Zeronian S.H., Pan N. & Yoon M.Y. (1994) Enzymatic hydrolysis of cotton, linen, ramie and viscose rayon fabrics. *Textile Research Journal* 64(5), pp. 270-279.

Cain R.B. (1992) Microbial degradation of synthetic polymers. In: *Microbial control of pollution.* Fry J.C., Gadd G.M., Herbert R.A., Jones C.W., Watson-Craik I.A. (Eds), pp.293-341, Cambridge University Press, England

Dastjerdi R., Montazer M. & Shaksavan S. (2009) A new method to stabilize nanoparticles on textile surfaces. *Colloids Surfaces A. Physicochemical Engeenering Aspects* 345, pp. 202-210.

Dastjerdi R. & Montazer M. (2010) A review on the application of inorganic nano-structured materials in the modification of textiles: focus on antimicrobial properties. *Colloids Surfaces B. Biointerfaces* 79, pp. 5-18.

Denizel T., Jarvis B., onions A.H.S., Rhodes A.C., Samson R.A., Simmons E.G., Smith M.Th. & Hueck-van der Plas E.H. (1974) Catalogue of potentially biodeteriogenic fungi held in the culture collection of the CBS, CMI and QM. *Interantional Biodeterioration Bulletin* 10 (1) pp. 3-23.

Dubas S.T., Kumlangdudsana P. & Potiyaraj P. (2006) Layer-by-layer deposition of antimicrobial silver nanoparticles on textiles fibers. *Colloids Surfaces A: Physicochem Engennering Aspects* 289, pp. 105-109.

Dymel M., Gutarowska B., Więckowska-Szakiel M. & Ciechańska D. (2008) Metody jakościowe oceny aktywności przeciwdrobnoustrojowej wyrobów włókienniczych. *Przegląd Włókienniczy Włókno-Odzież-Skóra* 11, pp. 27-31

Ennis D.M., Kramer A., Jameson C.W., Mazzocchi P.H. & Bailey W.J. (1978) Structural factors influencing in biodegradation of imides. *Applied Environmental Microbiology* 35 (1), pp. 51-53.

Evans E.T. (1996) Biodegradation of cellulose. *Biodeterioration Abstracts.* 10(30), pp. 275-285.

Flannigan B., Samson R.A. & Miller J.D. (2001) *Microorganisms in home and indoor work environments. Diversity, health impacts, investigation and control.* Taylor and Francis Publ., London, New York.

Forlani G., Seves A.M. & Ciferri O. (2000) A bacterial extracellular proteinase degrading silk fibroin. *International Biodeterioration and. Biodegradation* 46, pp. 271-275.

Friedrich J., Zalar P., Mohorcic M., Klun U. & Krzan A. (2007) Ability of fungi to degrade synthetic polymer nylon-6. *Chemosphere* 67, pp. 2089-2095.

Gochel M., Belly M. & Knott J. (1992) Biodeterioration of wool during storage. *International Biodeterioration and. Biodegradation* 30 (1), pp. 77-85.

Goetzendorf-Grabowska B., Królikowska H. & Gadzinowski M. (2004) Polymer microspheres as carriers of antibacterial properties of textiles: a preliminary study. *Fibers and Textiles in Eastern Europe* 12(4): 62-64.

Gorensek M. & Recelj P. (2007) Nanosilver functionalize cotton fabrics. *Textile Research Journal* 77(3), pp. 138-141.

Gupta B., Jain R., Anjum N., Revagade N. & Singh H. (2004) Antimicrobial properties of natural dyes against gram-negative bacteria. *Color Technology* 120, pp. 167-171.

Gupta B., Gulre S.K.H., Anjum N. & Singh H. (2007) Development of antimicrobial propylene sutures by graft copolymerization. II. Evaluation of physical properties, drug release and antimicrobial activity. *Journal Applied Polymers Science* 103, pp. 3534-3538.

Gupta B., Jain R. & Singh H. (2008) Preparation of antimicrobial sutures by preirradiation grafting onto polypropylene monofilament. *Polymers Advances Technology* 19, pp. 1698-1703.

Gutarowska B. & Michalski A. (2009) Antimicrobial activity of filtrating meltblown nonwoven's with addition of silver ions. *Fibers and Textiles in Eastern Europe* 17 (74), pp. 23-28.

Gutarowska B., Dymel M., Więckowska-Szakiel M. & Ciechańska D. (2009) Metody ilościowe oceny aktywności przeciwdrobnoustrojowej wyrobów włókienniczych. *Przegląd Włókienniczy Włókno-Odzież-Skóra* 3, pp. 34-37.

Gutarowska B., Brycki B., Majchrzycka K. & Brochocka A. (2010) Antimicrobial properties of filtering polypropylene nonwovens containing alkylammonium microbiocides on a perlit carrier. *Polimery* 7(8), pp. 568-574.

Halim El-Sayed A.H.M.M., Mahmoud W.M., Davis E.M. & Coughlin R.W. (1996) Biodegradation of polyurethane coatings by hydrocarbon-degrading bacteria. *International Biodeterioration and. Biodegradation* 37, pp. 69-79.

Han S. & Yang Y. (2005) Antimicrobial activity of wool fabric treated with curcumin. *Dyes and pigments* 64, pp. 157-161.

Hipler U.Ch., Elsner P. & Fluhr J.W. (2006) Antifungal and antibacterial properties of a silver –loaded cellulosic fiber. *Journal Biomedical Materials Research Part B. Applied Biomaterials* 77 B(1), pp. 156-163.

Hoare J.L. (1968) A review of chemical aspects of the yellowing of wool. *Wool Research Organisation of New Zealand Communication* 2, pp. 5-13.

Horan R.L., Antle K., Collette A.L., Wang Y., Huang J., Moreau J.E., Volloch V., Kaplan D.L. & Altman G.H. (2005) In vitro degradation of silk fibroin. *Biomaterials* 26, pp. 3385-3393.

Howard J.W. & Mc Cord F.A. (1960) Cotton Quality Study: IV: Resistance to Weathering *Textile Research Journal* 30: 75-117.

Howard G.T. (2002) Biodegradation of polyurethane: a review. *International Biodeterioration and. Biodegradation* 49, pp. 245-252.

Ishiguro Y. & Miyashita M. (1996) Deterioration of silk caused by propagation of microbes. *Proceedings of the Third International Silk Conference.* pp. 201-208, Suzhou, China.

Jain P. & Pradeep T. (2005) Potential of silver nanoparticle-coated polyurethane foam as an antibacterial water filter. *Biotechnology Bioengineering* 90 (1), pp. 59-63.

Jantas R. & Górna K. (2006) Antibacterial finishing of cotton fabrics. *Fibers and Textiles in Eastern Europe* 14 (1), pp. 88-91.

Jeffries T.W. (1987) Physical, chemical and biochemical considerations in the biological degradation of wood. In: *Wood and cellulosics: industrial utilization, biotechnology, structure and properties.* Kennedy J.F., Phillips G.O., Williams P.A. (Eds), pp. 213, Ellis Horwood Publ., Chichester.

Joshi M., Ali W. & Rejendran S. (2007) Antibacterial finishing of poliester/cotton blend fabrics using neem (Azadirachta indica): a natural bioactive agent. *Journal Applied Polymers Science.* 106, pp. 793-800.

Kanazawa A., Ikeda T. & Endo T. (1994) Polymeric phosphonium salts as a novel class of cationic biocides. *Journal Applied Polymers Science* 54 (9), pp. 1305-1310.

Kaplan D., Adams W.W., Farmer B. & Viney Ch. (1994) Silk: Biology, Structure, Properties and Genetics. In: *Silk Polymers Materials Science and Biotechnology.* Kaplan D., Adams W.W., Farmer b., Viney Ch. (Eds) pp. 3-16, American Chemical Society, Washington.

Kenawy E-R., Abdel-Hay F. I., El-Shanshoury A.E-R.R. & El-Newehy M.H. (2002) Biologically active polymers. V. Synthesis and antimicrobial activity of modified poly(glycidylmethacrylate-co-2-hydroxyethyl methacrylate) derivatives with quaternary ammonium and phosphonium salts. *Journal Polymer Science Part A: Polymers Chemistry* 40, pp. 2384-2393.

Kenawy E-R. & Mahmoud Y.A.-G. (2003) Synthesis and antimicrobial activity of some linear copolymers with quaternary ammonium and phosphonium groups. *Macromolecules Science* 3, pp. 107-116.

Khan M.I., Ahmad A., Khan S.A., Yusuf M., Shahid M., Manzoor N. & Mohammad F. (2011) Assessment of antimicrobial activity of Catechu an its dyed substrate. *Journal Cleaner Products* 19, pp. 1385-1394.

Kim H.W., Kim B.R. & Rhee Y.H. (2010) Imparting durable antimicrobial properties to cotton fabrics using alginate-quaternary ammonium complex nanoparticles. *Carbohydrate Polymers* 79, pp. 1057-1062.

Kowalik R.B. (1980) Microdecomposition of basic organic library materials. Part II. *Restaurator* 4: 135-219.

Kubicek C.P., Munhlbauer G., Klotz M., John E. & Kubicek-Pranz E. (1988) Properties of conidial-bound cellulose enzyme system from trichoderma reesei. *Journal General Microbiology* 134, pp. 1215-1222.

Kunert J. (1989) Biochemical mechanism of keratin degradation by the Actinomycete Streptomyces fradiae and the fugus Microsporum gypseum: a comparision. *Journal Basic Microbiology* 29(9), pp. 597-604.

Kunert J. (1992) Effect of reducing agents on proteolytic and keratinolytic activity of enzymes of Microsporum gypseum. *Mycoses* 35, pp. 343-348.

Kunert J. (2000) Phisiology of keratinophilic fungi. In: *Biology of dermatophytes and other keratinophilic fungi*. Kushwaha R.K.S., Guarro J. (Eds), pp. 77-85, Revista Iberoamericana de Micologia, Bilbao.

Lala N.L., Rammaseshan R., Boun L., Sundarrajan S., Barhate R.S., Ying-jun L. & Ramakrishna S. (2007) Fabrication of nanofibres with antimicrobial functionality used as filters: protection against bacterial contaminants. *Biotechnology Bioengineering* 97(6), pp. 1357-1365.

Lewis J. (1981) Wool. In: *Economic microbiology. Microbial degradation*. Rose A.H.(Ed), pp. 81-130, Academic Press, London.

Li S. & Vert M.(1995) Biodegradation of aliphatic polyesters. In: *Degradable polymers. Principles and applications*. Scott G., Gilead D. (Eds), pp. 43-87, Chapman and Hall, London.

Lim S.H., Hudson S.M. (2004) Application of a fiber –reactive chitosan derivative to cotton fabric as an antimicrobial textile finish. *Carbohydrate Polymers* 56, pp. 227-234.

Lucas N., Bienaime Ch., Belloy Ch., Queneudec M., Silvestre F. & Nava-Saucedo J.E. (2008) Polymer biodegradation: mechanisms and estimation techniques. *Chemosphere* 73, pp. 429-442.

Majchrzycka K., Gutarowska B. & Brochocka A. (2010): Aspects of tests assessment of filtering materials used for respiratory protection against bioaerozol. Part II– sweat in environment, microorganisms in the form of bioaerozol. *International Journal of Occupational Safety and Ergonomics* 16 (2), pp. 275-280.

McCarthy B.J., Greaves P.H. (1988) Mildew-causes. Detection Methods and Prevention. *Wool Science Review* 85, pp. 27-48.

McCarthy B.J. (1995) Biocides for use in textile industry. In: *Handbook of biocide and preservative use*. Rossmoore H.W. (Ed), 238-253, Blackie Academic and Professional, London.

Montazer M., Behzadnia A., Pakdel E., Rahimi M.K. & Moghadam M.B. (2011) Photo-induced silver on Nano titanium dioxide as an enhanced antimicrobial agent for wool. *Journal Photochemistry Photobiology B. Biology* 103, pp. 207-214.

Munoz-Bonilla A. & Fernandez-Garcia M. (2011) Polymeric materials with antimicrobial activity. *Progress Polymers Science* DOI: 10.1016/j.progpolymsci.2011.08.005, in press.

Nelson G.(2002) Application of microencapsulation in textiles. *International Journal Pharmaceutics* 242, pp. 55-62.

Nigam S.S., Agarwal P.N. & Tandan R.N. (1972) Fungi responsible for degradation of service materials in India. *Journal Science and Technology* 10-b(1), pp. 1

Nigam N. & Kushwaha R.K.S. (1992) Biodegradation of wool by Chrysosporium keratynophilum acting singly or in combination with other fungi. Trans. *Mycology Association Japan* 33, pp. 481-486.

Nurdin N., Helary G. & Sauvet G. J. (1993) Biocidal polymers active by contact. II biological evaluation of polyurethane coatings with pendant quaternary ammonium salts. *Applied Polymers Science* 50, pp. 663-670.

Pedersen G.L., Screws G.A.Jr. & Credoni D.M. (1992) Biopolishing of cellulosic fabrics. *Canadian Textile Journal* 109: 31-35.

Prijambada I.D., Negoro S. Yomo T. & Urabe I. (1995) Emergence of nylon oligomer degradation enzymes in *Pseudomonas aeruginosa* PAO through experimental evolution. *Applied Environmental Microbiology* 61(5), pp. 2020-2022.

Prusty A.K., As T., Nayak A. & Das N.B. (2010) Colourimetric analysis and antimicrobial study of natural dyes an dyed silk. *Journal Cleaner Products* 18, pp. 1750-1756.

Radetic M., ilic V., Vodnik V., Dimitrijevic S., Jovancic P., Saponjic Z., & Nedeljkovic J.M. (2008) Anibacterial effect of silver nanoparticles deposited on corona-treated polyester and polyamide fabrics. *Polymers Advances Technology* 19, pp. 1816-1821.

Ren X., Akdag A., Kocer H., Worley S.D., Broughton R.M. & Huang T.S. (2009) N-halamine-coated cotton for antimicrobial and detoxification applications. *Carbohydrate Polymers* 78, pp. 220-226.

Ruiz C., Main T., Hilliard N., Howard G.T. (1999) Puricication and characterization of two polyurethanase enzymes from pseudomonas chlororaphis. *International Biodeterioration and. Biodegradation* 43, pp. 43-47.

Sachinvala N., Parikh D.V., Sawhney P., Chang S., Mirzawa J., Jarrett W. & Joiner B. (2007) Silver (I) antimicrobial cotton nonwovens and printcloth. *Polymers Advances Technology* 18, pp. 620-628.

Safranek W.W. & Goos R.D. (1982) Degradation of wool by saprophytic fungi. *Canadian Journal Microbiology* 28, pp. 137-140.

Salerno-Kochan R. & Szostak-Kotowa J. (1997) Biodegradation poliester fibres. *Proceedings of 11th IGWT Symposium*, pp. 314-316, Vienna 1997.

Salerno-Kochan R. & Szostak-Kotowa J. (2001) Microbiological degradation of textiles. Part I. Biodegradation of cellulose textiles. *Fibers and Textiles in Eastern Europe* 9(3), pp. 69-72.

Sato M. (1976) The effects of molds on fibres and their products. VIII. Scanning electron microscopic study on the destruction of silk yarns damaged by molds. *Kyoto-Furitsu Daigaku Gakuju Hokoku: Rigaku, Seikatsu Kagaku* 27, pp. 59-64.

Seves A., Romano M., Maifreni T., Sora S. & Ciferri O. (1998) The microbial degradation of silk: a laboratory investigation. *International Biodeterioration and. Biodegradation* 42(4), pp. 203-211.

Silver S. (2003) Bacterial silver resistance: molecular biology and misuses of silver compounds. *FEMS Microbiology Reviews* 27, pp. 341-353.

Simpson W. (1987) The influence of pH on the reflectance and photostability of wool to sun light. *Journal Textiles Institute* 5, pp. 430-438.

Singh R., Jain A., Panwar S., Gupta D. & Khare S.K. (2005) Antimicrobial activity of some natural dyes. *Dyes and Pigments*, pp. 99-102.

Sionkowska A. & Planecka A. (2011) The influence of radiation on silk fibroin. *Polymers Degradation Stability* 96, pp. 523-528.

Sousa F., Guebitz G.M. & Kokol V. (2009) Antimicrobial and antioxidant properties of chitosan enzymatically functionalize with flavonoids. *Process Biochemistry* 44, pp. 749-756.

Struszczyk H., Lebioda J., Twarowska-Schmidt K. & Niekraszewicz A. (2003)New bioactive synthetic fibres developed in the Institute of Chemical Fibres. *Fibers and Textiles in Eastern Europe* 11 (2), pp. 96-98.

Szostak-Kotowa J. (2004) Biodeterioration of textiles. *International Biodeterioration and. Biodegradation* 53, pp. 165-170.

Szostak-Kot J. (2005) Fibres and nonwovens In: *Microbiology of materials* Zyska B., Żakowska Z.(Eds), pp. 89-136, Technical University of Lodz Publ., (in polish)

Szostak-Kot J. (2009) Biodeterioration of cultural heritage artefacts. Microbiological aspects of conservation. *Proceedings of V Symposium of Microbial Corrosion of Technical Materials*, pp. 75-84 Łódź 2009.

Tan S., Li G., Shen J., Liu Y., Zong M. (2000) Study of modified polypropylene nonwoven cloth. II Antibacterial activity of modified polypropylene nonwoven cloths. Journal Applied Polymers Sciences 77, pp. 1869-1876.

Targoński Z. & Bujak S. (1991) Mikrobiologiczna degradacja ligniny. *Postępy Mikrobiologii* 30 (1), pp. 89-106.

Tyndal R.M. (1992) Improving the softness and surface appearance of cotton fabrics and garments by treatment with cellulose enzymes. *Textile Chemist and Colorist* 24(6), pp. 23-26.

Wales D.S. & Sagar B.F. (1988) Mechanistic aspects of polyurethane biodeterioration. In: *Biodeterioration 7.* Houghton D.R., Smith R.N., Eggins H.O.W. (Eds) Elsevier Applied Science Publ.. Oxford, Melbourne

Yamada H., Asano Y., Hino T. & Tani Y. (1979) Microbial utilization of acrylonitryle. *Jouranl Fermentaion Technology* 57, pp. 8-14.

Yi E., Hong J.Y. & Yoo E.S. (2010) A novel bioactive fabric dyed with unripe Citrus grandis Osbeck extract part 2: effects of the Citrus extract and dyed fabric on skin irritancy and atopic dermatitis. *Textiles Research Journal* 80 (20), pp. 2124-2131.

Yu D.G., Teng M.Y., Chou W.L. & Yang M.C. (2003) Characterization and inhibitory effect of antibacterial PAN-based hollow fibe loaded with silver nitrate. *Journal Membrane Science* 225, pp. 115-123.

Zyska B. (1977) Nonwoven and textiles (Włókna i tkaniny) In: *Microbial corrosion of technical materials (Mikrobiologiczna korozja materiałów technicznych)* Zyska B (Ed) pp. 46-104, NT Publ., Warszawa (in polish)

Zyska B. (2001) Textile industry (Przemysł włókienniczy) In: *Disasters, accidents and microbiological threats in industry and building* (Katastrofy, awarie i zagrożenia mikrobiologiczne w przemyśle i budownictwie). Zyska B. (Ed), pp. 48-59, Technical University of Lodz Publ., Lodz (in polish).

Central Statistical Office Yearbooks – Roczniki GUS, Poland 2008, Zakład Wydawnictw Statystycznych, Warszawa, 2008

Permissions

The contributors of this book come from diverse backgrounds, making this book a truly international effort. This book will bring forth new frontiers with its revolutionizing research information and detailed analysis of the nascent developments around the world.

We would like to thank Han-Yong Jeon, for lending his expertise to make the book truly unique. He has played a crucial role in the development of this book. Without his invaluable contribution this book wouldn't have been possible. He has made vital efforts to compile up to date information on the varied aspects of this subject to make this book a valuable addition to the collection of many professionals and students.

This book was conceptualized with the vision of imparting up-to-date information and advanced data in this field. To ensure the same, a matchless editorial board was set up. Every individual on the board went through rigorous rounds of assessment to prove their worth. After which they invested a large part of their time researching and compiling the most relevant data for our readers. Conferences and sessions were held from time to time between the editorial board and the contributing authors to present the data in the most comprehensible form. The editorial team has worked tirelessly to provide valuable and valid information to help people across the globe.

Every chapter published in this book has been scrutinized by our experts. Their significance has been extensively debated. The topics covered herein carry significant findings which will fuel the growth of the discipline. They may even be implemented as practical applications or may be referred to as a beginning point for another development. Chapters in this book were first published by InTech; hereby published with permission under the Creative Commons Attribution License or equivalent.

The editorial board has been involved in producing this book since its inception. They have spent rigorous hours researching and exploring the diverse topics which have resulted in the successful publishing of this book. They have passed on their knowledge of decades through this book. To expedite this challenging task, the publisher supported the team at every step. A small team of assistant editors was also appointed to further simplify the editing procedure and attain best results for the readers.

Our editorial team has been hand-picked from every corner of the world. Their multi-ethnicity adds dynamic inputs to the discussions which result in innovative outcomes. These outcomes are then further discussed with the researchers and contributors who give their valuable feedback and opinion regarding the same. The feedback is then collaborated with the researches and they are edited in a comprehensive manner to aid the understanding of the subject.

Apart from the editorial board, the designing team has also invested a significant amount of their time in understanding the subject and creating the most relevant covers. They scrutinized every image to scout for the most suitable representation of the subject and create an appropriate cover for the book.

The publishing team has been involved in this book since its early stages. They were actively engaged in every process, be it collecting the data, connecting with the contributors or procuring relevant information. The team has been an ardent support to the editorial, designing and production team. Their endless efforts to recruit the best for this project, has resulted in the accomplishment of this book. They are a veteran in the field of academics and their pool of knowledge is as vast as their experience in printing. Their expertise and guidance has proved useful at every step. Their uncompromising quality standards have made this book an exceptional effort. Their encouragement from time to time has been an inspiration for everyone.

The publisher and the editorial board hope that this book will prove to be a valuable piece of knowledge for researchers, students, practitioners and scholars across the globe.

List of Contributors

B. K. Behera
Department of Textile Technology, Indian Institute of Technology, Delhi, India

Jiri Militky, Rajesh Mishra and Dana Kremenakova
Faculty of Textile Engineering, Technical University of Liberec, Czech Republic

Pelin Gurkan Unal
Namık Kemal University, Department of Textile Engineering, Turkey

Seung Jin Kim and Hyun Ah Kim
School of Textiles, Yeungnam University, Gyeongsan, Korea Institute for Knit Industry, Iksan, Korea

Jeng-Jong Lin
Department of Information Management, Vanung University, Taiwan, R.O.C.

Hatice Kübra Kaynak
Gaziantep University, Textile Engineering Department, Turkey

Osman Babaarslan
Çukurova University, Textile Engineering Department, Turkey

Jiří Militký
Faculty of Textile Engineering, Technical University of Liberec, LIBEREC, Czech Republic

Ján Široký, Barbora Široká and Thomas Bechtold
University of Innsbruck, Research Institute for Textile Chemistry and Physics, Austria

Gonca Özçelik Kayseri and Gamze Süpüren Mengüç
Ege University, Emel Akın Vocational Training School, Izmir, Turkey

Nilgün Özdil
Ege University, Textile Engineering Department, Izmir, Turkey

Nikola Jakši´c
Turbo inštitute, Slovenia

Danilo Jakši´c
University of Ljubljana, Slovenia

Beata Gutarowska
Technical University of Lodz, Institute of Technology Fermentation and Microbiology, Lodz, Poland

Andrzej Michalski
Spółdzielnia Inwalidów ZGODA, Konstantynów Łódzki, Poland